Applied Molecular Genetics of Filamentous Fungi

Applied Molecular Genetics of Filamentous Fungi

Edited by

JAMES R. KINGHORN
Plant Molecular Genetics Unit
University of St Andrews

and

GEOFFREY TURNER
Department of Molecular Biology and Biotechnology
University of Sheffield

SPRINGER-SCIENCE+BUSINESS MEDIA, B.V.

First edition 1992

Originally published by Chapman & Hall in 1992
Softcover reprint of the hardcover 1st edition 1992

Typeset in 10/12pt Times New Roman by Thomson Press (India) Ltd, New Delhi

ISBN 978-94-010-5301-3 ISBN 978-94-011-2930-5 (eBook)
DOI 10.1007/978-94-011-2930-5

A catalogue record for this book is available from the British Library

Library of Congress Cataloguing-in-Publication data available

∞ Printed on permanent acid-free paper, manufactured in accordance with the proposed ANSI/NISO Z 39.48-199X and ANSI Z 39.48-1984

Preface

The filamentous fungi are perhaps unique in the diversity of their metabolic activities. This includes biosynthetic as well as degradative activities, many of which are of industrial interest. The objective of this text is to provide an up-to-date and broad review which emphasizes the genetic and molecular biological contribution in the field of fungal biotechnology.

This text begins with an overview of the tools and methodologies involved which, to a large extent, have been developed in the model filamentous fungus *Aspergillus nidulans* and subsequently have been extended to commercially important fungi. This is followed by a chapter which provides a compilation of genes isolated from commercial fungi and their present status with respect to structure, function and regulation.

Chapters 3 and 4 highlight the degradative powers of filamentous fungi. First, a discussion of what is known regarding the molecular genetics of fungi and the genes and enzymes involved in the beverage and food industries. This has an oriental flavour, reflecting the tremendous importance of fungi in traditional Chinese and Japanese food production. An account of lignocellulose degradation by filamentous fungi follows, illustrating the potential of fungi to utilize this substance as a renewable energy source.

The ability of fungi to produce high-value foreign proteins is reviewed in chapters 5 and 6. Chymosin production, in particular, represents a good example of high-level yields being obtained, such as to warrant commercial production.

Filamentous fungi are equipped with the ability to synthesize biological molecules which enable such organisms to compete successfully in natural biosystems. These molecules are often useful to man in various ways. For instance, secondary metabolite production (chapter 10) has been at the heart of fungal biotechnology for decades. In particular, β-lactam antibiotic production has had a profound effect on the general well-being of not only humans, but domestic animals. Further classes of fungal molecules have been used for biological control of, for example, insect growth (chapter 7).

Finally, filamentous fungi, as exemplified by the mushrooms, can be used as a source of human food material. Significant differences in fungal species and cultural practices (fungal and human!) are seen in different areas of the world and their molecular genetics are covered in chapters 8 and 9.

J.R. Kinghorn
G. Turner

Contents

Contributors

Gale L. Armstrong Genencor International, 180 Kimball Way, South San Francisco, CA 94080, USA

Elizabeth A. Bodie Genencor International, 180 Kimball Way, South San Francisco, CA 94080, USA

Gina L. Carter Genencor International, 180 Kimball Way, South San Francisco, CA 94080, USA

John M. Clarkson School of Biological Sciences, Claverton Down, University of Bath, Bath BA2 7AY, UK

Dan Cullen Forest Products Laboratory, US Department of Agriculture, 1 Gifford Pinchot Drive, Madison, Wisconsin 53705–2398, USA

Nigel S. Dunn-Coleman Genencor International, 180 Kimball Way, South San Francisco, CA 94080, USA

Katsuya Gomi National Institute of Brewery Sciences, 2-6-30 Takinokawa, Kita, Tokyo, Japan 114

Santiago Gutierrez Department of Ecology, Genetics and Microbiology, Section of Microbiology, Faculty of Biology, University of León, 24071 León, Spain

David I. Gwynne Allelix Biopharmaceuticals, Mississauga, Ontario L4V 1P1, Canada

Paul A. Horgen Centre for Plant Biotechnology, University of Toronto, Erindale Campus, Mississauga, Ontario L5L 1C6, Canada

Hiroyuki Horiuchi Faculty of Agricultural Chemistry, Department of Agriculture, University of Tokyo, 1-1-1 Yayoi, Bunkyo, Tokyo, Japan 113

Peter Kersten Forest Products Laboratory, US Department of Agriculture, 1 Gifford Pinchot Drive, Madison, Wisconsin 53705–2398, USA

James R. Kinghorn Plant Molecular Genetics Unit, Sir Harold Mitchell Building, University of St Andrews, St Andrews, Fife KY16 9TH, UK

Juan F. Martin Department of Ecology, Genetics and Microbiology, Section of Microbiology, Faculty of Biology, University of León, 24071 León, Spain

Gregory May Department of Cell Biology, Baylor College of Medicine, 1 Baylor Plaza, Houston, TX 77030, USA

Kenji Sakaguchi BioProjects Inc., 1-11-18 Daizawa, Setagaya. Tokyo, Japan 155

Kazuo Shishido Department of Life Science, Tokyo Institute of Technology, Nagatsuta, Yokohama 227, Japan

Masamichi Takagi Faculty of Agricultural Chemistry, Department of Agriculture, University of Tokyo, 1-1-1 Yayoi, Bunkyo, Tokyo, Japan 113

Geoffrey Turner Department of Molecular Biology and Biotechnology, University of Sheffield, Sheffield S10 2TN, UK

Sheila E. Unkles Plant Molecular Genetics Unit, Sir Harold Mitchell Building, University of St Andrews, St Andrews, Fife KY16 9TH, UK

1 Fungal technology

G. MAY

1.1 Introduction

It has been nearly a decade since the first reports of the transformation of *Aspergillus nidulans* and during that time the number of selective markers has grown significantly [1–4]. The development of the *A. nidulans* transformation system was preceded by the development of a transformation system for another filamentous fungus, *Neurospora crassa* [5]. The success with *N. crassa* and the yeast *Saccharomyces cerevisiae* fostered continued efforts with *A. nidulans*. In this chapter the development of transformation systems for *A. nidulans*, the fate of the transforming DNA and the application of transformation in *A. nidulans* to questions of biological interest are reviewed. The focus is primarily on *A. nidulans* as a model system but mention is also made of other systems if specific examples do not exist in *A. nidulans* or other systems have exploited a particular approach more effectively. Examples are given of the technology applied to organisms of commercial importance.

Since the first reports of the transformation of *A. nidulans* there has been a substantial increase in the number of available selective markers. In addition the number of fungal species for which transformation systems are now available has grown significantly. These additional species include economically important industrial fungi and plant pathogens. Simultaneously there has been a significant growth in understanding the fate of transforming DNA in filamentous fungi in general, and particularly in *A. nidulans*. A conservative estimate of the markers available for selection in *A. nidulans*, based on the published literature, is 13. The nutritional selective markers for *A. nidulans* transformation are listed in Table 1.1, markers for which there is both forward and reverse selection are listed in Table 1.2 and the dominant selective markers are given in Table 1.3. For simplicity only the *A. nidulans* gene designation has been used but reference is included to the homologous gene from other systems, such as the *pyr4* gene to complement the *pyrG89* mutation of *A. nidulans* [1]. The utility in the design of certain types of molecular genetic manipulations of the host genome in those cases which have both a forward and a reverse selection will be discussed in detail later. In some cases the dominant selective markers listed have not been used routinely for the transformation of *A. nidulans* but a description of their usefulness is justified in that for some filamentous fungi they have been the only available marker for transformation and in other systems they have been used extensively. A good example of a frequently used dominant marker is a benomyl resistant β-tubulin mutant gene in *N. crassa* [6, 7]. Most of these selection systems have also been used for commercially important fungi.

1.2 Selective markers

1.2.1 *Nutritional selective markers and markers with forward and reverse selection*

The first group of selective markers, and the most widely used in the transformation of *A. nidulans*, are the nutritional selective markers. This group of markers represents the largest single group and also contains those for which there is both forward and reverse selection. Those nutritional markers commonly used in *A. nidulans* are listed in Tables 1.1 and 1.2. The most widely used among these nutritionally selective markers are *pyrG*, *argB*, *trpC* and *amdS* [1–4, 8, 9]. This can be most easily explained by the fact that these were also the first selective markers used in *A. nidulans* and have been in use long enough now to have become widely distributed. Transformation with nutritional markers is based on having auxotrophic mutant strains that can be transformed to prototrophy for the selective marker being used. This is relatively straightforward for *A. nidulans* in that numerous auxotrophic mutations have been previously characterized at both the genetic and biochemical level [10].

The greatest barrier to transformation of *A. nidulans* is obtaining cloned genes that can be used to transform strains to prototrophy. There are two approaches for obtaining such genes: the first, used to isolate the *amdS* gene, is the direct physical cloning of the gene [11] and the second is the complementing of auxotrophic mutants of *E. coli* or yeast with libraries made from *A. nidulans* genomic DNA. The latter strategy was employed to isolate the *trpC* and *argB* genes of *A. nidulans* [4, 12]. In an alternative approach, a cognate gene from another filamentous fungus was used to complement an

Table 1.1 Nutritional selective markers used to transform *Aspergillus nidulans*

Marker	Selection	Reference
amdS	acetamide utilization	[3, 121]
argB	arginine prototrophy	[2]
prn	proline utilization	[31]
pyroA	pyridoxine prototrophy	[32]
riboB	riboflavin prototrophy	[33]
trpC	tryptophan prototrophy	[4]

auxotrophic mutant of *A. nidulans*. The use of the *N. crassa pyr4* gene to complement the *pyrG89* mutation of *A. nidulans* is the best example of this approach and it has now been used in the development of transformation systems for numerous fungi as described below [1].

The development of a transformation system for *A. nidulans* has provided the springboard for the development of transformation systems for other filamentous fungi; in some cases by providing selective markers but also by supplying the basic methodology. This cross fertilization between different fungal systems is broadly based and includes industrial fungi and phytopathogenic fungi. A prime example of this is the utilization of the *amdS* gene of *A. nidulans* to transform *A. niger*, *Penicillium chrysogenum*, *Trichoderma reesei*, *Cochliobolus heterostrophus* and *Glomerella cingulata* allowing these species to utilize acetamide as the sole nitrogen source [13–16]. A particular advantage of *amdS* in these examples is that the host strain grows poorly on acetamide and it is not necessary to obtain a mutant. Similarly, the direct selection of auxotrophic mutations in industrial fungi and the use of genes from *A. nidulans* has become routine. The use of the *pyr4* gene of *N. crassa* to complement the *pyrG89* mutation of *A. nidulans* became the corner-stone for the transformation of the industrially important fungi, *A. niger* and *A. oryzae*, because it was possible to select for loss-of-function mutations in the orotidine-5′-phosphate decarboxylase gene by growth on media containing 5-fluoroorotic acid [17, 18]. This approach was adapted from methods employed in *S. cerevisiae* [19].

Additional selection systems have now been established for these and other industrially important fungi using similar strategies. These other systems include the *sC* gene, sulfate non-utilization, the *argB* gene, arginine auxotrophy, *trpC*, trytophan auxotrophy, and the *niaD* gene, nitrate non-utilization [14, 20–29]. The selection systems for *sC* and *niaD* are particularly useful in this application in that loss-of-function mutations in these genes are readily selected for on the appropriate medium. Loss-of-function mutations in the *niaD* gene are selected as being resistant to chlorate [24, 26–28]. Similarly, mutations in the *sC* genes are selected as being resistant to selenate [30]. One disadvantage of using these selection systems in other fungal systems is that the auxotrophic mutations are frequently selected following mutagenesis. Mutagenesis of the parental strain could also introduce other apparently silent mutations into the strain that would later interfere with the process being examined. When the mutant selections are sufficiently tight selection can be made for spontaneous mutants in which the likelihood that additional alterations in the genome have occurred is low. It is possible to select for spontaneous mutations in the *sC* and *niaD* genes thus eliminating this potential problem.

There are other nutritional selective markers that are utilized in the transformation of *A. nidulans*. These include the proline utilization cluster *prn*, the *pyroA* gene and the *riboB* gene [31–33]. The *prn* cluster is made up of four

Table 1.2 Nutritional selective markers with forward and reverse selection used to transform *Aspergillus nidulans*

Marker	Selection	Reference
niaD	nitrate utilization	[24–29]
pyrG	uridine prototrophy	[1, 8]
sC	sulfate utilization	[30]

Table 1.3 Dominant selective markers used to transform *Aspergillus nidulans*

Marker	Selection	Reference
benA	benomyl resistance	[7, 57]
oliC	oligomycin resistance	[51]
Bleomycin	bleomycin resistance	[43]
G418	G418 resistance	[120]
Hygromycin	hygromycin resistance	[44, 49]
Phleomycin	phleomycin resistance	[46]

genes, *prnA*, *prnB*, *prnC*, and *prnD*, one of which, *prnB* the permease, has been sequenced [31, 34]. There is little information about structure or expression of the *pyroA* and *riboB* genes but both provide added selective markers for laboratories working with *A. nidulans.*

Early accounts of successful transformation of *A. nidulans* reported relatively low frequencies. Using the *pyr4* gene of *N. crassa* to complement the *pyrG89* mutation of *A. nidulans*, reported rates of transformation were generally of the order of 20 transformants per microgram of plasmid DNA and similar rates of transformation were reported using *amdS* and *trpC* [1, 3, 4]. These frequencies of transformation were insufficient to clone genes by direct complementation using plasmid libraries, a goal that many investigators held. Several different approaches were taken to get around these low frequencies of transformation. One approach was to use cosmids that contained a selective marker. The advantage of this system is that large DNA fragments are cloned, typically 35 to 40 kilobase pairs, requiring fewer transformants to obtain a reasonable chance of cloning a gene of interest [35, 36]. Another approach was to look for sequences that increased the frequency of transformation, either because they maintain the transforming DNA as an autonomously replicating vector or because they increase the frequency in some unknown manner. Selecting for autonomously replicating sequences (ARS) in *S. cerevisiae* from *A. nidulans* did not produce an autonomously replicating vector for *A. nidulans* but did identify a sequence that increased the frequency of stable integrative transformation by 100-fold. This sequence, called *ans1*, is a dispersed repetitive

DNA element that also increases the frequency of stable transformants with *argB* and *trpC*, and is an AT-rich DNA sequence [37, 38]. The mechanism by which the *ans1* sequence increases the number of stable transformants is not known but it is not simply the fact that there are multiple sites in the genome where it can integrate [38]. The final approach used to increase transformation frequency was to alter the conditions for the preparation, handling, transformation and plating of protoplasts [8, 37, 39, 40]. Modifications in the preparation and transformation of protoplasts are dealt with in a later section in which transformation procedures are described.

Autonomously replicating extrachromosomally maintained plasmids have now been described for *A. nidulans* and the basidiomycete *Phanerochaete chrysosporium* [41, 42]. The sequence that confers autonomous replication properties on an *argB* plasmid in *A. nidulans* has been designated *AMA1* [41]. The *AMA1* sequence is an inverted repeat of 6.1 kilobases which increases the frequency of transformation by approximately 250-fold and allows the plasmid to exist at 10 to 30 copies per haploid genome. The plasmid used in these studies was not able to be transmitted from one nucleus to another in heterokaryons and exhibited mitotic instability, being lost from 65% of transformant conidiospores even when growth was on nutritionally selective media. Genomic blot hybridization analysis with the *AMA1* sequence suggests that it is present in several copies in the genome of *A. nidulans*. Finally, this sequence could also act in a similar manner in both *A. niger* and *A. oryzae*. In contrast to the *A. nidulans* autonomously replicating vector, that described for *P. chrysosporium* was derived from a naturally occurring endogenous plasmid of this fungus. In addition the plasmid containing the sequence that confers autonomous replication, ME-1, has a low copy number and is stably maintained under non-selective growth, even for extended periods in the absence of selection [42].

1.2.2 *Dominant selective markers*

The dominant selective markers used to transform *Aspergillus nidulans* and other filamentous fungi are resistance genes for the antibiotics (oligomycin, bleomycin, G418, hygromycin and phleomycin) and mutant β-tubulin genes that give resistance to the antimicrotubule compound benomyl [15, 43–52]. The mechanism of action of each of these selective markers is well understood and they will not be discussed here except to mention potential problems that may be encountered in their use. One particular problem which has been observed is that the selection of G418 resistance in *A. nidulans* was not possible using 0.6 M KCl as an osmotic buffer, even though on standard media the tested strains were very sensitive to this antibiotic.

One of the primary advantages of using dominant selective markers is that the recipient strain's genotype need not be known. It is this feature that has resulted in the application of these markers to fungal species for which there are few or no auxotrophic mutants. This includes fungal species of economic importance either as industrial species for the production of metabolites, or proteins and plant pathogens [15,45,46]. Although these selective markers have been available for several years they are not widely used with *A. nidulans*. This is probably because there are multiple nutritional markers that can be used in this system. A disadvantage of many of these selective markers is that the compounds used to select for transformants are often very expensive. This abrogates their use as general selective markers because of the high cost of preparing the selective media and may be another reason why they are not in general use.

One of the dominant selective markers that is widely used is the benomyl resistant β-tubulin mutant gene [6, 51, 53]. One of the advantages of this selective marker is that benomyl, or its tradename product benylate, is inexpensive and readily available. Additional compounds related to benomyl, such as nocodazole and thiabendazole are commercially available and in general benomyl-resistant strains are cross-resistant to these compounds [54]. It should be noted, however, that this cross-resistance is not absolute and should be tested for any given mutant allele before it is used with other compounds [55]. In addition, since benomyl is a widely used antifungal agent, naturally occurring resistant strains are available from which the resistance gene can be isolated. Isolation of β-tubulin genes from fungi has been made easier by the fact that DNA hybridization probes for the isolation of these genes are readily obtainable [6, 56].

In *N. crassa* a benomyl resistant β-tubulin mutant gene has gained wide use as a selective marker for transformation [6, 7]. In this system the semi-dominant benomyl resistance gene was placed on a cosmid vector and used to construct a genomic library of *N. crassa* DNA. The combination of the semi-dominant marker and the large size of cosmids has greatly facilitated the cloning of genes from *N. crassa*. Individual clones are maintained in the wells of microtiter plates and pools of these are grown for transformation. These cosmid pools are then used to transform the mutant for which the gene is to be cloned and transforming pools are identified. The identified pool is further subdivided until the individual clone transforming the mutant phenotype is identified within the pool. This general approach is known as sibselection and was first described for the transformation of *N. crassa* with plasmid libraries [57]. It should be stressed that the selection for benomyl resistance is not a true dominant but is semi-dominant. This is best understood by considering that there is already a pool of wild-type β-tubulin protein being synthesized and the microtubules formed in these cells will be composed of a mixture of wild-type and mutant benomyl-resistant β-tubulin along with an equal mole of α-tubulin. Experiments describing this phenomenon have been carried out in *A. nidulans* [58, 59]. A similar argument can be made for the oligomycin-resistant gene where the level of resistance will depend upon the contribution of the wild-type and mutant genes to the pool protein in the cell [52].

1.3 Transformation procedures

1.3.1 *Preparation and storage of protoplasts*

Transformation of *A. nidulans* begins with the removal of the cell wall and the production of protoplasts. The removal of the wall is achieved by incubating mycelia or germlings (newly germinated conidiospores) in the presence of lytic enzymes and an osmotic stabilizer. The lytic enzymes used range from Novozyme 234 to a combination of this enzyme with β-glucuronidase and driselase [1, 3, 4, 8, 39]. The osmotic stability of the protoplasts has been maintained using salts, such as magnesium chloride, potassium chloride and ammonium sulfate, and sugars, including sorbitol and sucrose [1, 4, 39, 40]. Variations on these general methods have been employed by a number of groups and adapted for the transformation of other fungal species. Given the relatively large number of methods and variations upon them that are employed in the transformation of other fungal systems it would be difficult to describe them all. Instead the methods employed in the author's laboratory to transform *A. nidulans* and some of its variations are described. In addition some of the variables that have been found to influence the frequency of transformation will be described. These methods also have been successfully used to transform industrial strains of *A. niger* and *A. oryzae*.

Conidia which are to be converted to protoplasts are germinated for 5 to 6 h, or until the start of germtube emergence at 32°C, in media supplemented for optimal growth of the strain to be transformed (10^9 conidia in 50 ml). These cells are sedimented by centrifugation and resuspended in protoplasting mix which is composed of 20 ml each of double strength media and 0.1 M citric acid, and 0.8 M ammonium sulfate brought to pH 5.8 with potassium hydroxide, to which is added 0.4 ml of 1 M magnesium sulfate, 200 mg bovine serum albumin, 100 mg of Novozyme 234, 100 mg Driselase and 100 μl β-glucuronidase. This mixture of enzymes and buffer system is similar to those described previously [8, 40]. Germlings are incubated in this mixture for approximately 3 h at 32°C. The conversion of cells to protoplasts is followed by phase microscopy and when the conversion is near completion the protoplasts are sedimented and washed twice in 50 ml of 0.05 M citric acid, 0.4 M ammonium sulfate and 1% (w/v) sucrose brought to pH 5.8 with potassium hydroxide. The final pellets of protoplasts are resuspended in 1 ml of 0.6 M potassium chloride, 0.1 M calcium chloride and 10 mM Tris-HCl (pH 7.5), and stored on ice. Protoplasts, 0.1 ml, are transformed by the addition of DNA, up to 10 μg in 20 μl, followed by the addition of 50 μl of PEG solution (25% (w/v) polyethylene glycol 8000, 0.1 M calcium chloride, 0.6 M potassium chloride and 10 mM Tris-HCl, pH 7.5) and incubated on ice for 15 min. One milliliter of PEG solution is added after the incubation on ice and the protoplasts are then incubated for 15 min at room temperature. Portions of the transformed protoplasts are added to 3 ml of selective media containing

1 M sucrose and 1% (w/v) agar and plated onto selective media containing 0.2 M sucrose and 1.5% (w/v) agar. Plates are then incubated for two to three days at the appropriate temperature.

Using this transformation protocol no effort is made to remove cell wall materials or undigested cells at the end of the protoplasting, a feature common to other procedures. Protoplasts made by other methods are either filtered or made under conditions where they become buoyant and are floated to remove them from the undigested cells and cell walls [1, 3, 4]. We have determined that this is not necessary and in fact reduces relative transformation frequencies. In addition, it has been noted that for many strains the frequency of transformation increases if the protoplasts prepared in this manner are allowed to incubate overnight on ice prior to transformation. This cannot be generalized to all strains of *A. nidulans*, as it has also been observed that this pre-incubation does not lead to an increased transformation rate and in fact with some strains has resulted in lower transformation rates. Therefore, it must be empirically determined whether a particular strain will have a decrease, an increase or no change in its transformation efficiency following an overnight incubation on ice.

Another feature of this transformation procedure is that large quantities of protoplasts can be prepared by simply scaling up the amounts of starting material and the extra protoplasts can be stored frozen at −70 to −80°C for several months using a modification of a method used to freeze *N. crassa* protoplasts for transformation [7]. *A. nidulans* protoplasts, 1 ml prepared as described above, are mixed with 0.5 ml of PEG solution and 15 μl of DMSO, thoroughly mixed and stored frozen in 150 μl portions. Aliquots are removed from the freezer, thawed on ice, DNA is added and the tube incubated for 15 min on ice. Then 1 ml of PEG solution is added and the protoplasts are incubated for 15 min at room temperature. The transformed protoplasts are then plated as described previously. Frozen protoplasts have been used to test the ability of DNA subfragments from a larger clone to rescue various *A. nidulans* mutants in cotransformation experiments, greatly facilitating subcloning experiments. As with the pre-incubation of protoplasts overnight to obtain an increased transformation efficiency, not all strains appear to tolerate being frozen, thus the strains being utilized must be tested for this.

1.3.2 *Factors affecting transformation*

Some factors that have been observed to alter transformation efficiencies of *A. nidulans* using our procedures have been mentioned. Other aspects that can dramatically affect transformation of *A. nidulans* have also been noted, for example, similar to transformation in *N. crassa*, the particular batch of Novozyme 234 used can alter the overall transformation rate [57]. This can be overcome in part by first fractionating the Novozyme 234 with ammonium sulfate [40] and then testing each batch for the quantity necessary for protoplasing. Another modification of this protocol, relative to many, is that a lower concentration of osmotic stabilizer is utilized in the medium on which the transformants are plated; 0.2 M sucrose in the plates and 1 M sucrose in the top agar. This lower concentration of sucrose in the plates has been found to give increased numbers of transformants and permits a more rapid growth of the colonies than if 1 M sucrose is utilized in both the plates and the top agar. This is presumably because the high sucrose concentration is inhibitory to the regeneration and growth of the protoplasts. Finally, strains carrying mutations in the *chaA* gene have been found to transform less efficiently for reasons which have not been investigated in any detail and are not yet understood.

1.3.3 *Cotransformation*

Cotransformation of *A. nidulans* has been shown to occur at reasonably high rates, with up to 95% cotransformation reported in one study [60]. It was also shown in this study that cotransformation could be used to facilitate direct gene replacement, in this case replacement of *trpC* with a *trpC–lacZ* fusion. The high frequency of cotransformation reported suggests an alternative to cloning genes by complementation in *A. nidulans*. Cotransformation can be used to identify cosmid pools that complement a mutation in a gene of interest, using a modification of the procedures described for *N. crassa* [7]. The construction of two cosmid libraries, the assignment of these cosmids to linkage groups and the fact that they are readily available should facilitate the cloning of genes by complementation in *A. nidulans* [61]. Using cotransformation of a nutritional marker with cosmid pools specific to a particular linkage group from these two libraries, it has been possible to identify in a matter of days a single cosmid that complements a particular mutation. This approach has been used successfully by three groups to clone three genes from *A. nidulans*. The demonstration that this approach works should greatly facilitate the cloning of other genes from *A. nidulans*. To localize the complementing activity and to help define the limits of the gene, restriction fragments with complementing activity are then identified from the larger DNA clones using cotransformation [40, 62].

1.4 Fate of transforming DNA in *Aspergillus nidulans*

The introduction of DNA into cells does not necessarily imply that it will be generally useful as a research tool. If, for example, the fate of transforming DNA can be predicted, it can be a useful tool in studying questions of biological interest. This is the case for *S. cerevisiae* where nearly all transforming DNA on non-replicating plasmids integrates into the genome by homologous recombination. It is this simple fact that has made *S. cerevisiae* among the most widely used and powerful of eucaryotic experimental systems.

This is to be contrasted with *N. crassa* where the frequency of homologous recombination is low [63]. Homologous recombination does occur in *A. nidulans* at a frequency that is intermediate between these two systems [4, 53]. The types of recombination events that take place in *A. nidulans* and how they can be utilized as tools to examine questions of biological interest are described here.

In the initial reports of the transformation of *A. nidulans*, the transforming DNA was shown to integrate into the chromosomal DNA [1, 3, 4]. In those studies that used selective markers from *A. nidulans* it was demonstrated that in some cases the transforming DNA had integrated into the homologous chromosomal locus [3, 4]. Thus, these early experiments clearly demonstrated that integration had occurred in some of the transformants by homologous recombination. This then made it possible to investigate the utility of this integrative homologous recombination to experimentally manipulate the genome of *A. nidulans* in a manner similar to that used to manipulate the genome of *S. cerevisiae* [64–66].

1.4.1 *Homologous recombination in* Aspergillus nidulans

Homologous recombination has been utilized extensively in *S. cerevisiae* to manipulate the genome of this organism [64–67]. Homologous recombination in *A. nidulans* has also been utilized to alter its genome in well defined ways. In order to fully appreciate how this can be done the manner in which homologous recombination with both circular and linear DNA molecules takes place between transforming DNA and the homologous locus on the chromosome must be understood. The integration of a circular molecule into the chromosome by homologous recombination leads to a duplication of the target sequences with the duplicated sequences being separated by the vector sequences. This is illustrated in Figure 1.1 in which a sequence, A′ to Z′,

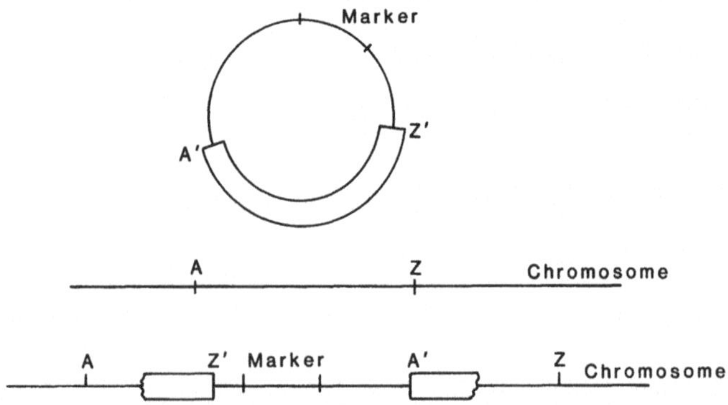

Figure 1.1 Integration of a circular DNA molecule into a chromosome by homologous recombination. The integration results in a tandem duplication of the target sequences A to Z′ and A′ to Z separated by the vector sequences containing the selective marker.

integrates at the homologous chromosomal sequences, A to Z. The integration leads to a tandem duplication of the homologous sequences separated by the vector sequences that also carry the selective marker. This type of integration event is potentially unstable because of the tandem duplication of sequences that exist in the genome. Reversal of this integration would result in the loss of the integrated vector sequences, along with the selective marker, and regenerate the original chromosome structure at this locus. Experiments that use variations on this type of integration and plasmid loss to manipulate the genome of *A. nidulans* are described in detail below.

Integration of a linear DNA molecule into the chromosome by homologous recombination involves two recombination events rather than the single event required for integration of a circular DNA molecule. The integration of a linear DNA molecule is depicted in Figure 1.2. A linear DNA molecule is interrupted by the insertion of a selective marker. Homologous recombination at both ends of the sequence, the intervals A′ to S′ and T′ to Z′, and the equivalent chromosomal sequences results in the transplacement of the chromosomal sequences with those of the linear DNA molecule. This type of integration does not result in the tandem duplication of target sequences that is observed with the integration of a circular DNA molecule, and the absence of the tandem duplication ensures the stability of the chromosomal structure generated by the integration of a linear DNA molecule. The structure of linear molecules lends itself well to the generation of null mutants either by the replacement of a gene coding region or by the insertional inactivation of a gene, as depicted in Figure 1.3. In the case of the replacement all, or nearly all, of the transcriptional unit is replaced by the nutritional marker. In contrast, insertional inactivation simply interrupts the transcriptional unit, presumably leading to inactivation of the gene.

1.4.2 *Direct and indirect gene replacement*

In *S. cerevisiae* homologous recombination between transforming DNA and the chromosome has been used to obtain direct and indirect gene replacements [65, 66]. Following the demonstration that transformation in *A. nidulans* involved homologous recombination it was soon shown that both direct and indirect gene replacements could be obtained in this system but at a lower efficiency than is obtained in *S. cerevisiae* [60, 68]. In these studies several different experiments were performed to test the feasibility of direct and indirect gene replacements in *A. nidulans*. The first experiment tested whether one *trpC* allele marked by a restriction polymorphism could replace a null chromosomal allele by direct gene replacement. Sixteen transformant strains were analysed and only one was found to have the restriction pattern, determined by genomic blots, for the expected gene replacement event. Of the remaining fifteen transformants analysed, five appeared to have repaired the null mutant by gene conversion, five were determined to have a tandem duplication of the two alleles and five had complex patterns that were not readily interpretable. The next experiment examined whether insertional inactivation of the *argB* gene by the *trpC* gene could transform as a linear molecule and lead to a strain disrupted for the *argB* gene, similar to that illustrated in Figures 1.2 and 1.3. The insertional inactivations tested had the *trpC* gene in the two possible orientations within the *argB* gene. Approximately 30% of the transformants with these constructs were found to be poorly conidiating and to be arginine auxotrophs, expected characteristics for *argB* mutants. Genomic blots of eight transformants, four for each of the constructs, showed that 6 of them had the expected pattern for the disruption of *argB* and two had complex patterns that could not be easily interpreted. Interestingly, the two transformants that exhibited the complex pattern of integration were

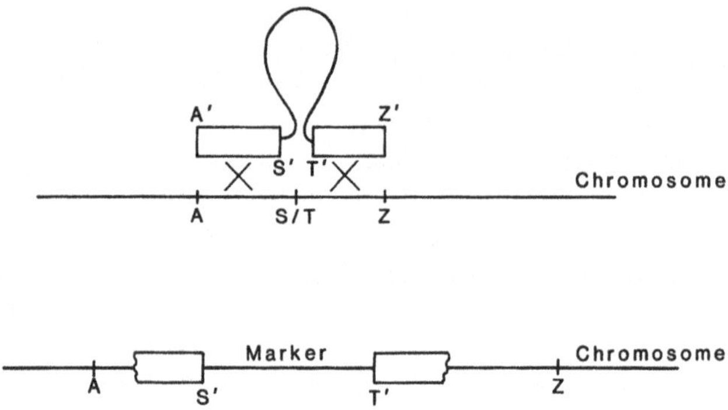

Figure 1.2 Integration of a linear DNA molecule into a chromosome by homologous recombination. The integration results in the separation of the contiguous chromosomal sequences by vector or non-contiguous chromosomal sequences including a selective marker in this illustration.

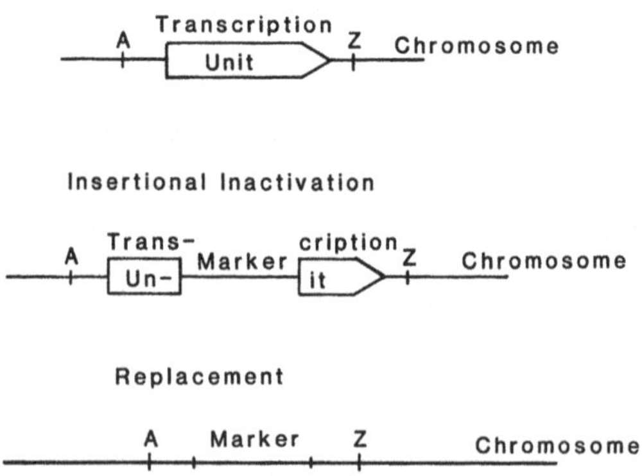

Figure 1.3 Structure of a chromosomal locus containing a transcription unit, top line, and the same chromosome after integration of a linear molecule with a selective marker disrupting the transcription unit, middle line, or a selective marker replacing the transcriptional unit, bottom line.

from one of the constructs. This could suggest an orientation-dependent influence on the frequency of the homologous integration event but this was not directly examined. From these two experiments it can be concluded that direct gene replacements occur in *A. nidulans* and insertional inactivations can be used to disrupt genes using linear DNA molecules as substrates to direct homologous recombination. Although these experiments have demonstrated that direct gene replacement and gene disruption by linear DNA molecules does occur in *A. nidulans*, they have not been widely applied in the study of various biological processes. In addition, the frequency with which these processes take place in *A. nidulans* is relatively low. Based upon this limited study it can be concluded that direct replacement will take place in 6% of the transformants and if the gene conversion events are included it occurs in 36% of the transformants. This latter number is consistent with the frequency with which a linear disruption of the *argB* gene could be obtained, i.e. 30% of the transformants. Although these frequencies are low relative to what has been observed in *S. cerevisiae*, they are useful in terms of providing a tool with which to manipulate the genome of *A. nidulans*. In other experiments it has been shown that cotransformation can also be used to obtain gene replacement strains, though at very low frequencies [60]. In these experiments it was also determined that only linear DNA molecules produced the gene replacement strains. It is unlikely that this method will be widely used as the frequency with which replacement takes place is very low.

Direct gene replacement has also been used in manipulating the genome of at least one industrially important fungus, *A. awamori* [69]. The aspergillopepsin gene was cloned and used to construct a linear gene disruption for the gene of this protease. This was done by replacing the aspergillopepsin structural gene with an *argB* gene and transforming the linear DNA fragment into the appropriate host. They determined that 16 to 40% of the prototrophic transformants were deficient for production of aspergillopepsin. Genomic blot hybridization analysis was used to confirm that the loss of protease activity resulted from gene replacement. This study confirms that for those industrially important fungi where homologous recombination has been demonstrated it should be possible to develop production strains that have desirable characteristics, e.g. in this case a reduction in the secretion of an extracellular protease.

In contrast to linear disruptions or direct gene replacements, indirect or two-step gene replacements have been used more extensively in *A. nidulans* [32, 40, 68]. Indirect gene replacements are obtained using circular DNA molecules and exploit the fact that homologous integration of these molecules produces tandem duplications, as shown in Figure 1.1. The basis for a two-step gene replacement is that integration by homologous recombination can be reversed and result in loss of the integrated plasmid. Plasmid loss takes place at a moderate frequency in *A. nidulans* when it is permitted to self-fertilize under non-selective conditions. The plasmid loss can result in either the original

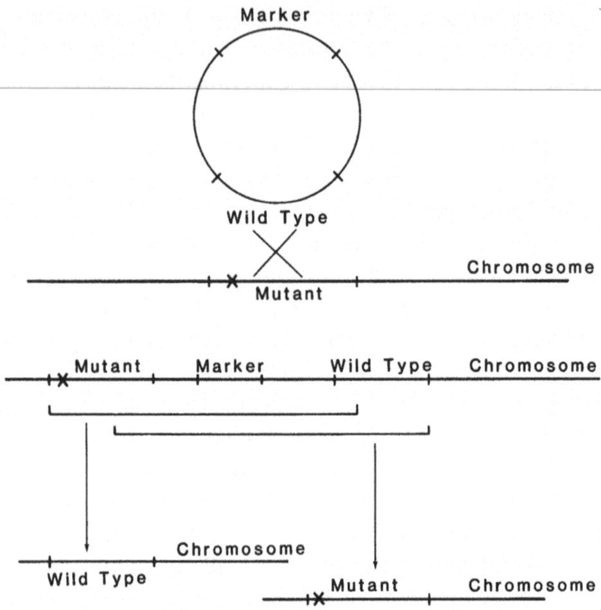

Figure 1.4 Indirect or two-step gene replacement. Integration of a circular plasmid containing a wild-type copy of a gene at the mutant chromosomal locus results in a tandem duplication of the target sequences separated by the vector sequences and the selective marker. Loss of the selective marker and plasmid sequences by homologous recombination between the tandemly duplicated sequences regenerates the original mutant chromosome or results in the wild-type information remaining in the chromosome.

chromosomal DNA sequences being retained or those that were carried in on the plasmid during transformation remaining. Figure 1.4 illustrates the transformation of a mutant strain by a wild-type gene producing a tandem duplication following integration of the plasmid by homologous recombination and subsequent loss by intrachromosomal recombination. Whether the mutant or wild-type information is retained within the chromosome depends on the interval where recombination takes place during plasmid loss (Figure 1.4). Two-step gene replacements can also be achieved without passing through the sexual cycle. This is done by utilizing one of the selective markers for which there are both forward and reverse selections, such as *pyrG*. Rare plasmid eviction events can be selected by vegetative growth of transformant strains on media containing the growth inhibitor, e.g. 5-fluoroorotic acid in the case of *pyrG*. Sectors of robust wild-type growth and conidiation are observed from point inoculations on non-selective media containing the inhibitor. These are subsequently tested for their phenotype and genotype to determine the type of plasmid loss event that has taken place [32]. The advantages of this approach are that it is not necessary to wait for the completion of the sexual cycle, it can be performed with many transformant strains simultaneously and fewer colonies need to be analysed in order to obtain the desired genotype. There is also the added advantage that new mutations can be rapidly introduced into any gene-facilitating structure and function studies of that gene.

1.4.3 *Gene disruption*

The way in which linear DNA molecules can be used to disrupt a gene and generate null or loss of function mutations has been described. Circular DNA molecules can also be used to generate gene disruptions [53, 70–74]. As illustrated in Figure 1.5, an internal DNA fragment of a gene, the segment D to T in this example, is placed on a circular plasmid with a selective marker. Upon integration by homologous recombination, this plasmid will produce a tandem duplication of the target sequences resulting in truncation of one gene at the 5′ end, or amino terminal portion of the polypeptide, producing the segment D to Z in the illustration, and truncation of the other at the carboxyl terminal portion of the polypeptide, or 3′ end, resulting in the segment A to T. The copy of the gene that has the normal promoter presumably produces a defective polypeptide that either lacks function or is degraded. It should be pointed out that it is only an assumption that the polypeptide lacks function as it is possible that a truncated polypeptide may retain some function and does not represent a true null mutation. It is also possible that, by interfering with the normal cellular process, a truncated polypeptide may be poisonous to the cell. Therefore caution must be used in interpreting results from this type of gene disruption and other lines of evidence must be obtained to show that the presumed disruption leads to a loss of function. These same arguments apply

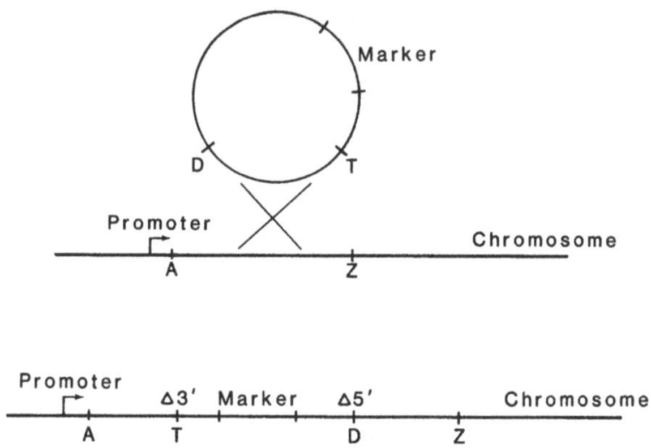

Figure 1.5 Gene disruption by a circular DNA molecule. An internal segment of a gene, D to T, on a circular DNA molecule containing a selective marker integrates at the homologous site in the chromosome. This integration produces two truncated genes, a 3′ truncated gene producing segment A to T and indicated by Δ3′, and a 5′ truncated gene producing segment D to Z and indicated by Δ5′.

to gene disruptions using insertional inactivation as it is also possible that the gene with the insertion may still be producing a truncated, but functional, polypeptide. Certainly a loss of function for a nutritional marker such as the example of the disruption of the *argB* gene described earlier [68] or antibiotic production [74] can readily be tested on the appropriate media. This is not so for a gene that functions as part of a cellular structure or participates in a function for which the null mutations phenotype cannot be easily rescued by supplementing the growth media.

Disruption of genes involved in complex structures or processes requires a different approach because loss-of-function mutations in this class of genes may result in cell death. In *S. cerevisiae* and *Schizosaccharomyces pombe* diploid strains are transformed with a gene-disrupting plasmid and transformants heterozygous for the disruption are identified and sporulated. The resulting tetrad is dissected, the products are germinated and the genotype of the surviving products is determined. If the gene under investigation is essential then two of the products will not be viable and the two viable products will be auxotrophic for the marker used to generate the disruption. In *A. nidulans* this is not possible because diploids are not fertile and mature ascospores are not found in tetrads. In contrast, a different approach has been taken to analyse whether a gene is essential in *A. nidulans*. This different approach takes advantage of the fact that *A. nidulans* can grow as a heterokaryon, and has been termed heterokaryon disruption [70–73, 75]. In a heterokaryon disruption multinucleate protoplasts are transformed with the disrupting plasmid and form heterokaryon colonies on the selective media. These heterokaryons are able to grow because the non-transformed nuclei lacking the nutritional marker of the transforming DNA provide the essential function of the gene under investigation and the transformed nuclei provide the nutritional marker selected for in the transformation (Figure 1.6). The appropriate heterokaryons are identified by testing the growth of conidiospores on selective media, where no growth would be expected, and non-selective media, where growth should occur because of the presence of conidia that are auxotrophic for the transformed marker but have the essential gene function being studied. Once the heterokaryon disrupted colonies are identified they can be propagated by transfer of part of the mycelial growth, part of which is also used to prepare genomic DNA to demonstrate the correct genotype by restriction endonuclease digestion and gel blot hybridization analysis. Conidia from these heterokaryons can then be germinated in liquid media under selective and non-selective conditions to determine the terminal phenotype of the conidia that carry the gene disruption.

There are several possible problems with this approach that need to be considered. First, the determination of the terminal phenotype may not be all that informative. For example, there may not be a single-terminal phenotype making any conclusion about the function of an unknown gene less certain. Second, it is also possible that the gene in question may not complement a

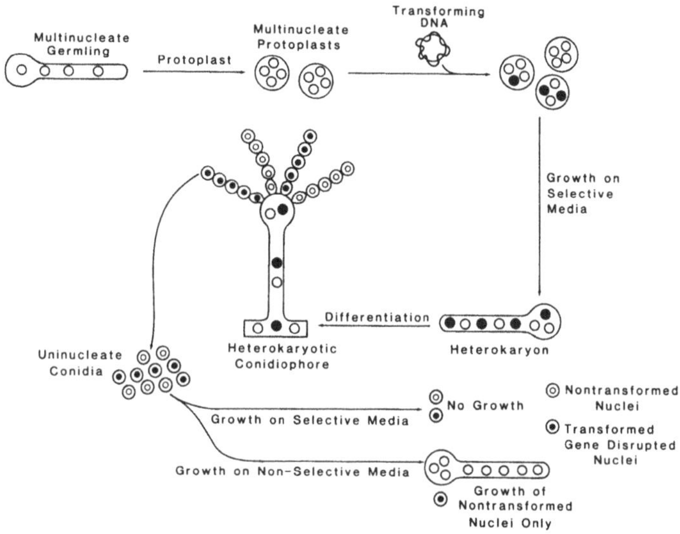

Figure 1.6 Heterokaryon disruption in *Aspergillus nidulans*. Filled nuclei contain the integrated nutritional marker and have the disrupted essential gene. The open nuclei are auxotrophic for the selective marker and contain a functional copy of the essential gene. See text for a full description.

mutation in a different nucleus, for which no example is known in *A. nidulans*. This would prevent the formation of heterokaryons and it might be concluded that a gene is therefore not essential. In cases where the heterokaryon disruption phenotype is not obtained, it must be demonstrated by analysis of genomic DNA from many transformants that it is possible to detect the predicted integration for the disruption. An additional proof would be to demonstrate that a conditionally lethal mutation within the gene in question could be disrupted with a concomitant loss of the conditional phenotype. Finally, it is possible that the gene product does not function or cross-feed during conidial development and the resulting conidiospores for gene disruption do not make it through to the point where they can be examined. In this case the heterokaryon disruption phenotype would still be identifiable and the genomic analysis could take place but there would be no further study of the null mutant phenotype.

1.5 Expression vectors

The development of expression vectors for *A. nidulans* and other filamentous fungi has been relatively slow with the greatest effort and progress being made in industrially important fungi. The emphasis in this area is due to the fact that expression vectors can be used for the production of heterologous proteins [76–84 and see chapters 5 and 6]. There are three expression systems in general use at this time based on the α-amylase promoter of *A. oryzae*, the glucoamylase promoters of *A. niger* or *A. awamori*, and alcohol dehydrogenase promoters of *A. nidulans*. These various promoters have been combined with terminator sequences from the same or different genes and introduced into all of these species as well as others. In addition to these expression systems for the different species of *Aspergillus*, the cellobiohydrolase I, *cbh1*, gene of *T. reesei*, has also been used to express a heterologous secreted protein, calf chymosin, in this system [81].

In contrast to the variety of expression systems used in the industrial fungi only the *alcA* system is widely used in the study of biological processes in *A. nidulans*. The *alcA* alcohol dehydrogenase gene is an ideal expression system with which to examine biological processes in that it is a conditional expression system. The *alcA* promoter is subject to carbon catabolite repression, is inducible by alcohols and is de-repressed on some carbon sources [58, 85–87]. These features allow the manipulation of the level of expression from this promoter by manipulating the carbon source used for growth. The level of expression can be manipulated from essentially no expression on glucose-containing media, to low level expression on non-repressing media such as glycerol, to very high levels of expression on media containing threonine or ethanol. Thus by simply manipulating the carbon source it is possible to obtain variable levels of expression from this promoter.

Several laboratories have used the *alcA* promoter to investigate the consequence of expressing genes out of context [58, 88–90]. Expression of *brlA*, a gene involved in conidiophore development, from the *alcA* promoter induced cultures to initiate the conidiophore development pathway when submerged in liquid medium, a condition that would normally block this developmental pathway [89]. Expression of the *brlA* gene out of context produced cellular structures characteristic of those found during conidiophore development, including the development of conidiophores and spores in these cultures. In similar set of experiments it was also shown that expression of *abaA*, another gene required for conidiophore development, from the *alcA* promoter activated the expression of both the *brlA* and *wetA* genes [90]. In addition, expression of *abaA* from the *alcA* promoter produced cells in submerged culture that exhibited significantly altered morphology, including vacuolization and abnormal septation. These two examples used the *alcA* promoter to express out of context genes that participate in a common developmental pathway to investigate the consequences of this altered expression. These studies led to a model for the interaction between some of the genes in this developmental program and should allow the direct testing of the model that has been developed.

The *alcA* promoter has also been used to study the role of the *nimA* protein kinase in initiating mitotic events in the cell cycle [88]. In this study it was shown that overexpression of the *nimA* protein kinase would induce morphological changes in cells that were characteristic of mitosis. These changes included chromatin condensation and reorganization of cytoplasmic microtubules into mitotic-like microtubule arrays. Thus it was shown that the *nimA* protein kinase could initiate mitotic events in cells that would normally not exhibit such mitotic events. This indicated that the *nimA* protein kinase was responsible for regulating some of the changes in cellular organization associated with a normal mitosis. This study, as well as those previously discussed, demonstrates the utility of a conditional expression system, such as that developed using the *alcA* promoter, to investigate regulation of development or other cellular processes.

The *alcA* promoter can also be used to develop conditional null mutant strains. As illustrated in Figure 1.7, the *alcA* promoter can be fused to part of the coding region of a gene such that upon integration it will disrupt the chromosomal copy of the gene. In this example the gene sequences A to F are fused to the *alcA* promoter and integration of the plasmid at the homologous chromosomal locus generates a disruption of the gene sequences. If the gene in question is essential then the disruption will produce a strain that requires non-repressing or inducing media for viability. Conditional null mutants of this type should be very useful in the isolation of bypass suppressor mutants by simply selecting for growth of suppressors on repressing media. This approach has not yet appeared in the literature, although one study in which it has proved useful is known, but it should provide a useful tool for future studies. This

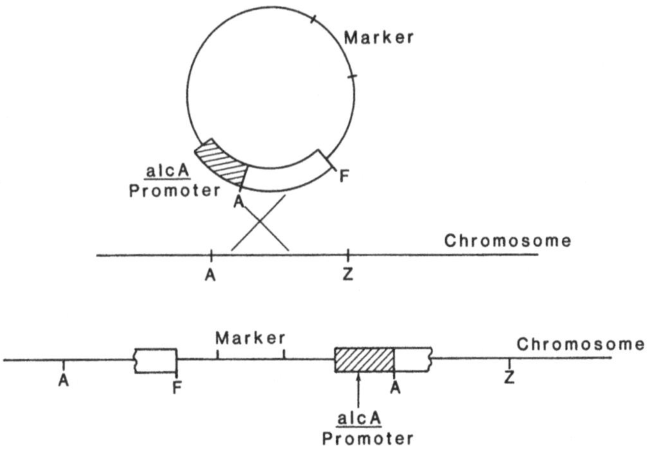

Figure 1.7 Integration of a circular molecule containing the *alcA* promoter fused to an amino terminal region of a gene that is truncated at the 3′ end, segment A to F, at the homologous chromosomal locus leading to a gene disruption and producing an *alcA* driven copy of a gene in the chromosome, segment A to Z. Integrations of this type into an essential gene produce strains dependent for cellular viability by growth on either non-repressing or inducing media for the *alcA* promoter. See text for a full description.

approach was used to generate a conditional null in the *tubC* gene but since *tubC* is not essential for growth this aspect of the conditional null could not be investigated [58].

1.6 Karyotyping

Karyotypes for organisms have traditionally been based on the microscopic examination of mitotic or meiotic chromosome preparations. Karyotypes for fungi have not usually been developed using this approach because of the small size of their genomes. Classical genetics has been more commonly used but this approach requires a collection of genetic markers that can be followed using either Mendelian genetics or by the parasexual cycle. These approaches also have their limitations. Mendelian genetic analysis requires that the sexual cycle for the organism be known. Parasexual analysis requires the organism to be haploid and competent to form diploids. Thus, these traditional methods for developing fungal karyotypes are limited in their utility, and for fungi that are obligate pathogens neither of the methods is useful. In contrast, modern electrophoretic approaches do not have the same limitations and thus provide an alternative in the development of karyotypes for fungi, though this approach has its own limitations which are discussed below.

The development of pulsed-field gel electrophoresis to separate megabase-sized DNA molecules has led to the rapid development of electrophoretic karyotypes for numerous fungal species [91–93]. Improvements in the original methods, which have permitted the separation of DNA molecules of up to 10 megabases in size, are primarily responsible for the development of electrophoretic karyotypes for the additional fungal species [94–97]. The size of DNA molecules that can be separated by pulsed-field gel electrophoresis is the main limitation of this approach. Another severe limitation that exists for this method is resolving DNA molecules that are of similar size. Resolution of doublets is aided in some cases by the existence of translocation strains [96, 98].

The first electrophoretic karyotypes for fungi were developed for the yeast *S. cerevisiae* [91–93]. These early results and improvements in the electrophoretic methods soon led to karyotypes for additional fungal species, including *S. pombe*, *Candida albicans*, *Ustilago maydis*, *U. hordei*, *Cryptococcus neoformans*, *Cephalosporium acremonium*, *A. nidulans*, *N. crassa* and *A. niger* [96, 98–107 and see chapter 6]. The ease with which karyotype data can be obtained using pulsed-field gel electrophoresis will lead to the analysis of additional fungal species. Using cloned genes from other fungal species as hybridization probes has provided a rapid means of assigning cloned genes to a chromosome, even in the absence of a mutation for classical genetic analysis.

Finally it is not necessary to have a formal genetic system in order to develop a karyotype using electrophoretic methods and the application of cloned genes can begin to provide the basis for a molecular linkage map that can be developed and used in the absence of any formal genetic system.

Pulsed-field gel electrophoresis methods are also used as a means to classify field isolates of some phytopathogenic fungi because of the chromosome polymorphisms that exist in some species, such as *U. hordei.* Similarly, chromosomal polymorphisms are potentially useful in the molecular diagnostics of human fungal pathogens. For example, the four different serotypes, A B, C and D, of the human pathogen *C. neoformans* can be separated into three groups based on chromosomal polymorphisms [104]. In addition, the serotype D group exhibited a high degree of chromosomal polymorphism, even among individual isolates of this serotype. If the chromosomal polymorphisms that are associated with pathogenicity can be identified, this could be used as a diagnostic tool in the classification of field isolates for phytopathogenic fungi and clinical isolates of human pathogens.

The development of electrophoretic karyotypes for fungi has other applications. Studies of genome organization will be facilitated by the existence of these karyotypes. One particular application is the development of chromosomal restriction maps for restriction enzymes that have rare recognition sequences. This has already been accomplished for the yeast *S. pombe* for which a *NotI* restriction map has been constructed for the entire genome [108]. For industrial fungal species this information could be used to develop rational approaches to strain improvement using integrative recombination [105]. These types of applications are likely to be extended to other fungal species in the future. The development of large scale restriction maps for species lacking any formal genetics provides an alternative to classical genetics. In addition there is no requirement for a sexual phase, thus a physical map can be developed for any species.

The use of pulsed-field gel electrophoresis is relatively recent but it has already seen wide application in the development of electrophoretic karyotypes for many fungal species, as discussed already. It is likely that physical maps will be developed for many species in the future as the technology improves and the methods for their development are simplified and made more rapid. Cloned genes as hybridization probes could be used to develop such linkage maps first as described above for individual chromosomes and ultimately for large individual restriction fragments. In this way it should be possible to obtain a molecular linkage map for any fungal species of interest. However, it is unlikely that such maps will be developed for most species as a complete molecular map is not essential in the study of most organisms at this time. Instead, efforts are being made in the development of restriction fragment length polymorphism maps for some fungal species and this will be discussed in the next section.

1.7 Restriction fragment length polymorphism mapping

An alternative to a complete molecular map for species that have a sexual phase in their life cycle is a map based upon restriction fragment length polymorphisms (RFLP). The requirements for the development of an RFLP map are two individuals with different genetic backgrounds, and the progeny from a cross of these two individuals. Thus it is only possible to develop an RFLP map for organisms for which the sexual life cycle is known. The requirement that an organism should have a sexual phase from which one can obtain the products for analysis only holds for the construction of detailed RFLP maps, RFLP differences have other applications in the analysis of fungi such as the classification of sibling species of a particular group or the analysis of field isolates of phytopathogenic fungi. These applications, in addition to those of the more classical development of RFLP maps for specific fungal species, will be discussed.

RFLP genetic maps have been developed for several fungal species including *N. crassa*, *P. chrysosporium* and *Bremia lactucae* [109, 113, 114]. The development of the *N. crassa* RFLP map began as part of an investigation of the 5S ribosomal RNA (rRNA) genes of this fungus. Progeny from a cross of a standard wild-type laboratory strain, Oak Ridge, and an exotic strain, Mauriceville, were tested for RFLPs for several 5S rRNA genes and linkage to standard genetic markers. This preliminary study formed the foundation for what has become a generally accepted method for mapping cloned genes to locus in *N. crassa* [57, 109–111]. Using this approach in *N. crassa* it is now possible to obtain genetic linkage for a cloned DNA fragment for genes that do not have mutations. Cloned DNA fragments are used to identify RFLPs between the two strains and the segregation of the RFLP is subsequently followed in the progeny of the cross. Linkage between the polymorphism and known genetic markers or other RFLPs are then determined to drive a map.

DNA polymorphisms other than RFLPs can also be used to generate genetic maps. A relatively new method to generate polymorphism maps is based on the amplification of random DNA segments using oligonucleotide primers to generate DNA segments using the polymerase chain reaction (PCR) [112]. This new method has been given the name RAPD, for random amplified polymorphic DNA. A set of random 10 base pair oligonucleotides is used individually to prime PCR reactions. The DNA segments amplified from the two parents have been shown to segregate in a simple Mendelian manner and, like RFLPs, can be used to construct a genetic map. The advantage of this method is that no previous knowledge of the genome is required, segments of DNA need not already be cloned and the set of random primers can be used with any organism. In addition RAPD markers behave as dominant markers in that they are usually present in one parent but not the other. It has been reported, for example, that of 88 RAPD markers used to construct a map for

N. crassa, 84 were dominant and the remaining 4 were co-dominant. The application of this technology to fungal species for which it is possible to derive the products of a sexual cross should allow for the rapid development of genetic maps. It should also be possible to use RAPD markers in the generation of mitotic maps for those species in which it is possible to construct diploids, such as *A. niger* [115]. Finally, this approach could also be used in the development of physical maps when combined with the formation of large-scale restriction maps using rare cutting restriction enzymes.

Another approach to produce restriction polymorphism maps has been based on the distribution of intermediate repetitive sequence elements within the genome. This application of repeated DNA sequences has been most extensively studied in the rice pathogen *Magnaporthe grisea*. These repeat sequences, named MGR for *M. grisea* repeat, are highly polymorphic and were shown to be distributed on all the chromosomes, based on genetic crosses and hybridization of chromosome-sized DNA separated by pulsed-field gel electrophoresis [116]. The polymorphic distribution of these sequences between strains was used to demonstrate that these sequences segregated as Mendelian markers in crosses between strains. It was also shown that polyadenylated messenger RNA contained sequences homologous to the MGR sequence illustrating that the repeat contains expressed sequences. These sequences were also determined to be rare in non-pathogens of rice. The polymorphic distribution of the MGR sequence and its rare occurrence in non-pathogens of rice was used to map the SMO gene in a strain of *M. grisea* that infects weeping lovegrass [117]. This study demonstrated the utility of using the MGR sequences as genetic markers in mapping experiments. It should also be possible to walk from the MGR sequences that are linked to the SMO locus as it was also demonstrated that the copy number of MGR sequences could be reduced by additional crosses of segregants to strains which are non-pathogenic to rice. Finally, it was shown in this study that the MGR sequences were meiotically stable making them suitable markers for RFLP mapping.

Two additional fungal species have had RFLP maps constructed using more classical approaches. These are *P. chrysosporium* and *Bremia lactucae*, known to cause white rot and lettuce powdery mildew respectively [113, 114]. *P. chrysosporium* has assumed biotechnological significance for the degradation of lignocellulose (see chapter 4). These maps have several potential applications but the most likely application is the cloning of genes by walking from the site of RFLPs. For example, the RFLP map of *B. lactucae* showed linkage between one of the RFLP markers and an avirulence gene. This linkage should provide a starting point for chromosome walking to clone this avirulence gene. Similar approaches should be applicable to the cloning of genes of interest from other fungi as well once maps are developed and linkage is detected between an RFLP marker and a gene of interest. The use of RAPD markers in such an analysis should not only speed the development of maps but also provide an initial probe for the walk independent of having cloned DNA sequences at the start of the analysis.

RFLP studies have applications other than the development of genetic maps. One such application is the analysis of the relationship of groups of morphologically closely related species [118, 119]. The black Aspergilli species are difficult to classify using traditional morphological criteria and this has led to confusion in their classification. Using RFLP analysis for ribosomal DNA and cloned DNA probes from *A. niger*, what had previously been classified as being at least five species subdivided into seven varieties were resolved into two distinct groups. This approach provides an unambiguous means to classify black *Aspergillus* isolates into a group [118]. Similarly, cloned genes have been used to subdivide the *A. flavus* group of species into three distinct categories. In addition RFLP variation was detected in this study for membranes within a group suggesting that subgroups may exist for the larger aggregates. Additional variations were also detected for isolates based upon geographic distribution. Such analyses should ultimately provide a foundation for identifying not only the species of a particular isolate but also the geographic origin of the isolate.

Although the development of traditional RFLP genetic maps is limited to organisms that have a sexual phase and for which the products can be recovered and analysed, it is clear that RFLP and, potentially, RAPD differences can provide useful information. As discussed here they can be used to help classify members of groups that are morphologically very similar. In the case of plant pathogens they could also be used to help identify field isolates. This has been done for the plant pathogen *Leptosphaeria maculans*. Isolates of this fungus were classified according to their aggressiveness when grown on different cultivars and were subsequently examined for RFLPs using two restriction enzymes and 42 DNA hybridization probes [120]. In this study it was noted that aggressive and non-aggressive isolates had different RFLP patterns and that they could be separated into two groups. This phylogenetic approach has potential application in diagnostic analysis of field isolates for this pathogen. Similar developments should be possible for other fungal pathogens and could provide a rapid, reliable and objective means of classification for field isolates. The application of RAPD analysis to such studies should speed any analysis of field isolates in that genomic DNA need not be isolated in large quantities, and restriction digest and transfers need not be done for the analysis to be completed. For these reasons it is probable that the application of this emerging technology will be seen in several aspects of the study of fungi in general, and particularly in the fields of biotechnology and plant pathology.

Acknowledgements

The author would like to thank Dr S. Osmani for comments on this chapter and helpful discussions. The writing of this chapter was supported in part by a grant (GM41626) to the author from the National Institutes of Health.

References

1. Ballance, D.J., Buxton, E.P. and Turner, G. (1983) *Biochem. Biophys. Res. Commun.* **112**: 284–289.
2. John, M.A. and Peberdy, J.F. (1984) *Enzyme Microbiol. Technol.* **6**: 386–389.
3. Tilburn, J., Scazzocchio, C., Taylor, G.G., Zabicky-Zissman, J.H., Lockington, R.A. and Davies, R.W. (1983) *Gene* **26**: 205–221.
4. Yelton, M.M., Hamer, J.E. and Timberlake, W.E. (1984) *Proc. Natl Acad. Sci. USA* **81**: 1470–1474.
5. Case, M.E., Schweizer, M., Kushner, S.R. and Giles, N.H. (1979) *Proc. Natl Acad. Sci. USA* **76**: 5259–5263.
6. Orbach, M.J., Porro, E.B. and Yanofsky, C. (1986) *Mol. Cell. Biol.* **6**: 2452–2461.
7. Vollmer, S.J. and Yanofsky, C. (1986) *Proc. Natl Acad. Sci. USA* **83**: 4869–4873.
8. Oakley, B.R., Rinehart, J.E., Mitchell, B.L., Oakley, C.E., Carmona, C., Gray, G.L. and May, G.S. (1987) *Gene* **61**: 385–399.
9. Upshall, A. (1986) *Curr. Genet.* **10**: 593–599.
10. Clutterbuck, A.J. (1984) *Aspergillus nidulans.* In: *Handbook of Genetics* Vol. 1, (King, R.C. ed.), Plenum Publishing Corp., New York, pp. 447–510.
11. Hynes, M.J., Corrick, C.M. and King, J.A. (1983) *Mol. Cell. Biol.* **3**: 1430–1439.
12. Berse, B., Dmochowska, T.R., Skrzypek, M., Weglenski, P., Bates, M.A. and Weiss, R.L. (1983) *Gene* **25**: 109–117.
13. Kelly, J.M. and Hynes, M.J. (1985) *The EMBO Journal* **4**: 475–479.
14. Penttila, M., Nevalainen, H., Ratto, M., Salminen, E. and Knowles, J. (1987) *Gene* **61**: 155–164.
15. Rodriguez, R.J. and Yoder, O.C. (1987) *Gene* **54**: 73–81.
16. Turgeon, B.G., Garber, R.C. and Yoder, O.C. (1985) *Mol. Gen. Genet.* **201**: 450–453.
17. Mattern, I.E., Unkles, S., Kinghorn, J.R., Pouwels, P.H. and van den Hondel, C.A.M.J.J. (1987) *Mol. Gen. Genet.* **210**: 460–461.
18. Van Hartingsveldt, W., Mattern, I.E., van Zeijl, C.M.J., Pouwels, P.H. and van den Hondel, C.A.M.J.J. (1987) *Mol. Gen. Genet.* **206**: 71–75.
19. Boeke, J.D., LaCroute, F. and Fink, G.R. (1984) *Mol. Gen. Genet.* **197**: 345–346.
20. Buxton, F.P., Gwynne, D.I. and Davies, R.W. (1985) *Gene* **37**: 207–214.
21. Buxton, F.P., Gwynne, D.I., Garven, S., Sibley, S. and Davies, R.W. (1987) *Gene* **60**: 255–266.
22. Gomi, K., Iimura, Y. and Hara, S. (1987) *Agric. Biol. Chem.* **51**: 2549–2555.
23. Kos, A., Kuijvenhoven, J., Wernars, K., Bos, C.J., van den Broek, H.W.J., Pouwels, P.H. and ven den Hondel, C.A.M.J.J. (1985) *Gene* **39**: 231–238.
24. Unkles, S.E., Campbell, E.I., Carrez, D., Grieve, C., Contreras, R., Fiers, W., van den Hondel, C.A.M.J.J. and Kinghorn, J.R. (1989) *Gene* **78**: 157–166.
25. Malardier, L., Daboussi, M.J., Julien, J., Roussel, F., Scazzocchio, C. and Brygoo, Y. (1989) *Gene* **78**: 147–156.
26. Whitehead, M.P., Unkles, S.E., Ramsden, M., Campbell, E.I., Gurr, S.J., Spence, D., van den Hondel, C., Contreras, R. and Kinghorn, J.R. (1989) *Mol. Gen. Genet.* **216**: 408–411.
27. Campbell, E.I., Unkles, S.E. and Kinghorn, J.R. (1989) *Curr. Genet.* **16**: 53–56.
28. Whitehead, M.P., Gurr, S.J., Grieve, C., Unkles, S.E., Spence, D., Ramsden., M. and Kinghorn, J.R. (1990) *Gene* **90**: 193–198.
29. Johnstone, I.L., McCabe, P.C., Greaves, P., Gurr, S.J., Cole, G.E., Brow, M.A.D., Unkles, S.E., Clutterbuck, A.J., Kinghorn, J.R. and Innis, M.A. (1990) *Gene* **90**: 181–192.
30. Buxton, F.P., Gwynne, D.I. and Davies, R.W. (1989) *Gene* **84**: 329–334.
31. Hull, E.P., Green, P.M., Arst Jr., H.N. and Scazzocchio, C. (1989) *Mol. Microbiol.* **3**: 553–559.
32. May, G.S., Waring, R.B., Osmani, S.A., Morris, N.R. and Denison, S.H. (1989) The Coming of Age of Molecular Biology in Aspergillus nidulans. In: *Proceedings of the EMBO-Alko Workshop on Molecular Biology of Filamentous Fungi*, Helsinki. (Nevalainen, H. and Penttila, M. eds). Foundation for Biotechnical and Industrial Research 6, 11–20.
33. Oakley, C.E., Weil, C.F., Kretz, P.L. and Oakley, B.R. (1987) *Gene* **53**: 293–298.
34. Sophianopoulou, V. and Scazzocchio, C. (1989) *Mol. Microbiol.* **3**: 705–714.
35. Boylan, M.T., Mirabito, P.M., Willett, C.E., Zimmerman, C.R. and Timberlake, W.E. (1987) *Mol. Cell Biol.* **7**: 3113–3118.
36. Yelton, M.M., Timberlake, W.E. and van den Hondel, C.A.M.J.J. (1985) *Proc. Natl Acad Sci. USA* **82**: 834–838.

37. Ballance, D. and Turner, G. (1985) *Gene* **36**: 321–331.
38. Cullen, D., Wilson, L.J., Grey, G.L., Henner, D.J., Turner, G. and Ballance, D.J. (1987) *Nuc. Acids Res.* **15**: 9163–9175.
39. Johnstone, I.L., Hughes, S.G. and Clutterbuck, A.J. (1985) *The EMBO Journal* **4**: 1307–1311.
40. Osmani, S.A., May, G.S. and Morris, N.R. (1987) *J. Cell Biol.* **104**: 1495–1504.
41. Gems, D., Johnstone, I.L. and Clutterbuck, A.J. (1991) *Gene* **98**: 61–67.
42. Randall, T., Reddy, C.A. and Boominathan, K. (1991) *J. Bacteriol.* **173**: 776–782.
43. Austin, B., Hall, R.M. and Tyler, B.M. (1990) *Gene* **93**: 157–162.
44. Cullen, D., Leong, S.A., Wilson, L.J. and Henner, D.J. (1987) *Gene* **57**: 21–26.
45. Herrera-Estrella, A., Goldman, G.H. and Montagu, M.V. (1990) *Mol. Microbiol.* **4**: 839–843.
46. Kolar, M., Punt, P.J., van den Hondel, C.A.M.J.J. and Schwab, H. (1988) *Gene* **62**: 127–134.
47. Marek, E.T., Schardl, C.L. and Smith, D.A. (1989) *Curr. Genet.* **15**: 421–428.
48. Picknett, T.M. and Saunders, G. (1989) *FEMS Microbiol. Lett.* **51**: 165–168.
49. Punt, P.J., Oliver, R.P., Dingemanse, M.A., Pouwels, P.H. and van den Hondel, C.A.M.J.J. (1987) *Gene* **56**: 117–124.
50. Randall, T., Rao, T.R. and Reddy, C.A. (1989) *Biochem. Biophys. Res. Commun.* **161**: 720–725.
51. Seip, E.R., Woloshuk, C.P., Payne, G.A. and Curtis, S.E. (1990) *Appl. Environ. Microbiol.* **56**: 3686–3692.
52. Ward, M., Wilkinson, B. and Turner, G. (1986) *Mol. Gen. Genet.* **202**: 265–270.
53. May, G.S., Gambino, J., Weatherbee, J.A. and Morris, N.R. (1985) *J. Cell Biol.* **101**: 712–719.
54. Sheir-Neiss, G., Lai, M.H. and Morris, N.R. (1978) *Cell* **15**: 639–647.
55. Jung, M.K. and Oakley, B.R. (1990) *Cell Mot. and the Cytoskeleton* **17**: 87–94.
56. May, G.S., Tsang, M.L.-S., Smith, H., Fidel, S. and Morris, N.R. (1987) *Gene* **55**: 231–243.
57. Akins, R.A. and Lambowitz, A.M. (1985) *Mol. Cell. Biol.* **5**: 2272–2278.
58. Waring, R.B., May, G.S. and Morris, N.R. (1989) *Genet.* **79**: 119–130.
59. May, G.S., Waring, R.B. and Morris, N.R. (1990) *Cell Mot. and the Cytoskeleton* **16**: 214–220.
60. Wernars, K., Goosen, T., Wennekes, B.M.J., Swart, K., van den Hondel, C.A.M.J.J. and van den Broek, H.W.J. (1987) *Mol. Gen. Genet.* **209**: 71–77.
61. Brody, H., Griffith, J., Cuticchia, A.J., Arnold, J. and Timberlake, W.E. (1991) *Nuc. Acids Res.* **19**: 3105–3109.
62. Timberlake, W.E., Boylan, M.T., Cooley, M.B., Mirabito, P.M., O'Hara, E.B. and Willett, C.E. (1985) *Exp. Mycol.* **9**: 351–355.
63. Dhawale, S.S. and Marzluf, G.A. (1985) *Curr. Genet.* **10**: 205–212.
64. Orr-Weaver, T.L., Szostak, J.W. and Rothstein, R.J. (1981) *Proc. Natl Acad. Sci. USA* **78**: 6354–6358.
65. Scherer, S. and Davis, R.W. (1979) *Proc. Natl Acad. Sci. USA* **76**: 4951–4955.
66. Rothstein, R.J. (1983) *Methods Enzymol.* **101**: 202–211.
67. Struhl, K. (1983) *Nature* (*London*) **305**: 391–397.
68. Miller, B.L., Miller, K.Y. and Timberlake, W.E. (1985) *Mol. Cell. Biol.* **5**: 1714–1721.
69. Berka, R.M., Ward, M., Wilson, L.J., Hayenga, K.J., Kodama, K.H., Carlomagno, L.P. and Thompson, S.A. (1990) *Gene* **86**: 152–162.
70. Osmani, S.A., Engle, D.B., Doonan, J.H. and Morris, N.R. (1988) *Cell* **52**: 241–251.
71. May, G.S. (1989) *J. Cell Biol.* **109**: 2267–2274.
72. Oakley, B.R., Oakley, C.E., Yoon, Y. and Jung, M.K. (1990). *Cell* **61**: 1289–1301.
73. Rasmussen, C.D., Means, R.L., Lu, K.P., May, G.S. and Means, A.R. (1990) *J. Biol. Chem.* **265**: 13767–13775.
74. Smith, D.J., Burnham, M.K.R., Bull, J.H., Hodgsm, J.E., Ward, J.M., Browne, P., Brown, J., Barton, B., Earl, A.J. and Turner, G. (1990) *The EMBO Journal* **9**: 741–747.
75. Doshi, P., Bossie, C.A., Doonan, J.H., May, G.S. and Morris, N.R. (1991) *Mol. Gen. Genet.* **225**: 129–141.
76. Cullen, D., Gray, G.L., Wilson, L.J., Hayenga, K.J., Lamsa, M.H., Rey, M.W., Norton, S. and Berka, R.M. (1987) *Bio/Technology* **5**: 369–375.
77. Gwynne, D.I., Buxton, F.P., Williams, S.A., Garven, S. and Davies, R.W. (1987) *Bio/Technology* **5**: 713–719.
78. Upshall, A., Kumar, A.A., Bailey, M.C., Parker, M.D., Favreau, M.A., Lewison, K.P., Josseph, M.L., Maraganore, J.M. and McKnight, G.L. (1987) *Bio/Technology* **5**: 1301–1304.
79. Christensen, T., Woeldike, H., Boel, E., Mortensen, S.B., Hjortshoej, K., Thim, L. and Hansen, M.T. (1988) *Bio/Technology* **6**: 1419–1422.

80. Gwynne, D.I., Buxton, F.P., Williams, S.A., Sills, A.M., Johnstone, J.A., Buch, J.K., Guo, Z-M., Drake, D., Westphal, M. and Davies, R.W. (1989) *Biochem. Soc. Trans.* **17**: 338–340.
81. Harkki, A., Uusitalo, J., Bailey, M., Penttila, M. and Knowles, J.K.C. (1989) *Bio/Technology* **7**: 596–603.
82. Huge-Jensen, B., Andreasen, F., Christensen, T., Christensen, M., Thim, L. and Boel, E. (1989) *LIPIDS* **24**: 781–785.
83. Ward, M., Wilson, L.J., Kodama, K.H., Rey, M.W. and Berka, R.M. (1990) *Bio/Technology* **8**: 435–440.
84. Carrez, D., Janssens, W., Degrave, P., van den Hondel, C.A.M.J.J., Kinghorn, J.R., Fiers, W. and Contreras, R. (1990) *Gene* **94**: 147–154.
85. Pateman, J.A., Doy, C.H., Olsen, J.E., Norris, U., Creaser, E.H. and Hynes, M. (1983) *Proc. R. Soc. Lond. B.* **217**: 243–264.
86. Sealy-Lewis, H.M. and Lockington, R.A. (1984) *Curr. Genet.* **8**: 253–259.
87. Lockington, R.A., Sealy-Lewis, H.M., Scazzocchio, C. and Davies, R.W. (1985) *Gene* **33**: 137–149.
88. Osmani, S.A., Pu, R.T. and Morris, N.R. (1988) *Cell* **53**: 237–244.
89. Adams, T.H., Boylan, M.T. and Timberlake, W.E. (1988) *Cell* **54**: 353–362.
90. Mirabito, P.M., Adams, T.H. and Timberlake, W.E. (1989) *Cell* **57**: 859–868.
91. Schwartz, D.C. and Cantor, C.R. (1984) *Cell* **37**: 67–75.
92. Carle, G.F. and Olson, M.V. (1985) *Proc. Natl Acad. Sci. USA* **82**: 3756–3760.
93. Carle, G.F. and Olsen, M.V. (1984) *Nuc. Acids Res.* **12**: 5647–5664.
94. Chu, G., Vollrath, D. and Davis, R.W. (1986) *Science* **234**: 1582–1585.
95. Clark, S.M., Lai, E., Birren, B.W. and Hood, L. (1988) *Science* **241**: 1203–1205.
96. Orbach, M.J., Vollrath, D., Davis, R.W. and Yanofsky, C. (1988) *Mol. Cell. Biol.* **8**: 1469–1473.
97. Vollrath, D. and Davis, R.W. (1987) *Nuc. Acids Res.* **15**: 7865–7876.
98. Brody, H. and Carbon, J. (1989) *Proc. Natl Acad. Sci. USA* **86**: 6260–6263.
99. Smith, C., Matsumoto, T., Niwa, O., Klco, S., Fan, J-B., Yanagida, M. and Cantor, C.R. (1987) *Nuc. Acids Res.* **15**: 4481–4489.
100. Magee, B.B., Koltin, Y., Corman, J.A. and Magee, P.T. (1988) *Mol. Cell. Biol.* **8**: 4721–4726.
101. Snell, R.G. and Wilkins, R.J. (1986) *Nuc. Acids Res.* **14**: 4401–4406.
102. Kinscherf, T.C. and Leong, S.A. (1988) *Chromosoma* (Berl.) **96**: 427–433.
103. McCluskey, K. and Mills, D. (1990) *Mol. Plant-Microbe Interact.* **3**: 366–373.
104. Polacheck, I. and Lebens, G. (1989) *J. Gen. Microbiol.* **135**: 65–71.
105. Skatrud, P.L. and Queener, S.W. (1989) *Gene* **78**: 331–338.
106. Ehinger, A., Denison, S.H. and May, G.S. (1990) *Mol. Gen. Genet.* **222**: 416–424.
107. Debets, A.J.M., Holub, E.F., Swart, K., van den Broek, H.W.J. and Bos, C.J. (1990) *Mol. Gen. Genet.* **224**: 264–268.
108. Fan, J-B., Chikashige, Y., Smith, C.L., Niwa, O., Yanagida, M. and Cantor, C.R. (1988) *Nuc. Acids Res.*, **17**: 2801–2818.
109. Metzenberg, R.L., Stevens, J.N., Selker, E.U. and Morzycka-Wroblewska, E. (1985) *Proc. Natl Acad. Sci. USA* **82**: 2067–2071.
110. Metzenberg, R.L. and Groteluschen, J. (1987) *Fungal Genet. Newsl.* **34**: 39–44.
111. Tyler, B.M. and Harrison, K. (1990) *Nuc. Acids Res.* **18**: 5759–5765.
112. Williams, J.G.K., Kubelik, A.R., Livak K.J., Rafalski, J.A. and Tingey, S.V. (1990) *Nuc. Acids Res.* **18**: 6531–6535.
113. Raeder, U., Thompson, W. and Broda, P. (1989) *Mol. Microbiol.* **3**(7): 911–918.
114. Hulbert, S.H., Ilott, T.W., Legg, E.J., Lincoln, S.E., Lander, E.S. and Michelmore, R.W. (1988) *Genetics* **120**: 947–958.
115. Debets, A.J.M., Swart, K. and Bos, C.J. (1989) *Can. J. Microbiol.* **35**: 982–988.
116. Hamer, J.E., Farrall, L., Orbach, M.J., Valent, B. and Chumley, F.G. (1989) *Proc. Natl Acad. Sci. USA* **86**: 9981–9985.
117. Hamer, J.E. and Givan, S. (1990) *Mol. Gen. Genet.* **223**(3): 487–495.
118. Kusters-van Someren, M.A., Samson, R.A. and Visser, J. (1991) *Curr. Genet.* **19**: 21–26.
119. Moody, S.F. and Tyler, B.M. (1990) *Appl. Environ. Microbiol.* **56**: 2453–2461.
120. Koch, E., Song, K., Osborn, T.C. and Williams, P.H. (1991) *Mol. Plant-Microbe Interact.* **4**: 341–349.
121. Beri, R.K. and Turner, G. (1987) *Curr. Genet.* **11**: 639–641.

2 Gene organization in industrial filamentous fungi

S.E. UNKLES

2.1 Introduction

This chapter provides a brief review of the gene structure of filamentous fungi currently used in industrial processes. However, this is not an exhaustive account of every gene which has been sequenced to date and there may be some omissions. Additionally, the definition of an industrial filamentous fungi can be equivocal—today's production strain may be tomorrow's laboratory model and vice versa. For example, the useful genes from the more intractable fungi may be expressed in a more amenable host while organisms like the classic laboratory *Aspergillus nidulans* strain could be commercially useful for heterologous expression.

By far the greatest amount of DNA sequence information available for filamentous fungi has been obtained from the model ascomycetes *A. nidulans* and *Neurospora crassa*. This bias is the result of a natural progression from the wealth of information provided by classical genetic analyses of these two organisms, to a detailed study of the genes involved using molecular genetic approaches (chapter 1). Several recent reviews deal with gene organization in these fungi [1–3]. In contrast, the relative number of genes which have been isolated and sequenced from filamentous fungi used in industrial processes is small. This lag is due to a lack of basic genetic information: few clearly defined mutants, which could be used for gene isolation by complementation, have been described in industrial fungi. Therefore, isolation of many genes requires the corresponding gene of *A. nidulans* or *N. crassa* to be isolated first and used as a probe in cross-hybridizations. A further possible reason for this lag is that industrial organizations have not been interested in isolating genes which are not directly involved in their particular process.

A similar gene compilation, published in 1987 [3], listed 59 genes, which at that time represented most of those sequenced from filamentous fungi, of which only 12 were genes of industrial species. Table 2.1, which lists 54 genes which have now been cloned and sequenced from filamentous fungi currently used in industry, illustrates the increased pace at which modern molecular methods can generate information on gene structure in genetically poorly characterized strains. There are three main reasons for the cloning and sequencing of these genes. First, the genes encode proteins which are directly relevant to the industrial process carried out by the fungus, for example, the penicillin biosynthetic genes from *Penicillium chrysogenum* [35–39 and chapter 10] and the cellulase genes of *Trichoderma reesei* [45–49 and chapter 4]. Second, the genes are of technological interest, such as *pyrG* or *niaD* of *A. niger* which can be used for selection of transformants [7, 19 and chapter 1], or *pepA* of *A. niger* var. *awamori* which can influence heterologous protein production [23 and chapters 5 and 6]. Third, genes can be considered of more general interest providing information on evolutionary relationships and on gene structure. This latter point may be important for molecular manipulation of strains in order to improve production. Examples of this third group could include *trpC* [4, 21, 32, 58], *pgkA* [34, 50, 51, 55] and also *pyrG* [7, 33, 53, 59] and *niaD* [19]. A possible further reason for obtaining sequence information from industrial strains is the subsequent availability of the gene promotor regions to alter the expression of homologous genes or to allow expression of heterologous genes either to achieve transformation (chapter 1) or to obtain valuable products (chapters 5 and 6).

Several methods have been used to isolate the genes listed in Table 2.1. Predominant among these have been initial isolation and sequencing of the corresponding protein followed by generation of mixed oligonucleotide probes, and cross-hybridization using heterologous probes derived from the corresponding yeast or filamentous fungal gene. Examples of genes cloned by the former method include *pgaI*, *pgaII*, *pelD*, *pmeA* of *A. niger*, *pepA* of *A. niger* var. *awamori*, *amy* of *A. oryzae*, *aat*, *phoA* of *P. chrysogenum*, *ipnA*, *pcbEF* of *Cephalosporium acremonium*, *aspA* of *Mucor miehei* and *M. pusillus* while cross-hybridization using other fungal genes as probes has allowed cloning of *A. niger pyrG*, *oliC*, *pelD*, *niaD*, *trpC*, *A. oryzae amdS*, *P. chrysogenum ipnA*, *pgk*, *T. reesei pgk*, *ura3*, *ura5*, *T. viride cbh1*, *Phycomyces blakesleeanus leu1* and *pyrG*. Other cloning methods include differential hybridization (*T. reesei cbh1*, *cbh2*, *egl1*, *egl3*), complementation of *E. coli* (*P. chrysogenum trpC*, *P. blakesleeanus trp1*) or *A. nidulans* mutants (*A. niger argB*, *pacA*), antibody recognition of a cDNA expression library (*A. oryzae alp*, *nepII*) and hybridization to a mammalian gene probe (*Lentinus edodes*, *ras*, *pril*).

2.2 Gene structure

The availability of over 50 gene sequences, the majority of which are from Ascomycete (or Deuteromycete) species but which also include five genes of *Mucor* and *Phycomyces* representing Zygomycetes and two genes of *Lentinus*, a Basidiomycete, allows a comparison between these genes of industrially important organisms and previous studies of gene structure in filamentous fungi.

Table 2.1 Industrial filamentous fungi from which genes have been cloned and sequenced.

Species	Main industrial uses	Cloned gene(s)		Reference
A. niger[a]	Extracellular enzyme production	*trpC*	(tryptophan synthetase)	[4]
	(glucoamylase, lipase, glucose oxidase, etc.)	*glaA*	(glucoamylase)	[5]
	Heterologous protein production	*argB*	(ornithine carbamoyl transferase)	[6]
	(chymosin, phytase)	*pyrG*	(orotidine-5′-monophosphate decarboxylase)	[7]
	Citric acid production	*aldA*	(aldehyde dehydrogenase)	[8]
		oliC	(mitochondrial ATP synthase subunit 9)	[9]
		pgaII	(polygalacturonase II)	[10]
		pgaI	(polygalacturonase I)	[11]
		pga	(polygalacturonase)	[12]
		pelA	(pectin lyase A)	[13]
		pelD	(pectin lyase D)	[14]
		pmeA	(pectin methyl esterase)	[15]
		god	(glucose oxidase)	[16, 17]
		pacA	(acid phosphatase)	[18]
		niaD	(nitrate reductase)	[19]
var. *awamori*		*glaA*	(glucoamylase)	[20]
		trpC	(trytophan synthetase)	[21]
		amyA	(α-amylase)	[22]
		amyB	(α-amylase)	[22]
		pepA	(acid protease)	[23]
A. tubigensis[a]	Pectinolytic enzyme production	*pgaII*	(polygalacturonase II)	[10]
A. shirousami[a]	Shochu (Japanese brandy) fermentation	*gla*	(glucoamylase)	[24]
A. oryzae[a]	Extracellular enzyme production	*alp*	(alkaline protease)[d]	[25]
	(α-amylase, protease, etc.)	*nepII*	(neutral protease II)[d]	[26]
	Heterologous protein production	*amyA, amy3, amyI*	(α-amylase)[f]	[27–29]
	(lipase, aspartic protease)	*amyB, amyC, amy1, amy2, amyII*	(α-amylase)[f]	[27–29]
	Oriental food fermentations	*glaA*	(glucoamylase)	[30]
		amdS	(acetamidase)	[31]

P. chrysogenum[a]	Penicillin production	*trpC*	(trytophan synthetase)[e]	[32]
		pyrG	(orotidine-5′-monophosphate decarboxylase)[d]	[33]
		pgk	(phosphoglycerate kinase)[e]	[34]
		ipnA	(isopenicillin *N* synthetase)	[35, 36]
		acvA	(α-aminoadipyl-cysteinyl-valine synthetase)	[37, 38]
		aat	(amino acyl transferase)	[39]
		phoA	(acid phosphatase)	[40]
C. acremonium[a]	Cephalosporin production	*pcbC*	(isopenicillin *N* synthetase)	[41, 42]
		pcbAB	(α-aminoadipyl-cysteinyl-valine synthetase)	[43]
		pcbEF	(expandase/hydrolase)	[44]
T. reesei[a]	Cellulase production	*cbh1*	(cellobiohydrolase I)	[45]
		cbh2	(cellobiohydrolase II)	[46]
		egl1	(endoglucanase I)	[47, 48]
		egl3	(endoglucanase III)	[49]
		pgk	(phosphoglycerate kinase)	[50, 51]
		ura3/pyr4	(orotidine-5′-monophosphate decarboxylase)	[52, 53]
		ura5	(orotidylic acid pyrophosphorylase)	[52]
T. viride[a]	Cellulase production	*cbh1*	(cellobiohydrolase)	[54]
		pgk	(phosphoglycerate kinase)	[55]
M. miehei[b]	Extracellular enzymes (aspartic protease, lipase)	*aspA*	(aspartic protease)	[56]
M. pusillus[b]	Extracellular aspartic protease	*aspA*	(aspartic protease)	[57]
P. blakesleeanus[b]	Carotene production	*trp1*	(trytophan synthetase)	[58]
		pyrG	(orotidine-5′-phosphate decarboxylase)	[59]
		leu1	(α-isopropylmalate isomerase)	[60]
L. edodes[c]	Edible mushroom	*ras*	(oncogene homologue)	[61]
		pri1	(putative zinc-finger protein)	[62]

[a] Ascomycete.
[b] Zygomycete.
[c] Basidiomycete.
[d] cDNA sequence.
[e] coding region sequence only.
[f] a family of amylase genes exists in *A. oryzae*, members of which have been sequenced by several groups. See chapter 3 for clarification of synonymous sequences.

Table 2.2 Upstream regions.

Species	Gene	CAAT	TATA	tsp	CT-rich region	ATG
A. niger	*trpC*	ACAAT(−283)	TCTAAT(−36)	−5, −6 TTTTCGTC ☆☆	−68 → −59	TCGTCATGGC
	glaA	GCAAT(−172)	TATAAAT(−109)	−44, −68 CAAC, GAAG ☆ ☆	−64 → −52 −149 → −139	CAGCAATGTC
	argB	—	TATTAT(−217)	−112, −113 GTGA ☆☆	−65 → −48 −196 → −181	CACAAATGCC
	pyrG	TCAAT(−295) CCAAT(−319)	TATAA(−142)	NA	−37 → −23 −183 → −174	ACACCATGTC
	aldA	CCACT(−138) CCACT(−152)	TATTAAT(−102)	−24, −29, −30 CTAACCAAC ☆☆ ☆	−89 → −30 −184 → −151	TCATCATGTC
	oliC	—	TATTTA(−140)	NA	−33 → −24 −80 → −57 −109 → −88	TCACAATG
	pgaI	CCAAT(−219)	TATAAAA(−152)	NA	−98 → −72	TCACCATGCA
	pgaII	GCAAT(−130) TCAAT(−213)	TATAA(−117)	NA	−38 → −18	TAATCATGCA
	pga	ACAAT(−216) ACAAT(−384)	TATAAGAA(−115)	NA	−37 → −17	TGACCATGCA
	pelA	CCAAT(−224)	TATAAA(−140)	NA	−89 → −61 −40 → −25	TCACCATGAA
	pelD	—	TATATAA(−148)	NA	−78 → −51	CCAGGATGAA
	pmeA	CCAAT(−238)	TATAA(−88)	NA	−57 → −50	CAATCATGGT
	god	CCAAT(−293)	TATAA(−81)	−38 CAAC ☆	−32 → −20	CCATCATGCA
	pacA	CCAAT(−193)	TATCTA(−176)	−145 TGCT ☆	—	CGACCATGAA

	niaD	—	—	−85, −60, −43 TCCT, CTGA, CTTA ☆ ☆ ☆	—	ATACCATGGC
var. *awamori*	*glaA*	GCAAT(−172)	TATAAAT(−104)	−51 → −73 several	−64 → −52 −149 → −139	CAGCAATGTC
	trpC	ACAAT(−285)	—	NA	—	TCGTCATGGC
	amyA/amyB	TCAAT(−192)	TATAAA(−100)	−63 AGCA ☆	−40 → −28	CATTTATGAT
	pepA	NA	TATAA(−123)	−50, −52, −53 TGTCTG ☆☆ ☆	—	TCAAAATGCG
A. tubigensis	*pgaII*	ACAAT(−217)	TATAA(−117)	NA	−37 → −17	TAATCATGCA
A. shirousami	*glaA*	GCAAT(−190)	TATAAA(−118)	−38 CAAC ☆	−71 → −59 −161 → −136	CCGCAATGTC
A. oryzae	*alp*	NA	NA	NA	NA	TCATCATGCA
	nepII	NA	NA	NA	NA	CCAGAATGCG
	amdS	CCAAT(−212)	TATAAA(−93)	NA	—	TCACTATGCC
	glaA	ACAAT(−88) TCAAT(−331)	TATAAAA(−72)	several −24 → −30 TTTCACA	—	GCAAGATGGT
	amyB/amyC/amy1/ amy2/amyII/amyI/ amyI/amy3/amyA	GCAAT(−192) CCAA$^{A}_{T}$T(−376)	TATAAA(−100)	−66, −64, −61 GATAGCAAC ☆ ☆ ☆ [reference 27, 29] −69 GGAGG ☆ [reference 28]	−40 → −28	CATTTATGAT
P. chrysogenum	*trpC*	NA	NA	NA	NA	TGGCCATGGC
	ipnA(pcbC)	—	TATAAT(−198)	−131, −132, −397 TCAG TAAG ☆☆ ☆	—	ACACCATGGC

Table 2.2 (*Contd.*)

Species	Gene	CAAT	TATA	tsp	CT-rich region	ATG
	acvA/pcbAB	—	TATATA(−228) [reference 37]	NA	—	GAAAAATGGG [reference 37]
		—	TATATA(−137) TATAA(−98) [reference 38]	NA	—	CAGACATGAC [reference 38]
	aat, (*acyA*, *pcbDE*)	CCAAT(−112) TCAAT(−219)	TATAAA(−108)	NA	—	CAGAAATGCT
	phoA	CCAAT(−288)	TATAA(−73)	−21, −24, −33 AGCT, TGCA, TACC ☆ ☆ ☆	—	CAAACATGCT
C. acremonium	*pcbC* (*ipnA*)	—	TATAAA(−139)	−56, −58, GTCA, TCGT, ☆ ☆ −72, −78 TTCT, CGAG ☆ ☆	−74 → −64	TCACCATGGG
	pcbAB (*acvA*)	—	TATAA(−341)	NA	—	GGTCCGTGGC
	pcbEF	NA	NA	NA	—	TCAACATGAC
T. reesei	*cbh1*	ACAAT(−195)	TATATAA(−131)	NA	−112 → −77	GCATCATGTA
	cbh2	ACAAT(−188)	TATAAAA(−96)	NA	—	GCACCATGAT
	egl1	—	TATAAA(−97)	NA	−41 → −27	CCAAAATGGC
	egl3	NA	TATAA(−98)[a]	NA	—	GCACAATGAA
	pgk	TCAAT(−438)	TATAA(−72)	−57 TCACG ☆ −60 TCATC ☆ −68 TAACT ☆ −73 CCATA ☆ −259 CCACC ☆	−101 → −74 −131 → −119	CCAAAATGTC

				−267 CATAT ☆ −276 CAGAG ☆ −284 AACTG ☆		
	pyr4(*ura3*)	CCAAT(−184)	—	NA	−102 → −77	CAGCCATGGC
	ura5	CCAAT(−83)	—	NA	—	ACAGAATGGC
T. viride	*cbh1*	—	TATATAT(−128)	NA	−104 → −86	GCATCATGTA
	pgk	NA	NA	NA	NA	CCAAAATGTC
M. miehei	*aspA*(*mmr*)	ACAAT(−149)	TATAAAA(−50)	NA	—	CGACCATGCT
M. pusillus	*aspA*(*mpr*)	TCAAT(−113) ACAAT(−195)	TATAAA(−49)	NA	—	TCAACATGCT
P. blakesleeanus	*trp1*	ACAAT(−70)	TATATAAATA(−51)	−13, −18 CTAA, TTTA ☆ ☆	−28 → −17	TTCTAATGGC
	pyrG	CCAAAT(−109)	TATAT(−93)	−23, −60 TACT, TTCA ☆ ☆	—	TCTTTATGAT
	leu1	ACAAT(−189) TCAAT(−504)	TATAA(−30) TATAT(−91) TATATATA(−352)	−55, −139, −305 CAAC, TCAC, CATC ☆ ☆ ☆	−25 → −10 −51 → −38	TACTAATGTC
L. edodes	*ras*	NA	NA	NA	NA	AAGAAATGGC
	pri1	NA	NA	NA	NA	TTAAAATGAC

Gene synonyms are given in parentheses. Where sequence of the same gene has been obtained by more than one group using different gene designations, all designations are given.

For references, see Table 2.1.

Positions of the 5′ end of sequence elements are given in bp from the A(+1) of the translation initiation codon.

Absence of a motif is shown by —, while NA indicates the information is not available.

Major tsp are underlined, as are consensus A at position −3 and ATG.

[a] after removal of intron in 5′ non-coding sequence.

☆ indicates nucleotides at tsp.

2.2.1 *Transcription: 5′ non-coding region*

The promoter regions of eukaryotes can be described as being composed of the so-called 'core' promoter elements, notably CAAT, TATA and in fungi the CT block, which act as the basic transcriptional signals to determine the start site and efficiency of transcription. Superimposed on these core elements may be other gene or pathway specific motifs which respond to various regulatory stimuli, causing activation or repression of gene transcription. Unfortunately, although core elements may be seen in the gene promoters of industrial filamentous fungi, no functional analysis has been done in these species to assess their relative importance for transcription. Comments can therefore only relate to their presence and position and to their possible role by analogy to better characterized systems. Table 2.2 gives information on these core elements within the promoters of industrial filamentous fungal genes. Because so few transcriptional start points (tsp) have been determined, positions are given in base pairs (bp) from the A(+1) of the translational start.

CAAT motif. In higher eukaryotes the conserved sequence GC$^{C}_{T}$CAATCT is often found around 80 bp upstream from the tsp [63] but in yeast this element is rare. In the filamentous fungi, sequences similar to this consensus are common (three times more genes have a CAAT sequence than do not in Table 2.2) but if these regions are significant, they occur at far greater distances from the tsp than in higher eukaryotic promoters. Most lie 100–200 bp from the tsp but some, such as those of *A. oryzae amy* and *T. reesei pgk*, are more than 300 bp distant.

TATA motif. The conserved sequence TATAAA common in eukaryotic promoters is thought to be involved in binding of the TFIID component of the general transcription machinery [63]. Similar motifs can be found in most of the genes in Table 2.2 with the exceptions of *A. niger niaD*, *A. niger* var. *awamori trpC* and *T. reesei ura3* and *ura5*. *A. niger niaD* is, in fact, the only gene in this compilation which does not have a recognizable CAAT, TATA or CT-rich sequence. Most fungal TATA motifs are found 50–150 bp from the translation initation site and 40–100 bp from the tsp. The TATA of *T. reesei pgk*, however, is very close to some tsp and between other tsp and the translational start, suggesting that in this case at least the sequence has questionable significance.

CT-rich region. Pyrimidine-rich tracts or CT boxes have been described in the promoters of yeast and filamentous fungal genes often immediately before the major tsp. Deletion analysis of CT boxes in *A. nidulans gpdA* and *oliC* genes has shown them to be important for determining the position of transcription initiation [64]. CT motifs from 10 up to 60 bp (average 20 bp) can be found in most genes in Table 2.2 but only in *A. niger glaA*, *aldA*, *A. niger* var.

awamori glaA, *T. reesei pgk* and *P. blakesleeanus trp1* has the position of these tracts been shown to occur immediately upstream from the tsp. However, CT-rich regions can also occur between the tsp and translation start as in the amylase genes of *A. niger* and *A. oryzae*, or between the CAAT and TATA as in *glaA* of *A. niger*, *A. niger* var. *awamori* and *A. shirousami* or even further upstream as in *A. niger aldA*. The significance of these CT boxes is therefore unclear.

Transcriptional starting points (tsp). Relatively few tsp have been determined for filamentous fungal genes, but the trend is for there to be two or more initiation sites. Only *A. niger godA*, *A. niger pacA*, *A. niger var. awamori amyA*, *amyB* and *A. shirousami glaA* have single tsp. There is no clear sequence preference around the tsp.

2.2.2 *Transcript processing and translation*

As with mammalian transcripts, any base can be used as a transcript cap site for filamentous fungal genes with a preference of A > U > G > C. The 5′ non-translated mRNA may range from 5 bp (*A. niger trpC*) to 305 bp (*P. blakesleeanus leu1*) but more commonly is around 30–70 bp.

Introns. Most filamentous fungal protein-encoding genes have been found to contain introns. For example, Gurr *et al.* [3] determined the amount to be 68%, a figure which is in good agreement with previous findings. Genes which do not contain introns are shown in Table 2.3. In the case of the *ipnA* and *acvA* genes of *P. chrysogenum* and their equivalents in *C. acremonium*, there is evolutionary evidence that these genes contain no introns because they have been transferred directly from prokaryotes [65].

Table 2.4. lists those genes with introns. All the introns are positioned within the coding sequences of the genes with the exception of the first intron of *T. reesei egl3* which occurs in the 5′ non-translated region. Multiple introns are common and up to 8 are found in the amylase genes of *A. niger* var. *awamori* and *A. oryzae*. Filamentous fungal introns are shorter than mammalian and

Table 2.3 Genes without introns.

Species	Genes
A. niger	*trpC*, *argB*, *oliC*, *god*
A. niger var. *awamori*	*trpC*
P. chrysogenum	*trpC*, *ipnA*, *acvA*
C. acremonium	*pcbC*, *pcbAB*, *pcbEF*
T. reesei	*ura3*, *ura5*
M. miehei	*aspA*
M. pusillus	*aspA*
P. blakesleeanus	*trpl*, *leul*

Table 2.4 Introns.

Species	Gene	Number of introns	Length (bp)	5′ splice site	3′ splice site	Putative internal consensus (bp from intron 3′ end)
A. niger	*glaA*	4–5	75	GTATGT	TAG	TGCTGAC (27)
			55	GTATGT	CAG	AGCTGAC (14)
			61	GTGTGT	CAG	AGCTAAC (19)
			58	GTAAGT	TAG	TACTAAC (15)
			169	GTACGT	CAG	TGCTGAC (21)
	pyrG	1	68	GTAGGC	GAG	GACTGAG (7) or AACTTAT (28)
	aldA	3	69	GTATGT	CAG	AACTAAC (24)
			57	GTGAGT	TAG	TGCTAAT (19)
			53	GTAAGC	TAG	CACTAAC (17)
	pgaI	2	52	GTATGT	TAG	CGCTAAC (16)
			62	GCACGA	CAG	CGCTAAC (21)
	pgaII	1	52	GTAAGC	TAG	TATTGAT (16)
	pga	1	52	GTAAGC	TAG	TGTTGAT (16)
	pelA	4	52	GTATGT	TAG	ATCTAAC (15)
			48	GTAGGT	CAG	CTCTAAC (14)
			64	GTACGT	CAG	AGTTAAC (16)
			54	GTAAGT	CAG	GACTAAT (18)
	pelD	4	65	GTGAGT	CAG	CGGTGAC (26)
			62	GTAAGT	TAG	TGCTGAC (17)
			63	GTATGC	CAG	AACTAAC (23)
			57	GTAAGT	CAG	CACTAAT (18)

	pmeA	6	63	GTAAGC	CAG	GGCTGAT (18)
			50	GTACGT	TAG	TACTAAA (17)
			50	GTCAGT	CAG	CGCTAAC (17)
			50	GTAGGG	TAG	GACTAAT (17)
			54	GTATGT	CAG	AGCTAAC (18)
			58	GTATGC	TAG	CACTAAC (18)
	pacA	2	201	GTATGT	CAG	CACTACC (24)
			256	GTCCGC	CAG	TCCTAAT (13)
	niaD	6	36	GTACTC	TAG	TGCTCAC (17)
			56	GTAAGG	CAG	ATCTGAC (20)
			52	GTAAGT	CAG	CCCTGAC (20)
			51	GTGAGA	CAG	GTCTAAC (17)
			51	GTACGA	TAG	AGCTAAC (17)
			54	GTACGT	CAG	TGCTGAC (17)
var. *awamori*	*glaA* as for *A. niger*					
	amyA/amyB	8	55	GTGTGT	TAG	AACTGAC (16)
			85	GTAAAT	TAG	TCCTAAC (17)
			69	GTAAGT	TAG	AACTAAC (20)
			68	GTTCGT	CAG	TACTGAC (22)
			58	GTAAGA	CAG	GGCTGAT (28)
			65	GTATGG	CAG	TCCTAAC (24)
			65	GTAAGT	AAG	ACCTAAC (17)
			79	GTAAGC	CAG	GACTAAT (21)
	pepA	3	51	GTAAGC	TAG	GACTAAC (16)
			52	GTGAGT	CAG	TGCTGAC (17)
			59	GTAAGA	TAG	TACTAAC (17)

Table 2.4 (*Contd.*)

Species	Gene	Number of introns	Length (bp)	5′ splice site	3′ splice site	Putative internal consensus (bp from intron 3′ end)
A. tubigensis	*pgaII*	1	52	GTAAGC	TAG	TGTTGAT (16)
A. shirousami	*glaA*	4	74	GTATGT	TAG	GACTGGC (27)
			55	GTATGT	CAG	AGCTGAC (14)
			62	GTGTGT	CAG	AGCTAAC (20)
			58	GTAAGT	TAG	TACTAAC (15)
A. oryzae	*amdS*	6	51	GTGGGT	TAG	AACTAAT (15)
			52	GTAAGA	TAG	TTCTAAC (16)
			52	GTAGCT	CAG	GACTAAT (16)
			48	GTAGGG	TAG	TACTGAT (15)
			53	GTAAGT	AAG	GCCTGAC (18)
			48	GTAAGT	TAG	TGCTAAC (16)
	glaA	4	49	GTACGT	TAG	GACTTAC (16)
			45	GTAAGT	TAG	AGCTTAC (14)
			50	GTGAGC	CAG	AGCTAAC (15)
			56	GTACGT	CAG	TGCTAAT (15)
	amy1,2,3, amyA,B,C	8	55	GTGTGT	TAG	AACTGAC (16)
	amyI,II		85	GTAAAT	TAG	TCCTAAC (17)
			69	GTAAGT	TAG	AACTAAC (20)
			68	GTTCGT	CAG	TACTGAC (22)
			58	GTAAGA	CAG	TACTTAC (15)
			65	GTATGG	CAG	TCCTAAC (24)
			65	GTAAGT	AAG	ACCTAAC (17)
			79	GTAAGC	CAG	GACTAAT (21)

P. chrysogenum	*pyrG*	1	55	GTAAGT	CAG	CACTAAC (16)
	pgk	2	55	GTAAGC	TAG	CCCTAAC (22)
			62	GTATGT	CAG	TTTTAAC (17)
	acyA	3	64	GTAAGT	CAG	TTCTGAC (16)
			68	GTGAGT	CAG	TGCTGAC (28)
			69	GTACGT	AAG	GACTAAT (20)
	phoA	1	52	GTAGGT	TAG	CGCTCAC (17)
T. reesei	*cbh1*	2	67	GTAAGT	AAG	AGCTGAC (20)
			63	GTGAGT	CAG	AGCTGAC (20)
	cbh2	3	49	GTAATT	AAG	TACTGAG (17)
			56	GTAGGT	CAG	TACTGAA (26)
			90	GTATGA	CAG	TGCTAAC (22)
	egl1	2	70	GTGAGC	CAG	TGCTGAC (19)
			57	GTTCGT	CAG	TGCTAAC (20)
	egl3	2	121	GTGAGT	TAG	CGCTGAA (10)
			174	GTGAGT	CAG	AACTGAG (29)
	pgk	2	219	GTGAGT	TAG	CGCTAAC (30)
			75	GTAAGT	TAG	GACTAAT (26)
T. viride	*cbh1*	2	66	GTAAGT	AAG	AGCTGAC (20)
			64	GTGAGT	TAG	GACTGAC (20)
	pgk	2	209[a]	GTGAGT[a]	CAG	GGCTGAC (28)
			73	GTAAGT	TAG	GGCTAAT (23)
P. blakesleeanus	*pyrG*	2	65	GTAGAT	AAG	ATCTCAA (20)
			67	GCAAGT	TAG	AGCTGAC (15)

Table 2.4 (*Contd.*)

Species	Gene	Number of introns	Length (bp)	5′ splice site	3′ splice site	Putative internal consensus (bp from intron 3′ end)
L. edodes	*ras*	6	54	GTTCGT	TAG	TTCTGAT (19)
			56	GCAAGT	CAG	GGTTAAT (21)
			52	GTCAGC	CAG	TTCTGAG (17)
			58	GTTAGT	TAG	GCCTAAA (19)
			63	GTGCGC	CAG	AACTCAC (21)
			63	GTGTGG	CAG	TCCTTAC (27)
	pri1	2	52	GTGGGT	TAG	ATCTGAA (23)
			56	GTATGT	TAG	TTCTGAT (20)

[a]corrected from published version.
For references see Table 2.1.
Consensus C, T and A at positions 3, 4, and 6 respectively of intron internal motif are underlined.

plant introns; the average length (from Table 2.4) is 69 bp, with a range of 36 bp (*A. niger niaD* intron 1) to 256 bp (*A. niger pacA* intron 2). Intron positions within corresponding genes of different species are generally conserved although the length and sequence of the intron is usually not. Examples of such position conservation are the amylase genes of *A. niger* var. *awamori* and *A. oryzae* and the *pgk* genes of *P. chrysogenum*, *T. reesei* and *T. viride*. The *glaA* genes of *A. niger*, *A. niger* var. *awamori*, *A. shirousami* and *A. oryzae* have four introns in common. It is unlikely that the fifth intron of *A. niger* and *A. niger* var. *awamori*, which is spliced out in a minority of transcripts, is removed in *A. shirousami* and *A. oryzae* as they lack the splice consensus sequences. The *pyrG* genes of *A. niger*, *P. chrysogenum* and *P. blakesleeanus* share the first intron position but only the *P. blakesleeanus* gene has a second intron.

Of the 5′ intron splice sites, 80% follow the consensus GTPuNGPy where Pu is a purine, N is any base and Py is a pyrimidine. Two exceptions are *A. niger pgaI* intron 2 and *L. edodes ras* intron 2 both of which have GC at the splice junction instead of GT. The majority of 3′ splice sites are PyAG with only 8% AAG and 1% GAG. An approximation to an internal site for lariat formation can be seen in all the introns of Table 2.4 on average 19 bp from the 3′ end of the intron. Few introns follow the yeast consensus TACTAAC exactly but most, 92%, have the CT at positions 4 and 5 of the element and the A at position 6, with a consensus being NNCTPuAPy.

3′ non-coding region. Following cleavage of the primary transcript, a poly(A) tail is added to most eukaryotic mRNAs. Several motifs have been found to be associated with cleavage (usually T or TG rich [66–68]) and in higher eukaryotes, the sequence AATAAA is recognized as the polyadenylation signal [69]. Perfect matches for the AATAAA sequence can be found in a few genes such as *A. oryzae alp* (Table 2.5) in which the sequence occurs at a position, 19 bp upstream from the polyadenylation site, expected from the mammalian example [69]. However, the same sequence occurs immediately around one of the major polyadenylation sites of *P. blakesleeanus trp1* and in other genes is truncated or absent. Putative mRNA cleavage signals have been recognized in several genes (Table 2.5) but often these do not coincide with the presence of an obvious AATAAA motif. Both cleavage and polyadenylation signals are required for cleavage in higher eukaryotes [70] and the lack of identifiable motifs in most fungal genes suggests differences in processing machinery.

Translational starts and stops. The translational starts listed in Table 2.2 follow the previously observed pattern in fungi [1, 3] showing a clear preference for A (64%) or a purine (84%) at position −3. Only one gene, *C. acremonium pcbAB*, does not use ATG as a start codon. Unusually, if not uniquely in the eukaryotes, it uses the codon GTG to encode methionine, a situation more often seen in the prokaryotes [71]. Translational termination

Table 2.5 3′ non-coding regions of genes with known polyadenylation sites.

Species	Gene	Distance (bp) from stop codon to polyadenylation site	Polyadenylation signal AATAAA and distance (bp) from stop codon	Distance (bp) from stop codon to reported putative mRNA cleavage signals
A. niger	*glaA*	122	—	NA
	pga	188	ATAA 87	TTTATCGC 145[a]
				CCAAGTTCC 208[b]
	pmeA	137	—	TAATCACTGC 98[a]
				CGTGATTG 151[b]
				TAGT 43, 94[c]
	aldA	110, 111, 156, 158	—	NA
	pacA	123	—	CGAGTTGA[b]
var. *awamori*	*glaA*	120	ATAA 85	NA
A. shirousami	*glaA*	108	—	NA
A. oryzae	*alp*	204	AATAAA 185	NA
	nepII	114	—	NA
	amyI	118, 133	AATAAA 127	TATG 14, 71, 78[c]
				TAGT 96[c]
	amyII	168, 232	—	TAGT 131[c]
				TATGTTTT 161[c]
	amy1	278	ATAA 180	NA
	glaA	161	AATAGA 120	NA
T. reesei	*egl3*	140	—	NA
P. blakesleeanus	*trp1*	48, 68 (major) 126, 156, 181 (minor)	AATAAA 45 AACTAAAA 64	NA

[a]similar to TTTCACTGC (66).
[b]similar to PyGTGTTPyPy (68).
[c]similar to TAGT...TTT...TATGT (67).
For references see Table 2.1.
—, no recognisable motif; NA, not available.

may have an influence on the regulation of protein synthesis in eukaryotes [72]. The fungal genes in this study show little bias in their use of stop codons (UAA 42%, UAG 28%, UGA 30%) and like other eukaryotes [72] have a preference (70%) for a purine at the position following the stop codon.

Codon usage. Several recent reviews have dealt in depth with codon usage in lower [73, 74] and higher [75] eukaryotes. The genes of the industrial fungi appear to follow the general pattern of filamentous fungal codon usage [3, 74]. Briefly, highly expressed genes have a preferred subset of sense codons, probably reflecting their corresponding tRNA species abundance. For example, *A. niger pgaI* and *pelD* use only 38 codons while *A. oryzae amdS*, which is probably not highly expressed, uses all the sense codons. In codons where there is a choice, those ending in C are usually preferred while there is a general bias against those ending in A. Interspecies variations may be superimposed on these basic rules although the small sample size for each species makes meaningful comparisons difficult. Nevertheless, comparison of codon usage in *glaA* genes (Table 2.6) of three different aspergilli would suggest a stronger bias in *A. niger* and *A. shirousami* than in *A. oryzae.*

2.2.3 *Chromosomal arrangement and location of genes*

Information on the effect of arrangement and location on gene expression may be valuable in the construction of recombinant strains containing multiple copies of homologous or heterologous genes but at present little is known except in a few cases. Conventional molecular methods have shown that the three structural genes of penicillin biosynthesis are clustered in *P. chrysogenum* and that the ability to produce antibiotics can be transferred *en bloc* by transformation to a non-producing species (*A. niger* [76]). Additionally, production strains selected over many years by classical genetic means have been found to contain multiple tandem copies of the cluster [77], a fact which encourages attempts to increase production of other useful metabolites by gene amplification in other species.

With the recent development of chromosome separation techniques, it has become possible by hybridization to assign chromosomal locations of genes in organisms in which the lack of classical genetic systems makes this otherwise impossible. Thus, the three *amyA*, *B* and *C* genes of *A. oryzae* have been demonstrated to be dispersed on different chromosomes [29] while the *glaA* gene is located on yet another chromosome [30]. It is possible also to determine the chromosomal locations of transforming DNA. In *C. acremonium*, multiple integration events occurred in a single chromosome although the target chromosome was random for each transformant [78].

Table 2.6 Codon usage in *glaA* genes.

		Number of times used		
Amino acid	Codon	*A. niger*	*A. shirousami*	*A. oryzae*
Phe	UUU	4	6	5
	UUC	18	16	18
Leu	UUA	0	0	2
	UUG	6	6	11
	CUU	3	7	6
	CUC	17	14	8
	CUA	2	0	3
	CUG	20	20	8
Ile	AUU	12	10	7
	AUC	11	13	17
	AUA	1	1	2
Met	AUG	3	2	4
Val	GUU	6	9	15
	GUC	15	13	15
	GUA	2	3	5
	GUG	19	17	12
Ser	UCU	16	20	9
	UCC	19	16	16
	UCA	4	2	7
	UCG	14	18	15
Pro	CCU	4	6	12
	CCC	10	8	7
	CCA	0	0	7
	CCG	8	9	4
Thr	ACU	20	18	11
	ACC	39	42	23
	ACA	5	4	7
	ACG	10	8	14
Ala	GCU	25	22	18
	GCC	19	20	18
	GCA	10	6	16
	GCG	11	13	16
Tyr	UAU	6	5	14
	UAC	21	22	13
Ter	UAA	0	0	0
	UAG	1	1	1
His	CAU	0	1	3
	CAC	4	4	1
Glu	CAA	4	2	11
	CAG	13	15	18
Asn	AAU	6	6	10
	AAC	19	19	13
Lys	AAA	0	0	8
	AAG	13	13	8
Asp	GAU	21	19	15
	GAC	23	23	17

Table 2.6 (*Contd.*)

Amino acid	Codon	Number of times used		
		A. niger	*A. shirousami*	*A. oryzae*
Glu	GAA	9	7	6
	GAG	17	20	12
Cys	UGU	3	3	5
	UGC	7	7	4
Ter	UGA	0	0	0
Trp	UGG	19	19	17
Arg	CGU	4	6	3
	CGC	7	7	5
	CGA	4	1	6
	CGG	3	4	6
Ser	AGU	12	11	10
	AGC	23	24	19
Arg	AGA	1	1	2
	AGG	1	1	1
Gly	GGU	14	19	11
	GGC	22	20	15
	GGA	7	6	9
	GGG	4	5	12

2.3 Gene regulation

While much work has been done to unravel the complex regulatory circuitry of systems in *A. nidulans* [79, 80] and *N. crassa* [81, 82], in contrast, relatively little is known about gene regulation in industrially important filamentous fungi. The regulation of gene expression in filamentous fungi is thought primarily to exist at the level of mRNA accumulation which in most cases is assumed to be the result of transcriptional control. Translational control may also be a component of the regulation of gene expression as demonstrated by cAMP-mediated catabolite expression of *glaA* in *A. niger* var. *awamori* [83]. Several investigators have demonstrated mRNA induction in response to the presence of substrate or substrate components (e.g. *A. niger* genes encoding pectinolytic enzymes [10,13]; *T. reesei* cellulase genes [45,47]; *A. oryzae* amylase genes [84]; and *amdS* [31]), repression (e.g. phosphate repression of acid phosphatase genes of *A. niger* and *P. chrysogenum* [18,40]; glucose repression of amylase and cellulase genes [84,85]) and differential regulation (e.g. *T. reesei pgk* [51]) but detailed study of the mechanisms involved is largely missing.

Coordinate expression of genes involved in a pathway such as penicillin biosynthesis or a process such as cellulose degradation as well as similarities in the regulation of genes introduced into heterologous hosts, for example *niaD* [86], *pgaII* [10] and *pacA* [18], might suggest that control mechanisms are

Table 2.7 Other sequence elements described in upstream regions.

Species	Gene	Motif	Position	Postulated role
A. niger	*aldA*	ACCATCAACCATCAAACCAACTTCTCT	−204	Carbon catabolite repression
		TGG direct repeat eight times	−659	Unknown
		$\mathrm{T}^{\mathrm{A}}_{\mathrm{T}}\mathrm{CCA}^{\mathrm{C}}_{\mathrm{G}}$ direct repeat six times	−763	Unknown
	pacA	CCTTG direct repeat three times	−159	Unknown
		$\mathrm{CTCTCC}^{\mathrm{A}}_{\mathrm{G}}\mathrm{CTGCC}$	−266, −306	Phosphate regulation
	niaD	TAGCTA	−156	Nitrate regulation
A. oryzae	*amy1, 2, 3*	AGGGGCGGAAA	−266	Similar to Spl transcription factor binding site
		TCCCCGCCCCT	−160	Complementary to Spl binding site
		TTTACGTAAA	−521	Unknown
		GGGCAACTCGCTT	−556, −573	Unknown
P. chrysogenum	*phoA*	TTCCAAGGTT	−560, −402	Phosphate regulation
		AACCTTGGAA	−103, −95	Phosphate regulation
T. reesei	*cbh2*	TATTTCCTCGCC	−414	Complementary to cAMP transcription signal sequence
	pgk	TCACGTCA	−387	cAMP transcription regulation
		CTTGCCTATCGTTGCAGCTT	−162	Upstream activating sequence
		ACGGAATGAACCTTGAAGGTTACAT	−346	Heat shock element

For references see Table 2.1. Positions are given in bp from the 5′ end of the element of the A(+1) of the translation initiation codon.

sufficiently alike as to allow recognition of motifs or structures at the DNA sequence level. Table 2.7 shows sequence elements in the 5′ non-coding regions of genes whose roles have been tentatively assigned by comparison with motifs known to exist in other eukaryotic genes. Additionally, there are other sequence motifs of unknown function recognized by their repetitive nature. In only one instance, however, has a functional analysis been undertaken. This is the putative heat shock element of *T. reesei pgk* which was shown by deletion studies to have no detectable effect on the site or efficiency of transcription under the conditions tested [51].

Another approach to the analysis of gene regulation has been the generation of promoter–reporter gene fusions to study gene transcription. Effects of specific deletions within the promoter on transcription and hence expression can be determined by assay of an easily detectable reporter product. Preliminary results using this technique have been reported in two industrial systems. The first is construction of *amy* promoter–*E. coli* β-glucuronidase gene (*uidA*) fusions to study regulation of α-amylase in *A. oryzae* [87 and chapter 3]. The work confirms that, while not discounting translational control, regulation occurs at the level of transcription, expression being induced by starch and repressed by glucose, and deletion studies are now underway. Gene fusions were also created between the *P. chrysogenum ipnA* (*pcbC*) promoter and *E. coli* β-galactosidase gene (*lacZ*). In the absence at that time of a suitable *Penicillium* transformation system, promoter deletions were studied in the heterologous, but penicillin-producing host, *A. nidulans* [88]. Data appeared to confirm carbon catabolite repression and indicated a region between nucleotides −759 and −394 possibly involved in the observed repression of penicillin biosynthesis by ammonium. These preliminary results require repetition in *P. chrysogenum.*

Promoter deletion studies have also been carried out directly on the *A. niger glaA* gene, the product of which is easily assayed [89]. The work revealed that the basal level of transcription was controlled by a region extending −214 bp from translation initiation, but high level expression in response to maltose was the responsibility of a region between nucleotides −562 and −318.

More recent methods of gene analysis such as DNA–protein binding assays and DNA footprinting are as yet untried in the industrial fungi. Likewise no regulatory genes have been characterized in detail, although a recent report described a putative zinc-finger encoding protein of unknown function in the mushroom *L. edodes* [62]. Clearly, investigation of the regulation of commercially important genes is a research area requiring further study.

2.4 Prospects

The accumulation of information at the DNA sequence level permits the hitherto unavailable option of industrial strain improvement and manipulation in a specific manner. It is now possible to consider the introduction of multiple copies of a desirable gene and to monitor the effect on final product formation. Indeed, classical strain improvement has fortuitously achieved this in the case of penicillin production although it must be assumed that genes encoding proteins involved in positive regulation of the pathway have similarly been amplified. Titration of regulatory protein is, of course, a potential problem in expression of large copy numbers, hence the need for further data on regulatory genes and the interaction of their products.

In addition to gene amplification, it should be possible to produce 'designer strains' which will produce different cocktails of enzymes to carry out different functions. Such technology is already in progress, for example in the manipulation of *T. reesei* cellulases (chapter 4). Strains can likewise be programmed to produce entirely novel products by introduction of completely new enzymes [90, 91]. Manipulation of promoter function to produce responses to alternative effector molecules with more economical substrates is also feasible.

Already valuable heterologous proteins may be expressed in industrial filamentous fungi (see chapters 5 and 6) but the use of these organisms is, in general, hindered at present by a dearth of knowledge of the molecular processes involved in post-transcriptional modification and secretion. This is an obvious area for intensive research to obtain the same level as that achieved with homologous proteins (20–30 g/l protein).

For any given microbial industrial process, be it production of secondary metabolites, extracellular enzymes or heterologous proteins, there has to be a limit to cellular capacity. However, general opinion agrees that this limit has not yet been met and so factors such as gene expression remain of major commercial importance. Therefore, obvious target areas must in future include basic molecular research into regulation at the transcriptional and translational level, regulatory genes and secretory pathways as well as investigation of general transcriptional and post-transcriptional signals.

References

1. Ballance, D.J. (1986) *Yeast* **2**: 229–236.
2. Rambosek, J. and Leach, J. (1987) *CRC Crit. Rev. Biotechnol.* **6**: 357–393.
3. Gurr, S.J., Unkles, S.E. and Kinghorn, J.R. (1987) The structure and organisation of nuclear genes of filamentous fungi. In: *Gene Structure in Lower Eucaryotes* (Kinghorn, J.L., ed.), IRL Press, pp. 93–193.
4. Kos, T., Kuijvenhoven, A., Hessing, H.G.M., Pouwels, P.H. and van den Hondel, C.A.M.J.J. (1988) *Curr. Genet.* **13**: 137–144.
5. Boel, E., Hansen, M.T., Hjort, I., Hoegh, I. and Fiji, N.P. (1984) *The EMBO Journal* **3**: 1581–1585.
6. Buxton, F.P., Gwynne, D.I., Garven, S., Sibley, S. and Davies, R.W. (1987) *Gene* **60**: 255–266.
7. Wilson, L.J., Carmona, C.L. and Ward, M. (1988) *Nuc. Acids Res.* **16**: 2339.
8. O'Connell, M.J. and Kelly, J.M. (1989) *Gene* **84**: 173–180.

9. Ward, M., Wilson, L.J., Carmona, C.L. and Turner, G. (1988) *Curr. Genet.* **14**: 37–42.
10. Bussink, H.J.D., Buxton, F.P. and Visser, J. (1991) *Curr. Genet.* **19**: 467–474.
11. Bussink, H.J.D., Brouwer, K.B., de Graaff, L.H., Kester, H.C.M. and Visser, J. (1991) *Curr. Genet.* **20**: 301–307.
12. Ruttkowski, E., Khanh, N.Q., Wientjes, F.-J. and Gottschalk, M. (1991) *Mol. Microbiol.* **5**: 1353–1361.
13. Kusters-van Someren, M.A., Harmsen, J.A.M., Kester, H.C.M. and Visser, J. (1991) *Curr. Genet.* **20**: 293–299.
14. Gysler, C., Harmsen, J.A.M., Kester, H.C.M., Visser, J. and Heim, J. (1991) *Gene* **89**: 101–108.
15. Khanh, N.Q., Ruttkowski, E., Leidinger, K., Albrecht, H. and Gottschalk, M. (1991) *Gene* **106**: 71–77.
16. Kriechbaum, M., Heilmann, H.J., Wientjes, F.J., Hahn, M., Jany, M., Gassen, H.G., Sharif, F. and Alaeddinoglu, G. (1989) *FEBS Lett.* **255**: 63–66.
17. Whittington, H., Kerry-Williams, S., Bidgood, K., Dodsworth, N., Peberdy, J., Dobson, M., Hinchliffe, E. and Ballance, D.J. (1990) *Curr. Genet.* **18**: 531–536.
18. MacRae, W.D., Buxton, F.P., Sibley, S., Garven, S., Gwynne, D.I., Davies, R.W. and Arst, H.N. Jr. (1988) *Gene* **71**: 339–348.
19. Unkles, S.E., Campbell, E.I., Punt, P.J., Hawker, K.L., Contreras, R., Hawkins, A.R., van den Hondel, C.A.M.J.J. and Kinghorn, J.R. (1992) *Gene* **111**: 149–155.
20. Nunberg, J.H., Meade, J.H., Cole, G., Lawyer, F.C., McCabe, P., Schweickart, V., Tal, R., Wittman, V.P., Flatgaard, J.E. and Innis, M.A. (1984) *Mol. Cell. Biol.* **4**: 2306–2315.
21. Adams, R.R. and Royer, T. (1990) *Nuc. Acids Res.* **18**: 4931.
22. Korman, D.R., Bayliss, F.T., Barnett, C.C., Carmona, C.L., Kodama, K.H. Royer, T.J., Thompson, S.A., Ward, M., Wilson, L.J. and Berka, R.N. (1990) *Curr. Genet.* **17**: 203–212.
23. Berka, R.M., Ward, M., Wilson, L.J., Hayenga, K.J., Kodama, K.H., Carlomagno, L.P. and Thomson, S.A. (1990) *Gene* **86**: 153–162.
24. Shibuya, I., Gomi, K., Iimura, Y., Takahashi, K., Tamura, G. and Hara, S. (1990) *Agric. Biol. Chem.* **54**: 1905–1914.
25. Tatsumi, H., Ogawa, Y., Murakami, S., Ishida, Y., Murakami, K., Masaki, A., Kawabe, H., Arimura, H., Nakano, E. and Motai, H. (1989) *Mol. Gen. Genet.* **219**: 33–38.
26. Tatsumi, H., Murakami, S., Tsuji, R.F., Ishida, Y., Murakami, K., Masaki, A., Kawabe, H., Arimura, H., Nakano, E. and Motai, H. (1989) *Mol. Gen. Genet.* **228**: 97–103.
27. Wirsel, S., Lachmund, A., Wildhardt, G., and Ruttkowski, E. (1989) *Mol. Microbiol.* **3**: 3–14.
28. Gines, M.J., Dove, M.J. and Seligy, V.L. (1989) *Gene* **79**: 107–117.
29. Gomi, K., Tada, S., Kitamoto, K. and Takahashi, K. *Gene* (in press).
30. Hata, Y., Tsuchiya, K., Kitamoto, K., Gomi, K., Kumagai, C., Tamura, G. and Hara, S. (1991) *Gene* **108**: 145–150.
31. Gomi, K., Kitamoto, K. and Kumagai, C. (1991) *Gene* **108**: 91–98.
32. Peñalva, M.A. and Sánchez, F. (1987) *Nuc. Acids Res.* **15**: 1874.
33. Cantoral, J.M., Barredo, J.L., Alvarez, E., Diez, B. and Martin, J.F. (1988) *Nuc. Acids Res.* **16**: 8177.
34. van Solingen, P., Muurling, H., Koekman, B. and van den Berg, J. (1988) *Nuc. Acids Res.* **16**: 11823.
35. Carr, L.G., Skatrud, P.L., Scheetez, M.E., II, Queener, S.W. and Ingolia, T.D. (1986) *Gene.* **48**: 257–266.
36. Barredo, J.L., Cantoral, J.M., Alvarez, E.A., Diez, B. and Martin, J.F. (1989) *Mol. Gen. Genet.* **216**: 91–98.
37. Smith, D.J., Earl, A.J. and Turner, G. (1990) *The EMBO Journal* **9**: 2743–2750.
38. Diez, B., Gutierrez, S., Barredo, J.L., van Solingen, P., van der Voort, L.H.M. and Martin, J.F. (1990) *J. Biol. Chem.* **265**: 16358–16365.
39. Barredo, J.L., van Solingen, P., Diez, B., Alvarez, E., Cantoral, J.M., Kattevilder, A., Smaal, E.B., Groenen, M.A.M., Veenstra, A.E. and Martin, J.F. (1989) *Gene* **83**: 291–300.
40. Haas, H., Redl, B., Friedlin, E. and Stöffler, G. *Gene* (in press).
41. Samson, S.M., Belagaje, R., Blankenship, D.T., Chapman, J.L., Perry, D., Skatrud, P.L., van Frank, R.M., Abraham, E.P., Baldwin, J.E., Queener, S.W. and Ingolia, T.D. (1985) *Nature* **318**: 191–194.
42. Smith, A.W., Ramsden, M., Dobson, M.J., Harford, S. and Peberdy, J.F. (1990) *Bio/Technology* **8**: 237–240.

43. Guitiérrez, S., Diez, B., Montenegro, E. and Martin, J.F. (1991) *J. Bacteriol.* **173**: 2354–2365.
44. Samson, S.M., Dotzlaf, J.E., Slisz, M.L., Backer, G.W., van Frank, R.M., Veal, L.E., Wu-Kuang Yeh, Miller, J.R. Queener, S.W. and Ingolia, T.D. (1987) *Bio/Technology* **5**: 1207–1214.
45. Shoemaker, S., Schweickart, V., Ladner, M., Gelfand, D., Kwok, S., Myambo, K. and Innis, M. (1983) *Bio/Technology* **1**: 691–696.
46. Chen, C.M., Gritzali, M. and Stafford, D.W. (1987) *Bio/Technology* **5**: 274–278.
47. Penttilä, M., Lehtovaara, P., Nevalainen, H., Bhikhabhai, R. and Knowles, J. (1986) *Gene* **45**: 253–263.
48. van Arsdell, J.N., Kwok, S., Schweickart, V.L., Lander, M.B., Gelfand, D.H. and Innis, M.A. (1987) *Bio/Technology* **5**: 60–64.
49. Saloheima, M., Lehtovaara, P., Penttilä, M., Teeri, T.T., Stählberg, J., Johansson, G., Pettersson, G., Claeyssens, M., Tomme, P. and Knowles, J.K.C. (1988) *Gene* **63**: 11–21.
50. Vanhanen, S., Penttilä, M., Lehtovaara, P. and Knowles, J. (1989) *Curr. Genet.* **15**: 181–186.
51. Vanhanen, S., Jokinen, A., Illmén, M., Knowles, J.K.C. and Penttilä, M. *Gene* (in press).
52. Berges, T., Perrot, M. and Barreau, C. (1990) *Nuc. Acids Res.* **18**: 7183.
53. Smith, J.L., Bayliss, F.T. and Ward, M. (1991) *Curr. Genet.* **19**: 27–33.
54. Cheng, C., Tsukagoshi, N. and Udaka, S. (1990) *Nuc. Acids Res.* **18**: 5559.
55. Goldman, G.H., Villarroel, R., Van Montagu, M. and Herrera-Estrella, A. (1990) *Nuc. Acids Res.* **18**: 6717.
56. Gray, G.L., Hayenga, K., Cullen, D., Wilson, L.J. and Norton, S. (1986) *Gene* **48**: 41–53.
57. Tonouchi, N., Shoun, H., Uozumi, T. and Beppu, T. (1986) *Nuc. Acids Res.* **14**: 7557–7568.
58. Choi, H.T., Revuelta, J.L., Sadhu, C. and Jayaram, M. (1988) *Gene* **71**: 85–95.
59. Diaz-Minguez, J.M., Iturriaga, E.A., Benito, E.P., Corrochano, L.M. and Eslava, A.P. (1990) *Mol. Gen. Genet.* **224**: 269–278.
60. Iturriaga, E.A., Diaz-Minguez, J.M., Benito, E.P., Alvarez, M.I. and Eslava, A.P. (1990) *Nuc. Acids Res.* **18**: 4612.
61. Hori, K., Kajiwara, S., Saito, T., Miyazawa, H., Katayose, Y. and Shishido, K. (1991) *Gene* **105**: 91–96.
62. Yamaoka, K., Kajiwara, S., Hori, K., Miyazawa, H., Saito, T., Kanno, T. and Shishido, K. *Gene* (submitted).
63. Nussinov, R. (1990) *Crit. Rev. Biochem. and Mol. Biol.* **25**: 185–221.
64. Punt, P.J. and van den Hondel, C.A.M.J.J. *Crit. Rev. Biochem.* (in press).
65. Ramon, D., Carramolino, L., Patino, C., Sanchez, F. and Penalva, M.A. (1987) *Gene* **57**: 171–181.
66. Benoist, C., O'Hare, K., Breathnach, R. and Chambon, P. (1980) *Nuc. Acids Res.* **8**: 127–142.
67. Zaret, K.S. and Sherman, F. (1982) *Cell* **28**: 563–573.
68. McLauchlan, J., Gaffney, D., Whitton, J.L. and Clements, J.B. (1985) *Nuc. Acids Res.* **13**: 1347–1368.
69. Humphrey, T. and Proudfoot, N.J. (1988) *Trends in Genetics* **4**: 243–245.
70. Wickens, M. (1990) *Trends in Biochemistry* **15**: 277–281.
71. Sharp, P.M., Higgins, D.G., Shields, D.C., Devine, K.M. and Hoch, J.A. (1990) *Genet. Biotechnol. Bacilli* **3**: 89–98.
72. Cavener, D.R. and Ray, S.C. (1991) *Nuc. Acids Res.* **19**: 3185–3192.
73. Andersson, S.G.E. and Kurland, C.G. (1990) *Microbiol. Rev.* **54**: 198–210.
74. Lloyd, A.T. and Sharp, P.M. (1992) *Mol. Gen. Genet.* (in press).
75. Campbell, W.H. and Gowri, G. (1990) *Plant Physiol.* **92**: 1–11.
76. Smith, D.J., Burnham, M.K.R., Edwards, J., Earl, A.J. and Turner, G. (1990) *Bio/Technology* **8**: 39–41.
77. Smith, D.J., Bull, J.H., Edwards, J. and Turner, G. (1989) *Mol. Gen. Genet.* **216**: 492–497.
78. Smith, A.W., Ramsden, M. and Peberdy, J.F. (1992) *Gene* (in press).
79. Punt, P.J., Dingemanse, M.A., Kuyvenhoven, A., Soede, R.D.M., Pouwels, P.H. and van den Hondel, C.A.M.J.J. (1990) *Gene* **93**: 101–109.
80. van Heeswijck, R. and Hynes, M.J. (1991) *Nuc. Acids Res.* **19**: 2655–2660.
81. Fu, Y.-H. and Marzluf, G.A. (1990) *Proc. Nat Acad. Sci. USA* **87**: 5331–5335.
82. Paluh, J.L., Orbach, M.J., Legerton, T.L. and Yanofsky, C. (1988) *Proc. Nat Acad. Sci. USA* **85**: 3728–3732.
83. Bhella, R.S. and Altosaar, I. (1988) *Curr. Genet.* **14**: 247–252.

84. Erratt, J.A., Douglas, P., Morannelli, F. and Seligy, V.L. (1984) *Can. J. Biochem. Cell Biol.* **62**: 678–690.
85. Teeri, T., Salovouri, I. and Knowles, J. (1983) *Bio/Technology* **1**: 696–699.
86. Unkles, S.E., Campbell, E.I., Carrez, D., Grieve, C., Contreras, R., van den Hodel, C.A.M.J.J. and Kinghorn, J.R. (1989) *Gene* **78**: 157–166.
87. Tada, S., Gomi, K., Kitamoto, K., Takahashi, K., Tamura, G. and Hara, S. (1991) *Mol. Gen. Genet.* **22**: 301–306.
88. Kolar, M., Holzmann, K., Weber, G., Leitner, E., and Schwab, H. (1991) *J. Biotechnol.* **17**: 67–80.
89. Fowler, T., Berka, R.M. and Ward, M. (1990) *Curr. Genet.* **18**: 537–545.
90. Gutiérrez, S., Diez, B., Alvarez, E., Barredo, J.L. and Martin, J.F. (1991) *Mol. Gen. Genet.* **225**: 56–64.
91. Isogai, T., Fukagawa, M., Aramori, I., Iwami, M., Kojo, H., Ono, T., Ueda, Y., Kohsaka, M. and Imanaka, H. (1991) *Bio/Technology* **9**: 188–191.

3 Fungal enzymes used in oriental food and beverage industries

K. SAKAGUCHI, M. TAKAGI, H. HORIUCHI and K. GOMI

3.1 Introduction

From ancient times, filamentous fungi have served in oriental countries as industrial catalytic sources to degrade starch to sugars and proteins to amino acids for the production of fermented foods such as soy sauce, miso and tempé. This is somewhat analogous to the role played by fungi in cheese production in the western hemisphere. In addition, fungi are used in the production of beverages, including Japanese saké and Chinese spirits (such as jiu), which, according to production levels, is a major industry.

Today, there are many fungal-based fermentation industries for the production of alcohol, amino acids, antibiotics and organic acids, which use glucose derived from starch as the common starting material. The saccharification of starch to glucose, maltose and other low molecular weight dextrins, is efficiently carried out by fungal glucoamylase and other amylases, which are initially produced in large-scale fermentation vessels.

Why is malt used as a substrate in the Western world, and fungi in the Oriental world? The difference may be due to eating habits. For instance, beer originated in Mesopotamia and continued to be used in Egypt where the main foods of the people were wheat and barley. Such ancient people perhaps encountered the sprouting of wet wheat and barley after floods or heavy rain were followed by dry conditions. This climate provided good natural conditions for furthering knowledge of malt and experimenting with it as a sweetening agent. One of the main Oriental foods, in contrast, is steamed rice and in the wet but warm climate of Japan, steamed rice can very easily succumb to a variety of fungal growth within several days.

3.2 Fungal industries in Japan

The water-field rice agriculture system is thought to have originated in the southern part of China in the second or third century BC. The two oldest history books of Japan (more than 2000 years old), which relate mythological and historical descriptions of the Imperial families and ancient Japan, relate stories of rice-saké drinking at the court and in private life between the first and fifth centuries AD. In addition, an orthodox Chinese history book of the late third century contains a description of Japan and its people which relates that saké was traditionally drunk after funerals and other social functions.

The filamentous fungi, *Aspergillus oryzae* and *Aspergillus sojae* have been almost solely utilized in Japan for saké and soy sauce manufacture. Steamed rice succumbs easily to growth of such aspergilli which have a natural selective advantage over other moulds. Rice is mixed with wood ash on its surface, giving it the necessary dryness, and potassium phosphates and alkali are added which provide selective conditions for fungi on the surface of rice. In medieval times, and also today, starter cultures of *Aspergillus* were mixed with ash. The ash of camellia branches and leaves harvested in summer is considered to be the best substrate.

There are descriptions of white and black saké consumption in ancient Japanese literature. White, slightly turbid saké was common even until the eighteenth century and its nature well understood, but black saké is not easy to define today. However, it is likely that it was the product of fermentation mixed with a herb ash. This compound was used for making sound Koji (steamed rice and other grains grown with aspergilli) and for the absorption of deteriorated flavour by carbon after fermentation. The latter technique is still frequently used for the clearance of flavour and colour from over-flavoured and coloured saké.

In the southern part of Japan, including Kagoshima and the Okinawa islands, black Koji moulds (so-called black aspergilli) such as *A. awamori*, *A. usamii*, *A. saitoi*, *A. luchuensis* and *A. niger* are commonly used for saccharification of steamed rice and steamed sweet potato. The black aspergilli produce a considerable amount of citric acid, and the pH of the digested starch materials is lowered to 3.2–3.5, much lower than that produced by yellow aspergilli, *A. oryzae* or *A. sojae* (pH 4.2–4.8). Of course, a lower pH mash, especially in southern and warmer areas, is very desirable for yeast fermentations which are consequently protected from bacterial infections. However, sour tastes caused by the black Koji moulds inspired distillation procedures to produce spirits; a similar situation to that in the Cognac and Armagnac areas where sour wine is used for the production of brandy.

3.2.1 *Shochu*

The white liquor produced from fermented mashes is generally called shochu and has an alcohol content of around 35%. Shochu is produced in the Okinawa islands from rice using black awamori aspergilli which gives a pleasant flavour after storage in porcelain pots for three to ten years. Shochu produced in the Kagoshima area of Japan is derived from steamed sweet potato saccharified by black Koji moulds previously cultured on rice. The fermented and distilled product is called imo-shochu or sweet potato shochu and contains various aromatic substances such as fatty acid ethyl esters.

3.2.2 *Saké*

The procedure for brewing saké is divided into three parts. In the first, Koji (moulded rice) is made by growing *A. oryzae* on polished steamed rice to provide high amylase activity. In the second part the starter (Moto) is made which promotes vigorously growing yeast. The third process is the major fermentation through stepwise supplementation of steamed rice and Koji.

Steamed rice is cooked, brought into a very clean, warm (28°C) chamber (the Koji chamber) and inoculated with *A. oryzae* spores. The growing temperature of *A. oryzae* on steamed rice is between 30 and 40°C (but is often below 20°C for good quality) and is very carefully controlled by crushing and breaking the rice heap where the mycelia of the fungus is growing. This process allows the

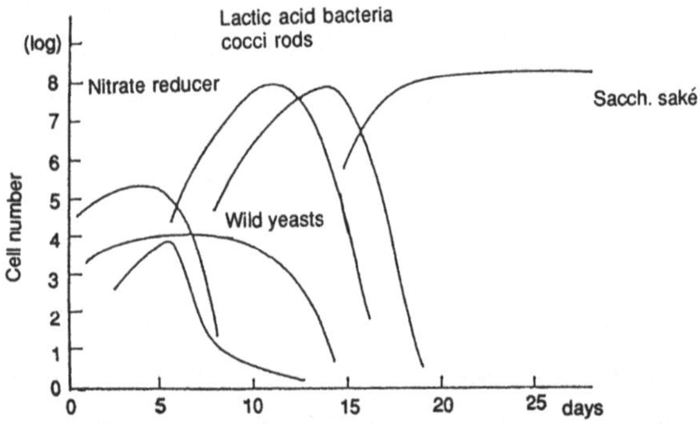

Figure 3.1 Transition of microbes in starter culture of saké.

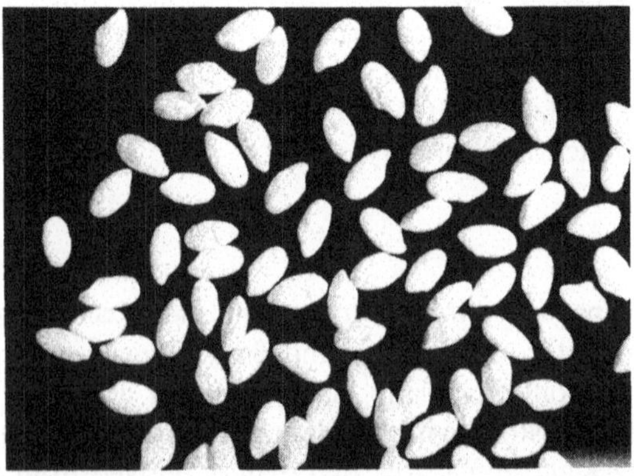

Figure 3.2 Rice Koji grown with *A. oryzae* (from Ozeki Saké Brewery).

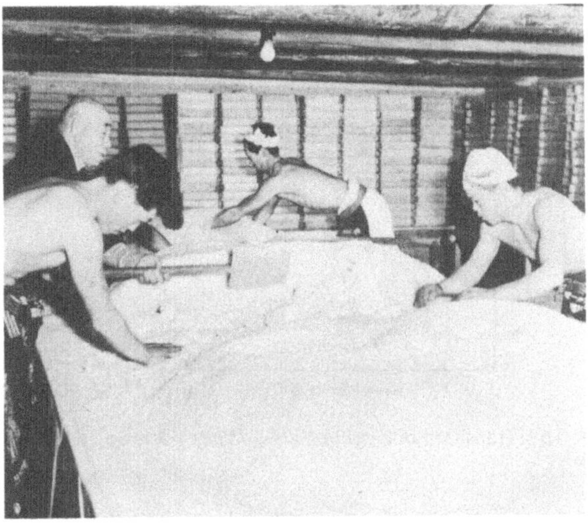

Figure 3.3 Traditional Koji chamber (from Hakushika Saké Brewery).

Figure 3.4 Major fermentation of saké (from Hakushika Saké Brewery).

growth of mycelia into the rice grain, which gives more efficient amylase production.

During the cold winter period, the rice Koji and double amounts of steamed rice are mixed with water for the preparation of vigorously growing yeast starter cultures (moto). With the aid of nitrate-reducing bacteria (*Pseudomonas*) and lactic acid bacteria (*Lactobacillus saké*), which lower the pH to 3.5, *Saccharomyces saké*, which is very similar to *S. cerevisiae*, grows abundantly.

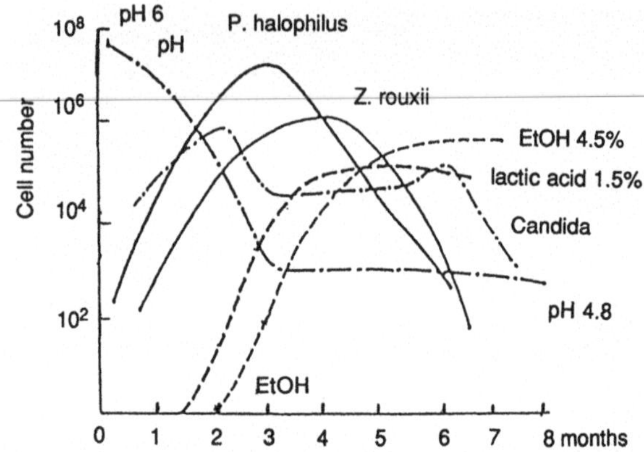

Figure 3.5 Transition of microbes, pH and alcohol in soy sauce brewing.

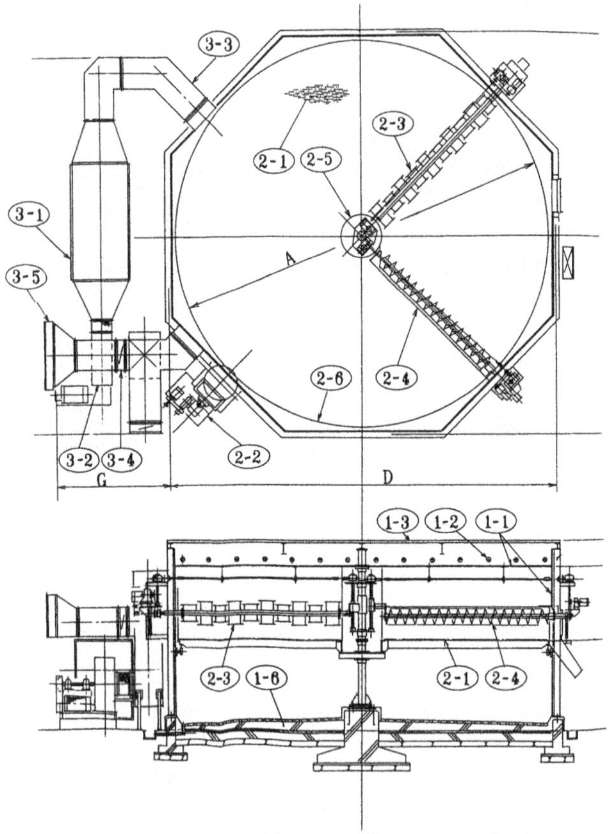

Figure 3.6 Koji-making machine for soy sauce and saké (from Fujiwara Joki Co.): 2-1 bottom plate; 2-2 motor for rotating; 2-3 crusher of crumped Koji; 2-4 feeder and harvester; 2-5 centre; 2-6 side wall; 3-1 air conditioner; 3-2 blower; 3-3 airduct; 3-4 dumper; 3-5 plate finn heater.

Nowadays synthetic lactic acid is artificially added to make a safer starter culture.

To the small amount of the starter culture, rice Koji, steamed rice and water are added stepwise three or four times. The alcohol content reaches 20% after fermentation for twenty days.

3.2.3 *Soy sauce*

Soy sauce is a major seasoning in Japan for almost all fish, vegetable and meat dishes, and its production reaches 1.2 million kilolitres per year. It is a fermented product in heavy brine of 18%, made by fermenting soy bean and wheat with *Aspergillus* proteinases and amylases, and then brewing salt tolerant lactic acid bacteria (*Pediococcus halophilus*) [1, 2] with yeasts (*Zygosaccharomyces rouxii* and others) [3]. Its alcohol content reaches 4–5% before pasteurization and its amino acid and glutamic acid contents are around 7.6% and 1.8% respectively, after brewing for 6 to 12 months. Its use for flavouring is now spreading throughout the world [4].

The major difference between soy and saké brewing is that all of the steamed soy bean and parched wheat are mixed and inoculated with *Aspergillus* spores allowing the full growth of this mould for two or three days. This causes coverage of mycelia and spores on the whole material, which maximizes proteinase and amylase activity. In the case of saké brewing, only one fifth of the total rice is grown with *Aspergillus*.

3.3 Fungal industries in China

China is the origin of culture and civilization in East Asia, just as Mesopotamia, Egypt, Greece and Rome are the origins of European culture. Every great civilization makes a good quality and large quantity of alcoholic beverages, because alcohol forms an important part of human culture.

The Chinese saccharifying fungus traditionally has been *Rhizopus*, in contrast to the aspergillis used in Japan. This species difference reflects the use of uncooked ground grains for the natural growth of the starter cultures. If such grains are cooked, aspergilli becomes dominant. In northern China, the ground grains of Kaoliang, millet, Dekkan grass grain, Indian millet, pea, soy bean, wheat and barley are added to 20–30% of water and the mixture used to make dumplings which are steamed and eaten as a main daily food. In order to provide the growth of saccharifying *Rhizopus*, the uncooked, brick-sized, or smaller, dumplings are kept at 28–30 °C for four or five days. Even without seed inoculation, the growing fungi is predominantly *Rhizopus* because the temperature of the material naturally rises to 45–65°C. The material produced is not unlike malt in that it has significant amounts of amylase, proteinases and other enzymes. This material, called Da qu (big ferment) serves as the enzymic source, after crushing and drying, for the saccharification of steamed Kaoliang, millet, wheat and other grains. The smaller sized flat ball of moulded material, produced in southern China, is called Xiao qu (small ferment) and serves as the enzymic source for saccharifying steamed rice mash [5].

3.3.1 *Chinese jiu*

Jiu is comparable with saké in Japan, since it is also made from rice and has an alcohol content of 14–18%, which is similar to that of saké. However, it is different from saké in that it must be stored for several years at low temperature in underground pots in order to mature. As a result, jiu has more acids and sugars than saké, which is a one-year-old product. In ancient China, spirits were not produced and several types of jiu were made from grains other than rice. These were popular also in northern China where such grains were predominant. Nowadays, jiu from rice, which has better quality, is produced in the southern area of the longest Chinese river, Yang-zu-jiang. In the north and other parts of China, the distilled spirits made from Da qu are more common because these are made using stills which were transferred from Arabian civilizations during the Yuan dynasty (twelfth century) [5].

3.3.2 *Bai jiu*

Da qu is produced from a mixture of ground barley, wheat and 20–40% peas. Moulded Da qu is dried and stored. *Rhizopus* predominates but there is a small amount of *Mucor, Aspergillus, Monascus, Penicillium, Absidia, Cladosporium* and *Neurospora*. In addition, there is a large amount of *Saccharomyces* and *Hansenula*, and *Pichia, Debaryomyces, Kluyveromyces* and other yeasts are also found. The existence of lactic acid bacteria, acetic acid bacteria and spore-forming bacilli has also been reported.

The ground Kaoliang and other grains are steamed with rice hull and stuffed with crushed Da qu and a small amount of water into holes in the ground. This solid is fermented by *Saccharomyces, Hansenula, Pichia* and others for several weeks or months. The fermented solid material is distilled, giving an alcohol content of 50% or more, and can be stored for years. The residue is recycled for fermentation [5].

3.3.3 *Huang jiu*

In contrast to bai jiu, huang jiu (yellow jiu) is brewed, rather than distilled, saké. The brewing method is rather complicated and varies according to location. However, the basic method starts with the saccharifying enzyme source, Xiao qu. Barley, wheat, rice grain and herbs are ground in 25% of water, and rounded to small balls of 2–3 cm in diameter. Without cooking, the balls are brought into a warm chamber at around 30°C and *Rhizopus* growth soon covers the grain balls, as in Da qu production, and the product is dried and stored.

In autumn and in winter (from October to March), amylopectin-rich rice is polished, with 10% loss of the rice grain surface, soaked in water for about a day, steamed and cooled by adding pasteurized water. The rice mash is put into large porcelain pots and Xiao qu is added (1 part Xiao qu to 80 parts rice mash). Within several days *Rhizopus* and yeasts from Xiao qu grow and bring about saccharification of rice starch which then allows alcohol fermentation to take place. Finally, Da qu (15–18%) produced from ground wheat is added with water to the rice mash and the fermentation continues for 20–25 days [5].

3.3.4 *Red jiu*

In southern parts of China and in Taiwan and Malaysia, the fungus *Monascus*, a red pigment producer, is used to make Koji (saccharifying ferment) from steamed rice. *Monascus* produces amylases, similar to aspergilli, which have a red colour and special flavours and are used in cooking to flavour meat and in Chinese medicine to lower the cholesterol value. *M. anka* and *M. purpureus* are the most common types and *M. anka* produces four types of amylases similar to *A. oryzae* [6, 7].

3.3.5 *Rice beer*

In south-eastern Asian countries a filamentous yeast forming a pseudo-mycellium (*Saccharomycopsis fibuligeri*) which produces α-amylases [8] and glucoamylase [9], in a similar way to fungi, is used for the saccharification of amylopectin-rich rice. The yeast starter culture which grows on rice is known as Ragi, Lukpang and Bubod in Indonesia, Thailand and the Philippines, respectively. This culture is used as the starter for brewing a type of rice beer. Its amylase genes have been studied recently, by transformation to *S. cerevisiae*, and show similarity with fungal amylase proteins [10–13].

3.3.6 *Tempé*

Tempé is soy bean fermented with *Rhizopus*. It is an important proteinous nutrient in Indonesia where its annual production reaches 600 000 tons. The United Nations Food and Agricultural Organization (FAO) recommends it as a nutritive and non-fatty proteinous food which is not easily digested in its original cooked form. In the United States there is a trend to consume tempé as a diet food. There is a classical review article on the oriental fermented foods by Hesseltine [14] and good descriptions of tempé by Kozaki [15], Shurtleff and Aoyagi [16] and Gandjar [17].

The seed culture of *Rhizopus* usually used is *R. oligosporus* (Saito) which is a proteinase-rich species. By sandwiching the de-skinned and cooked soy bean with Usar leaves, which contain *Rhizopus* mycelia and spores on their surface, a seed culture is made and the fungus grows abundantly in the cooked soy bean layer after only a few days.

3.4 Modern industries based on fungal enzymes

3.4.1 *Glucose production from starch*

Glucoamylase (EC 3.2.1.3) is very important for the production of glucose from various starch materials. So far glucoamylases have only been found in microorganisms such as fungi and yeasts. Glucoamylase is an exo-type enzyme which cleaves starch or glycogen from their non-reducing termini by breaking the α-1, 4 glucoside linkage, resulting in the production of monomer β-glucose. This enzyme also cleaves the branching point of the substrates, an α-1, 6 glucoside linkage, and therefore is indispensable for the production of glucose from starch. Starch from various sources is first liquefied at 105°C by addition of heat tolerant α-amylase produced from *Bacillus licheniformis*, which cleaves the amylase chain lowering the viscosity and avoiding the formation of non-soluble starch fragments [18]. Glucoamylase from *R. delemar*, *A. awamori* var. *kawachii* or *Endomyces* is then added and glucose is produced [19]. Most of the glucose is converted to isomerized sugar or starch sugar by glucose isomerase [20–22] which converts the glucose solution to a mixture of 42–45% fructose and glucose. The high fructose corn syrup of 55% fructose (HFCS) is believed to possess a better flavour than glucose alone and is widely accepted as a sweetener for soft drinks.

3.4.2 *Alcohol production*

Fermentation from starch and sugar materials produces 95% and 100% alcohol. The annual world production reaches 25 billion kilolitres per year. The alcohol is blended to various liquors, for example in Japan it is used to make saké. Other fractions are added to gasoline or used in the chemical industry. Fungal cultures or their enzymes are commonly used to saccharify starch material because they give high initial yields of alcohol as a result of the production of fungal glycoamylases.

The amylo process, developed in the 1930s, uses a submerged culture of *R. javanicus* to saccharify wheat, sweet potato and other starchy materials. It forms the basis of the modern fermentation industry which now uses an array of large tank fermenters of 200 kilolitres or more. The alcohol yield from the starch material is improved by cooperational enzymation with *Rhizopus* and the use of *A. awamori* var. *fumeus* culture broth as the source of various amylases. This shortens the main fermentation period from six to four days, and also gives a better yield of alcohol with more concentrated starch mash in the tank. The process completely dependent on commercial enzymes, such as *Bacillus* α-amylase and *Rhizopus* and *Aspergillus* glucoamylase preparations is currently practised in many factories, with yeast as the only living organism in the whole alcohol fermentation process [23].

3.5 Genetic transformation systems for *Rhizopus*

Rhizopus has been widely utilized in the Chinese fermentation industry as the major saccharifying fungus. *Rhizopus* belongs to the zygomycetes and secretes a large number of extracellular enzymes. Its culture supernatant from wheat bran is used as a commercial digestive in Japan. Therefore, *Rhizopus* is a promising organism to use as a host for the synthesis of valuable proteins because of its ability to secrete protein.

Another interesting characteristic of *Rhizopus* is that it grows very rapidly in certain directions with a filamentous morphology, generally called cell-tip growth. A genetic transformation system is a powerful tool for analysing these characteristics and utilizing them in industry for breeding more useful strains by molecular technology, especially when dealing with a species such as *R. niveus*, which has neither sexual cycle nor sexual mating system. Therefore a genetic transformation system for *R. niveus* is described here.

3.5.1 *Formation of protoplasts from* Rhizopus niveus [24]

Although transformation is successful in yeast using intact cells in the presence of specific cations [25], it was thought appropriate to develop a genetic transformation system of *R. niveus* using protoplasts. A number of factors affecting protoplast formation of *R. niveus* IFO 4810 were examined.

Sporangiospores are harvested by vigorous shaking in distilled water of sporangia with mycelia from cultures grown in RMM agar medium (2% glucose, 0.2% asparagine, 0.05% KH_2PO_4, 0.025% $MgSO_4 \cdot 7H_2O$). Cell debris is removed by filtration through a G-3 filter. Sporangiospores are resuspended at about 5×10^6 per millilitre in YPG medium (1% glucose, 2% polypepton and 2% yeast extract) containing 10 mM proline and germinated at 30°C, with shaking, for 4–6 h. Germinated spores (germlings) are filtered through a G-1 glass filter to remove overgrowth spores and washed twice with 0.3 M mannitol, 22 mM citric acid, and 55 mM sodium phosphate, pH 5.6. The germlings are added to a lytic enzyme solution dissolved in the same buffer to give a final concentration of about 5×10^6 per millilitre and incubated at 30°C for about 2 h, with shaking.

When the effects of mannitol, sorbitol, glucose or $MgSO_4$ are compared as osmotic stabilizers for protoplast formation, protoplasts are efficiently obtained when mannitol, sorbitol or glucose is used at a concentration of 0.3–0.5 M. In contrast to sugar and sugar alcohols, protoplasts are not obtained when 0.3 M $MgSO_4$ is used.

If Novozyme 234, chitinase T-1, and chitosanase are examined singly and in combination as to their ability to release protoplasts, no protoplasts are obtained when these lytic enzymes are used alone or when Novozyme 234 and chitinase T-1 are used in combination. However, chitosanase release protoplasts from mycelium when used in combination with Novozyme 234 or chitinase T-1, and the efficiency of protoplast formation is significantly increased when all three enzymes are used together.

The culture age of the mycelium affects the efficiency of protoplast formation. Protoplasts are obtained most efficiently from germlings which are 4–6 h old. Younger germlings are resistant to lytic enzymes, and the efficiency of protoplast formation from older ones is lower. Examination of protoplasts of intact germlings 4–6 h old prepared under the best conditions shows that the protoplasts are perfectly spherical with diameters of 10–20 mm. The regenerating frequency of the protoplasts thus prepared is approximately 4%, judging from their colony-forming ability.

3.5.2 *Transformation of* Rhizopus niveus *using drug-resistance genes as dominant selectable markers* [24, 26]

Initially, drug resistance was used as a selection marker for *R. niveus* transformation. The effects of various antibiotics such as kanamycin, neomycin, hygromycin B, and cycloheximide on the growth of *R. niveus* were examined. The antibiotic G418 turned out to inhibit growth effectively. A plasmid pGGR1 for transformation experiments was constructed as shown in Figure 3.7. Protoplasts of *R. niveus* were mixed with pGGR1 in the presence of PEG 4000 and Ca^{2+} and plated on the selection medium. Even without the addition of plasmid DNA, certain colonies (mock colonies) appeared. This phenomenon might be due to the incomplete effect of G418 on the growth of *R. niveus* on agar medium, as described previously for the cases of some other fungi. However, more colonies appeared upon the addition of pGGR1. To identify genuine transformants, the colonies were transferred one by one onto a plate with fresh selective medium, and the plates incubated at 30°C to induce sporulation. The spores thus formed were harvested and inoculated into liquid medium containing 200 μg/ml of G418. Under these selection conditions, the spores from the non-transformed colonies did not grow, indicating that only the true transformants could be selected by this method.

Southern blot analysis was carried out to ascertain the presence of the plasmid DNA sequence in one of the transformants. These results (Figure 3.8) suggested that the transformant had acquired the plasmid DNA sequence and that the recombination of the introduced DNA with the genomic DNA had occurred, since the restriction enzyme digestion patterns were different between the plasmid DNA and the total DNA of the transformant. The transformant described above was further cultivated for 96 h, and the total DNA prepared from the cells at intervals of 24 h. Southern blot analysis of

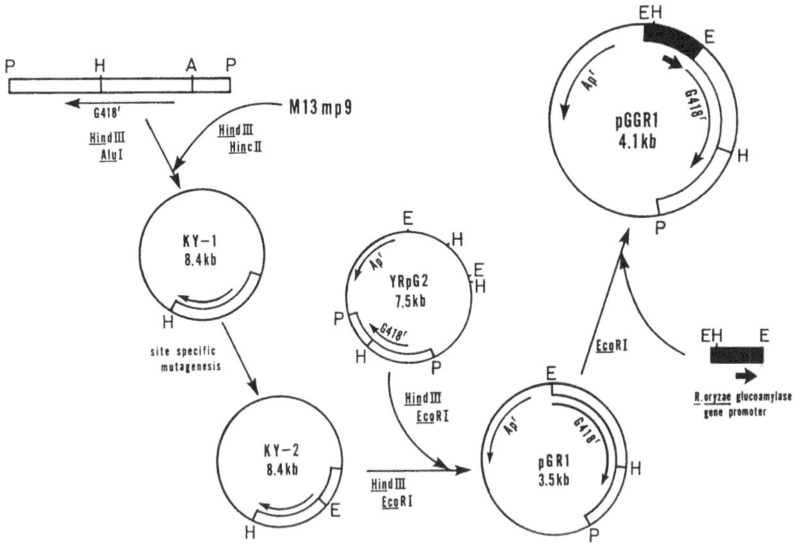

Figure 3.7 Construction of pGGR1. The 0.8 kbp *Hin*dIII-*Alu*I fragment derived from the 1.7 kbp *Pvu*II fragment containing the Tn*903* derived from G418 resistance gene was inserted between the *Hin*dIII and *Hin*cII site of M13mp9 to give rise to KY-1. KY-2 was obtained by introducing an *Eco*RI site just in front of the initiation codon of G418 resistance gene using site-specific mutagenesis. pGR1 was obtained by ligation of the 0.55 kbp *Hin*dIII-*Eco*RI fragment of KY-2 and the 2.95 kbp *Hin*dIII-*Eco*RI fragment of YRpG2. pGGR1 was constructed by insertion of the 0.6 kbp *Eco*RI fragment, carrying the *R. oryzae* glucoamylase gene promoter, into the *Eco*RI site of pGRI. M13mp9 and YRp7, Tn*903*-derived DNA and *R. oryzae* DNA are represented by the thin line, open box line, and solid box line, respectively. Restriction enzymes are denoted as follows: A, *Alu*I; E, *Eco*RI; H, *Hin*dIII; P. *Pvu*II.

these DNA samples indicated that rearrangement of DNA occurred rather frequently. Another genetic marker for a drug-resistant transformation is a bacterial blasticidin S resistance gene [26]. Blasticidin S is an amino-acylnucleoside antibiotic used to control rice blast disease. It specifically inhibits protein synthesis in both prokaryotic and eukaryotic cells. The blasticidin S resistance gene encoding blasticidin S deaminase, an inactivating enzyme, has been isolated from *B. cereus* [27]. It is noteworthy that *S. cerevisiae* and *Nicotiana tobacum* have been transformed to blasticidin S resistance [28].

The sensitivity of *R. niveus* to blasticidin S was tested by inoculating spores both onto PD agar medium (2.4% potato dextrose broth) and into PD liquid medium containing various concentrations. The growth of spores on agar medium and in liquid medium was completely inhibited at concentrations of more than 30 μg/ml and 1.5 μg/ml of blasticidin S, respectively. However, in the case of both agar and liquid medium, the sensitivity was lower when germinated spores or mycelia were inoculated. Spores could grow, although weakly, on PD agar medium containing as much as 300 μg/ml of blasticidin S

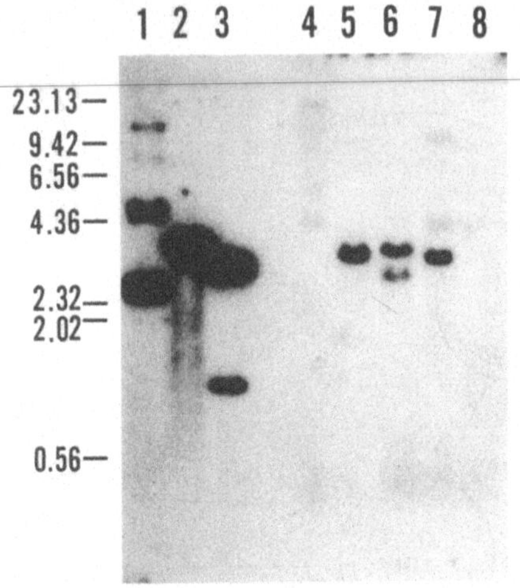

Figure 3.8 Southern blot analysis of total DNA prepared from *R. niveus* wild-type and transformant cells. Lane 1, undigested pGGR1 (0.1 ng); lanes 2 and 3, pGGR1 digested with *Eco*RI and *Hin*dIII, respectively; lane 4, undigested transformant DNA (15 μg); lanes 5, 6 and 7, transformant DNA digested with *Eco*RI, *Hin*dIII, and *Sal*I, respectively; lane 8, wild-type DNA digested with *Eco*RI. pGR1 was used as a probe. The position and size (kbp) of DNA markers used (λ-DNA digested with *Hin*dIII) are shown on the left.

when the pH of the medium was adjusted to pH 2.7 in order to make colonies of *R. niveus* compact. These results indicate that the sensitivity of *R. niveus* to blasticidin S is dependent on the conditions of the media and growth phases, and that blasticidin S is most effective for the termination of *R. niveus* spores.

The plasmid pRGDB1, constructed as shown in Figure 3.9, contains the blasticidin S resistance gene derived from *B. cereus* under the control of the *R. oryzae* glucoamylase gene promoter [29]. Spores of transformants selected as described above were inoculated onto PD agar medium containing blasticidin S. The untransformed wild-type cells were not able to grow in the presence of 30 μg/ml of blasticidin S, whereas transformant cells were able to grow in 50 μg/ml of the drug. It was suggested that the extent of balsticidin S resistance differed from one transformant to another, since the colony sizes of the transformants were different.

Southern analysis of total DNA from the transformants suggested that the introduced DNA was arranged differently in each transformant and present extrachromosomally in one of them. It also indicated that the introduced DNA was present extrachromosomally in a transformant selected using G418 resistance as a selectable marker. In the transformation systems of some other zygomycetous fungi, such as *Mucor circinelloides* [30, 31], *Phycomyces blakesleeanus* [31, 32] and *Absidia glauca* [33], plasmid DNA autonomously replicated. In integrative transformation of *A. glauca*, repetitive DNA had to be inserted into a vector DNA [34]. These results suggest that it may be a

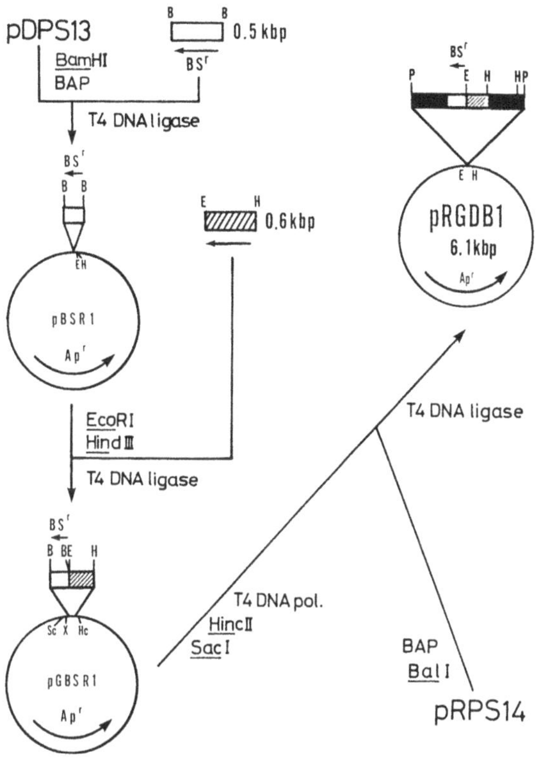

Figure 3.9 Construction of plasmid pRGDB1. Only the relevant restriction sites are shown. Symbols are: B, *Bam*HI; E, *Eco*RI; H, *Hin*dIII; Hc, *Hin*cII; P, *Pst*I; Sc, *Sac*I; X, *Xba*I, DNA:—, pUC119 or Bluescript KSM13+; ▭, *B. cereus* blasticidin S resistance gene; ▨, *R. oryzae* glucoamylase gene promoter; ■ *R. niveus* genomic DNA.

characteristic of zygomycetes that the frequency of integration is lower than in the other filamentous fungi.

The low intensity of the signals obtained by Southern blot analysis indicated that the copy number of the introduced DNA per haploid genome was low. The fact that the transformants are heterokaryones, in which nuclei harbouring the introduced DNA and those without it are present together, may be the cause of this low copy number. In order to increase the copy number of the introduced DNA, cultivation of a transformant under a more severe selection regime was attempted. However, the intensity of the signal in the sample from the cells cultivated in a medium with 50 μg/ml of blasticidin S was about the same as that with 20 μg/ml, indicating that the copy number could not be increased. This may be due to the decrease of selective pressure during cultivation, as a result of the inactivation of blasticidin S by the resistance gene product, or to the low sensitivity of germinated spores to blasticidin S.

This transformation system was used to express the *E. coli lacZ* gene in *R. niveus* but the activity was relatively low.

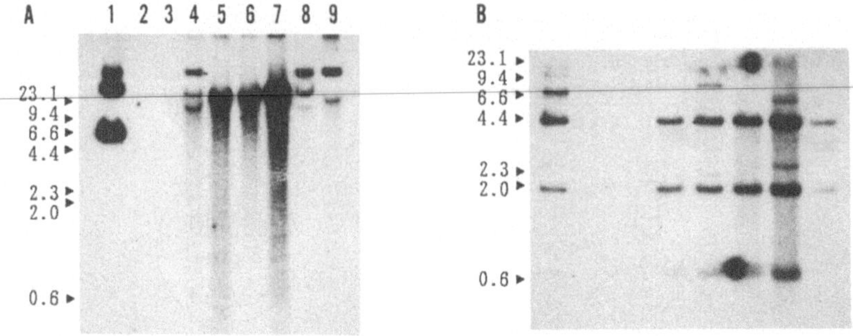

Figure 3.10 Southern blot analysis of total DNA prepared from a transformant obtained with pLeu4. (A) Analysis of undigested total DNA: lane 1, pLeu4; lane 2, λ-*Hin*dIII; lane 3, M-37 total DNA; lane 4, transformant No. 1 total DNA; lane 5, transformant No. 2; lane 6, transformant No. 3; lane 7, transformant No. 4; lane 8, transformant No. 7; lane 9, transformant No.8. (B) Analysis of *Hin*dIII digested total DNA. Lanes 1–8 are the same as in (A).

3.5.3 *Transformation of* Rhizopus niveus *using an auxotrophic selective marker*

To examine this type of transformation, several kinds of *R. niveus* leucine deficient mutants were generated and the *leuA* gene of *Mucor circinelloides* introduced to the mutants in an attempt to complement them [35]. An auxotrophy of a strain M-37 was found to be complemented by the plasmid pLeu4 which contains *leuA* and an adjacent sequence which functions as an ARS (autonomously replicating sequence) in *M. circinelloides.* Several transformants were cultured and their total DNAs extracted to perform Southern analysis. The result is shown in Figure 3.10. Plasmid rescue results in *E. coli* cells indicate that at least in the case of one transformant, plasmid DNA could be recovered in the original form. Therefore, it is suggested that in contrast to the ascomycetous fungi, transformation of *R. niveus* with pLeu4 plasmid results in integration of the introduced plasmid DNA chromosomes tandemly in multicopies or results in autonomous replication. In addition, when transformation with pLeu4 linearized by a restriction enzyme was performed, the transformation frequency was lower and introduced plasmid DNA tended to integrate into chromosomes tandemly, when compared with transformation with circular pLeu4 DNA.

3.6 Proteolytic enzymes

Proteases produced ad secreted by filamentous fungi play an important role in various production processes. For instance, aspartic proteases of *Rhizomucor pusillus* and *Rz. miehei* (*Mucor rennin*) have been used instead of bovine chymosin as the major catalytic agent for clotting milk cheese production. Thus, aspartic proteases are important examples of enzyme production and utilization in the food industry. Moreover, as fungal proteases can be obtained rather cheaply and in large amounts, their application has increased beyond their use in cheese production. For example, fungal proteases are utilized (i) as digestives to soften meat, (ii) as enzyme preparations and (iii) in detergents. Additionally, as fungal proteases can be purified in large amounts, some have been selected as targets of enzymatic characterization and higher conformational analysis as protease model systems.

3.6.1 *Aspartic proteases from* Rhizopus niveus

Rhizopus niveus secretes several classes of aspartic proteases (RNAPs) and many of these have been utilized as industrial enzyme preparations. The nucleotide sequence analysis of the aspartic protease I (RNAP-I) suggests that it encodes a polypeptide with 389 amino acid residues, of which 66 residues consist of a prepro-sequence. There is an intron of 64 bp in the coding region [36]. The mature protein has a sequence homology of 76% with rhizopepsin (an aspartic protease of *R. chinensis*) [37], 42% with penicillo-pepsin (*Penicillium janthinellum*) [38] and 41% with human pepsin [39]. Using the *R. niveus* RNAP-I gene as a probe, Southern hybridization was carried out with genomic DNA of *R. niveus* and several hybridizing bands were observed. Colony hybridization of cDNA and genomic libraries revealed clones of four related genes. Sequence analyses of these genes confirmed that there are at least four genes for RNAPs other than that for RNAP-I, and they were designated RNAP-II, -III, -IV and -V. All these RNAPs have a prepro-sequence and their sequence homology is 70–90%. All RNAP genes studied contain a single intron at the same position, although intron length is different. Their structure is summarized in Table 3.1. Comparing the expression levels of all four genes, the results carried out so far suggest that each of these genes is regulated differently according to the culture conditions.

The RNAP-I gene (after removal of its intron by site-directed mutagenesis) was inserted between the promoter and the terminator of the gene for glyceraldehyde-3-phosphate dehydrogenase (GAPDH) of *S. cerevisiae* and

Table 3.1 Summary of the RNAP genes.

	RNAP				
	I	II	III	IV	V
TATA motif[a]	+	+	+	+	+
CAAT motif[a]	+	CAAAT	–	+	+
Intron length (bp)	64	83	57	67	52
Poly(A) signal	AATAAA	ATTAAA	ATTAAA	AATAAA	ATTAAA
Amino acid residues	389	391	391	398	392

[a] + denotes presence of motif; – denotes motif is absent.

introduced into *S. cerevisiae* using a yeast vector. It was found that RNAP-I was efficiently secreted at the level of 40 mg/l [40].

3.6.2 *Mucor rennin*

Aspartic proteases produced by *Rz. pusillus* and *Rz. miehei* have been widely used instead of chymosin as a milk-clotting factor in cheese production . The essential characteristics of protease in cheese production are (i) a high substrate specificity (to cut at only one site in K-casein, a major protein component of milk) and (ii) a low activity of general protease. In this regard, *Rz. pusillus* rennin (MPR) and *Rz. miehei* rennin (MMR) are very similar to each other. The sequence homology of these two proteases is 83% [41, 42]. MPR consists of a pre-sequence of 22 amino acid residues and a pro-sequence of 44 amino acid residues. When the gene for MPR was inserted between the yeast *GAL7* promoter and the *GAL10* terminator, and the resulting construct introduced into *S. cerevisiae*, about 150 mg/l of the enzyme was found to be secreted into the medium [43]. Analysis of the N-terminus of the secreted enzyme indicated that the processing of the prepro-sequence occurs at the same site in *S. cerevisiae* as in its native *Rz. pusillus* and that the processing of the pro-sequence is carried out autocatalytically [44]. It was also demonstrated that two Asn-linked oligosaccharides are added to MPR when produced by *S. cerevisiae*. Such a modified enzyme has a different C/P to that of the native MPR (C/P is a ratio between milk-clotting activity and proteolytic activity (Table 3.2) [45]). MPR, produced in large amounts by *S. cerevisiae*, was purified and its structure–function relationships studied by site-directed mutagenesis. The pre-sequence and/or pro-sequence of MPR have been utilized as expression cassettes for extracellular production in *S. cerevisiae* of human growth hormone [46] and human pro-urokinase [47]. MMR has been produced by *A. nidulans* [42], *A. oryzae* [48] and *M. circinelloides* [49] as heterologous hosts. When MPR cDNA was inserted between the *A. oryzae* α-amylase gene promoter and the *A. niger* glucoamylase gene terminator and introduced into *A. oryzae*, more than 3 g/l of the enzyme was found to be secreted into the culture medium [48].

Table 3.2 Clotting and proteolytic activities of *Rz. pusillus* rennin[a].

	No. of glycosylated residues	Clotting activity (units/μg)	Proteolytic activity (units/μg)	Clotting activity/ proteolytic activity ratio
Yeast *Rz. pusillus* rennin				
Non-mutated (− endo H)	2	3.04 (45.1)	3.85 (198)	0.790 (22.8)
Non-mutated (+ endo H)	0	5.61 (83.2)	2.46 (127)	2.28 (65.7)
Commercial *Rz. pusillus* rennin	0	6.74 (100)	1.94 (100)	3.47 (100)

[a]Values in parentheses are expressed percentages of activity.

3.6.3 *Rhizopuspepsin*

This aspartic protease is secreted by *R. chinensis* in relatively large amounts and has been purified. This aspect can be useful for enzymatic characterization and X-ray diffraction analysis [50]. A cDNA copy for the mature enzyme has been isolated and its primary structure determined [51]. The primary structure of the purified enzyme has also been determined [37].

3.6.4 *Aspergillopepsin A*

This aspartic protease is secreted by *A. awamori*. The gene consists of 4 exons and 3 introns, and the coding region has a pre-sequence of 20 amino acid residues, a pro-sequence of 49 amino acid residues and a mature protein sequence of 325 amino acid residues [52]. The enzyme has a sequence homology of 62% with penicillopepsin, 37% with rhizopuspepsin, and 32% with swine pepsin. When the gene was disrupted by *in vivo* insertion of the *argB* gene into the coding sequence, the total activity of aspartic protease in the culture medium was found to decrease dramatically. This engineered mutant strain of *A. awamori* was proved to be useful for extracellular production of heterologous protein such as bovine chymosin [53].

3.6.5 *Alkaline protease*

In the production of soy sauce it is important to degrade raw materials in soy bean and this is done by two types of secreted *A. oryzae* enzymatic activities; high proteolytic activity and low amylolytic activity. Alkaline protease (serine protease) is one of the more important proteases synthesized by *A. oryzae* and used in high proteolytic activity. A cDNA clone encoding this enzyme has been isolated and sequenced [54]. It was found from the DNA sequence that this protease also has a prepro-sequence; a pre-sequence of 21 amino acid residues, a pro-sequence of 100 amino acid residues and a mature protein of 282 amino acid residues. It has sequence homology of 44% with protease B from *S. cerevisiae*, and 29% with subtilisin BPN′ from *B. amyloliquefacience*, but its sequence similarity with mammalian serine proteases, such as trypsin and chymotrypsin, is low. Ogawa and colleagues [55] reported that when the cDNA of this protease was inserted between the promoter and terminator of the GAPDH gene of the yeast *Z. rouxii*, about 120 mg/l was secreted into YPD medium and 275 mg/l into SOY medium after 7 days of cultivation [55]. They also showed that it is possible to create, by site-directed mutagenesis, mutant enzymes with higher specific activity or with higher thermoresistance (I. Ikegaya unpublished).

3.6.6 *Neutral protease*

It is known that the neutral proteases I and II (metallo-proteases) of *A. oryzae* also play important roles in soy sauce production. A cDNA clone of neutral protease II (Np II) was isolated recently and sequenced. The coding region consists of a pre-sequence of 19 amino acids, a pro-sequence of 15 amino acids and a mature sequence of 177 amino acids. No protease is reported thus far which shares any homology with Np II. Np II requires Zn for its activity and is thermostable [56].

3.7 Pectinolytic enzymes and RNases

3.7.1 *Pectinases*

Pectin is composed of the heteropolysaccharides of galacturonic acid residues and rhamnose residues, with side chains of neutral sugar residues and methyl esters. Pectin is a major component of higher plant cell wall and middle lamella. Pectin-degrading enzymes include pectin lyase, which cleaves glycosidic α-1,4 linkages of the main chain by trans-elimination, pectin esterase, which hydrolyses the methyl ester bond between the main and side chain, and polygalacturonase which hydrolyses glycosidic α-1,4 linkages of the main chain. Pectin-degrading enzymes are used commercially in the food industry, for clarification of fruit juices and wine, and in the confectionery industry.

The gene *pelD*, encoding pectin lyase D (PLD), has been cloned from *A. niger* [57]. The PLD deduced amino acid sequence shows that it is composed of 373 amino acid residues including a signal peptide of 19 amino acids at the N-terminus. The open reading frame of the *pelD* gene is interrupted by four introns. Five further genes (*pelA*, *B*, *C*, *E* and *F*) have been isolated from an *A. niger* genomic library by Southern hybridization under the low stringency hybridization condition using the *pelD* gene as the probe [58]. Moreover an *A. niger* transformant with multicopy *pelA* genes produces more extracellular pectin lyase activity (PLA) than the recipient strain when both are grown with presence of pectin as the sole carbon source. The *pelA* gene was sequenced recently and the deduced amino acid sequence of the PLA product shares 69% homology with that of PLD [59].

The gene (*pgaII*) encoding one of polygalacturonases, has been isolated from the genomic and cDNA libraries of *A. niger* [60, 61]. The nucleotide sequence of the *pgaII* gene and deduced amino acid sequence have been inferred. The protein has 362 amino acids including a prepropeptide of 27 amino acids. The open reading frame of the *pgaII* gene is interrupted by an intron of 53 bp, and an *A. niger* transformant with the *pgaII* gene is found which has elevated levels of polygalacturonase II [60].

The gene (*pgaI*), encoding polygalacturonase I (PGI) which has the second most abundant activity in commercial enzyme preparations, was cloned and sequenced. The amino acid homology between mature PGII and mature PGI is *c.* 60% [62].

The isolation of the cDNA encoding pectinesterase has been reported by

Khanh and colleagues [63]. The deduced amino acid sequence shows that it comprises 314 amino acids, including a signal peptide of 17 amino acids at the N-terminus.

cDNA and genomic clones encoding pectate lyase have also been isolated from *A. nidulans*. Although the nucleotide sequence of this cDNA clone has not yet been reported, the regulation of the expression of this gene has been investigated. The expression of this gene is induced in the presence of polygalacturonic acid as a carbon source, but is repressed by glucose or acetate as a carbon source. Gene disruption experiments have demonstrated that pectate lyase is not essential for growth on polygalacturonic acid [64].

3.7.2 *Ribonucleases*

The enzymology of ribonucleases (RNases) has been studied for a relatively long time because they are reasonably convenient to handle. For instance, their molecular weights are relatively small and they have thermotolerant structures. In particular, various RNases of filamentous fungi have been investigated extensively, since considerable yields are possible. RNases can be classified into two groups according to their specificities and molecular sizes.

The first group comprises RNases with smaller molecular weights (*c.* 11 000) which are guanine base-specific (including guanine-base-preferential, base-non-specific and purine-specific RNases). The primary structures of many RNases belonging to this type have been determined, e.g. RNase T_1 [65] from *A. oryzae* and RNase Ms [66] from *A. saitoi*. Such enzymes have high structural homology on the basis of primary structure analysis as well as crystallographic studies [67]. RNase T_1 structure has been characterized by means of NMR spectroscopy [68], chemical modification [69], and recombinant DNA technology [70]. RNase T_1 has been expressed in *E. coli* using the chemically synthesized form of the gene [70] and has been studied in order to establish relationships between structure and function [71, 72].

The second group of RNases contains those with larger molecular weights (*c.* 24 000–34 000). These enzymes are base non-specific and adenine base-preferential, and have been studied less extensively than small RNases. Primary structures of only three RNases have been elucidated: RNase Rh [73] from *R. niveus*, RNase T_2 from *A. oryzae* [74] and RNase M from *A. saitoi* [75]. The amino acid homology between RNase Rh and RNase T_2 is around 40% and that between RNase M and RNase T_2 is 70% (Figure 3.11). Similarity is observed with T_1 type RNases only at the active sites (Figure 3.12). The structure–function relation of RNase Rh and RNase M has been studied by means of chemical modification. It was found that in RNase M and RNase Rh at least two histidine residues and a carboxylic acid are involved at the active site [76]. Further information regarding the structure–function relation of this group of fungal RNases may be obtained by means of DNA recombinant technology using RNase Rh as a model system. This is a protein with a

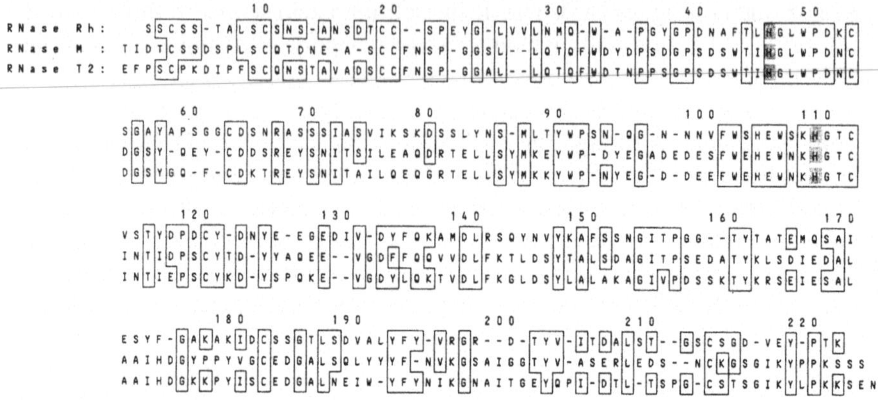

Figure 3.11 Alignment of *Rhizopus* RNase Rh amino acid sequences with RNase M from *A. saitoi* and RNase T_2 from *A. oryzae*.

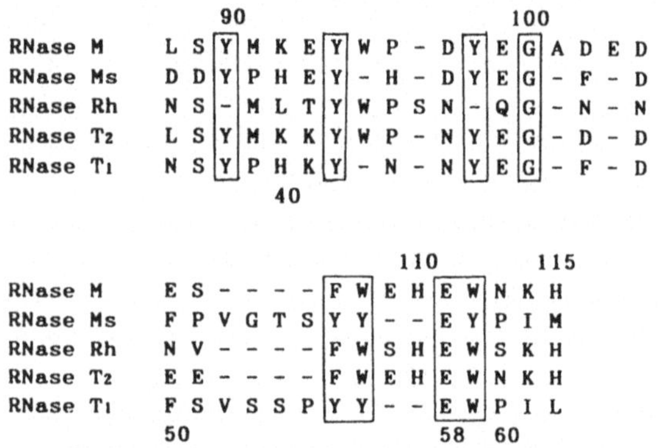

Figure 3.12 Composition of RNase amino acid sequence at the active sites. Residues which are identical or similar in all five RNases are boxed. The numbers on the upper and lower sides of the lines refer to RNase M and RNase T_1, respectively.

molecular weight of 24000, the smallest RNase of this type, and the gene has already been cloned and sequenced [73].

In addition a cDNA clone encoding RNase Rh, which is secreted extracellularly by *R. niveus*, has been isolated. When cDNA clones encoding both a pre-sequence and a mature polypeptide for this RNase Rh were introduced into *S. cerevisiae* under the control of the promoter of the glyceraldehyde 3-phosphate dehydrogenase gene of *S. cerevisiae*, the amount of RNase Rh detected in the culture medium was found to be only approximately 70 μg/l. In contrast, when the cDNA sequence for the mature enzyme of RNase Rh was fused with a nucleotide stretch encoding the prepro-sequence of RNAP-I and introduced into *S. cerevisiae* (using the same expression plasmid), mature RNase Rh was secreted into the medium at a level

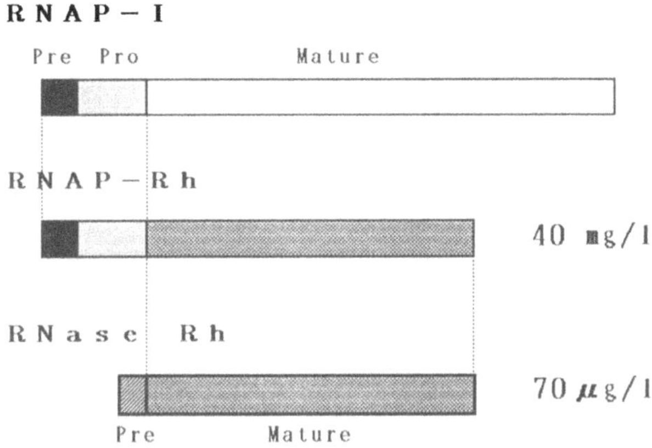

Figure 3.13 The pre-sequence (signal peptide) of *Rhizopus* RNase Rh is replaced by the prepro-sequence of RNAP-I. The total amount of secreted RNase Rh was determined by the activity of RNase in the culture supernatant.

of 40 μg/l (Figure 3.13) [77]. These results suggest that the prepro-sequence of RNAP-I may be useful to support efficient secretion of a polypeptide constructed downstream of the prepro-sequence. Thus, other mature polypeptides may be secreted from *S. cerevisiae* using the RNAP-I prepro-sequence.

Structure–function relationship studies of RNase Rh are now underway utilizing this expression system. The gene encoding RNase T_2 has also been cloned from the *A. oryzae* genomic library and its nucleotide sequence has been determined [78].

3.8 Lipases

Lipases are commonly found in all living organisms including humans, animals, plants, fungi and bacteria. Notwithstanding their ubiquity, their specificities are very different. Even within the fungi, lipases have very different specificities depending on the species. Lipases are very important industrially and are produced more abundantly from fungi than from other sources. The main uses of industrial fungal lipases are as additives to washing detergents and in the food industry. Lipases for washing powder are partly produced by recombinant strains of *A. oryzae*. With regard to food industries, lipases are utilized for dairy, bakery, brewing, mayonnaise and confectionery industries. Two representative examples are (i) the production of formulated cacao butter from palm oil for chocolate production through a lipase trans-esterification reaction and (ii) the addition of lipases to butter and milk flavours, margarines and other dairy products where milk cream requires partial lipase digestion [79] (Figure 3.14).

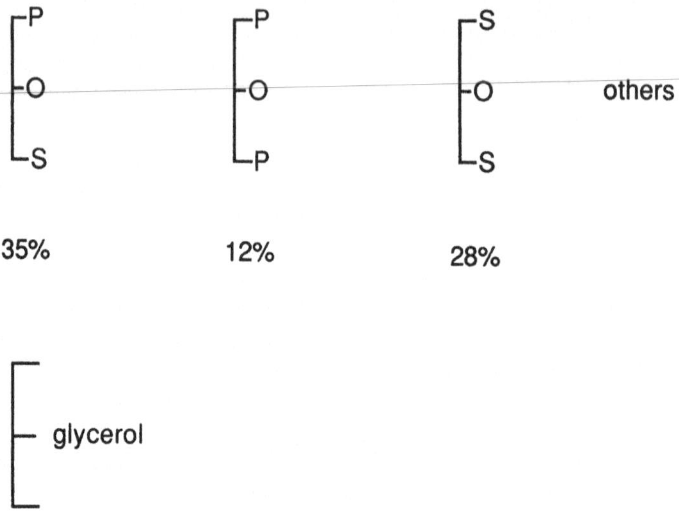

Figure 3.14 Glycerides in cacao fat. P: palmitic acid, O: oleic acid, S: stearic acid.

Table 3.3 illustrates the diverse specificities of lipases with regard to their molecular weights, stability, optimal pH and temperature. Supplementation of phospholipids (i.e. phosphatidylcholine) to *R. delemar* C-lipase causes a conformational change, resulting in a shift of isoelectric point to acid and an increase of hydrolysing activities to lipoproteins and other esters [96]. Rat spleen lipase was reported to shift its optimal pH from 8 to 6.5 with the addition to taurocholic acid (0.1%) and to pH 9.3 with 1.0% of the same detergent [97]. Other synthetic and natural detergents and emulsifiers, e.g. *Acacia* gum, give similar effects [98].

A. niger lipase is severely inhibited, reversibly at first, and then irreversibly, by the presence of even trace amounts of iron ions in tap water [99]. Certain lipases are readily degraded by proteinases, either added or simultaneously produced during growth of the organism [100, 101]. The substrate specificity with regard to various glycerides, is also different by their constituent fatty acids and cleaving points, i.e. 1,3 specific or non-specific for the cleavage of the three ester bonds of glycerides (Figure 3.14).

3.8.1 *Specific lipases*

Lipases are classified into two groups with regard to their specificity to cleave and to trans-esterify the position of the glyceride linkage with fatty acids (Table 3.4). Interestingly, 1,3 specific lipases are more important industrially: for the flavouring of dairy products *R. delemar* lipase is most useful and for washing powder additives, genetically engineered *Humicola lanuginosa* and *M.*

Table 3.3 Specificities of fungal lipases.

	Aspergillus niger [81]	*Geotrichum cylindracea* [82]	*Humicola lanuginosa S-38* [86]	*Humicola lanuginosa No. 3* [87]	*Candida cyclindracea* [82]
MW	38 000	54 000	27 500	39 000	—
Isoelectric point	4.3	4.33	—	6.6	—
pH optimum	5.6	6.3	8.0	7.0	5.2, 7.2
Temperature optimum (°C)	25	40	60	45	30
pH stability	2.2–6.8	4.2–9.8	6–10	5–9	2.0–8.5
Temperature optimum stability (°C)	50	56	60	60	37

	Mucor lipolytica [85]		*Penicillium cyclopium* [88]		*Phycomyces nitens* [89]
	(F3A)	(F3B)	(A)	(B)	
MW	25 400	29 000	27 000	36 000	26 500
Isoelectric point	9.7	10.2	4.96	4.15	5.9
pH optimum	9.0	8.0	7.0	6.0	6.0–7.0
Temperature optimum (°C)	37	30	35	40	37–40
pH stability	3.3–10.0	3.3–10.0	6.5–9.0	4.0–6.5	4.0–10.0
Temperature optimum stability (°C)	40	40	30	30	45

	Rhizopus chinensis [92]	*Rhizopus delamar* [93]		*Rhizopus japonicus* [91]	*Torulopsis ernobii* [95]
		(A)	(C)	(I) (II)	
MW	30 000	44 000	41 000	42 000	42 000
Isoelectric point	9.2	7.3	8.2	7.4(I), 7.9(II)	2.95
pH optimum	6.5	5.6	5.6	7.0–8.5	6.5
Temperature optimum (°C)	40–60	35	35	35–40	45
pH stability	2.5–11	3–8	4–7	5.0–7.0	3.0–9.0
Temperature optimum stability (°C)	75	65	45	45	65

Table 3.4 Specificities of fungal lipases.

1,3-specific	Non-specific
Rhizopus delemar[a]	*Candida cylindracea*[a]
Rhizopus arrhizus[a]	*Geotrichum candidum*[a]
Rhizopus chinensis	*Penicillium cyclopium*[a]
Rhizopus javanicus	*Penicillium roquefortii*
Rhizopus nivesus[a]	
Mucor miehei[a]	
Aspergillus niger[a]	
Humicola lanuginosa No. 3[a]	
Leptospirae pomana	

[a]Currently in industrial production.

miehei (known as *Rhizomucor miehei*) lipases are widely utilized. A noteworthy advantage of this specificity is in the production of formulated cacao butter from palm oil. Cacao butter has a sharp melting point at human body temperature. It is therefore, an indispensable additive to chocolate, cakes and suppositories [103].

More than seventeen lipases, from human, animal, bacterial and fungal sources, have been sequenced so far [173]. Although they vary in their characteristics and in their number of amino acids (from 135 of *Pseudomonas fragi* lipase through 475 of human hepatic lipase to 690 of *Staphylococcus aureus*), they share the common amino acid sequence Gly-X-Ser-X-Gly (Table 3.3) and, in many cases X-Ser- $(X)_6$-Ser. Both motifs are considered to be binding sites for lipid substrates. DNA similarity is also observed between

Table 3.5 A conserved sequence and number of amino acid residues among lipolytic enzymes.

Source	Substrate binding region	Amino acid
Rat hepatic [190]	^{140}K V H L I G Y S L G A^{150}	472
Human hepatic [179]	^{138}H V H L I G Y S L G A^{148}	476
Human lipoprotein [207]	^{125}N V H L L G Y S L G A^{135}	448
Mouse lipoprotein [189]	^{125}N V H L L G Y S L G A^{135}	447
Bovine lipoprotein [199]	^{127}N V H L L G Y S L G A^{137}	450
Canine pancreatic [188]	^{147}Q V Q L I G H S L G A^{157}	467
Porcine pancreatic [180]	^{145}N V H V I G H S L G S^{155}	449
Human lecithine cholesterol acyltransferase [193]	^{174}P V F L I G H S L G C^{184}	416
Rat lingual [181]	^{146}K I H Y V G H S Q G T^{156}	377
Staphylococcus hyicus [183]	^{362}P V H F I G H S M G G^{372}	603
Staphylococcus aureus [203]	^{405}K V H L V G H S M G G^{415}	690
Pseudomonas fragi [191]	^{76}R V N L I G H S Q G A^{86}	135
Rhizomucor miehei [175]	^{137}K V A V T G H S L G G^{147}	269
Humicola lanuginosa [182]	^{140}R V V F T G H S L G G^{150}	269
Rhizopus niveus [192]	^{170}K V I V T G H S L G G^{180}	297
Rhizopus delemar [202]	^{170}K V I V T G H S L G G^{180}	297
Penicillium camembertii [210]	^{138}E L V V V G H S L G A^{148}	305

lipases from related origins. For instance, there is 65% homology between *R. miehei* and *R. niveus* lipases [210] (Table 3.5).

Penicillium camembertii strain U-150 lipase possesses a peculiar specificity, cleaving mono- and diglycerides but not triglycerides [208, 209]. Therefore this lipase is useful for the synthesis of monoglycerides from glycerol and fatty acids or from glycerol and fat. The monoglycerides are potent emulsifiers for foods and cosmetics. The nucleotide homology of *P. camembertii* U-150 lipase with other fungal lipases is 29% with *R. miehei* lipase, 25% with *R. niveus* lipase, and 40% with *H. lanuginosa* lipase [210]. The *R. miehei* lipase active centre consists of Ser-His-Asp, which is also found in other fungal lipases.

The lipase gene of *R. miehei* and its function have been analysed extensively [175]. Two forms of triglyceride lipases were produced from this fungus, namely RML-A of molecular weight (MW) 42.5 kDa and RML-B of MW 27.5 kDa. Lipase RML-B was found to be a partially deglycosylated form of RML-A after treatment at the lower pH (4.5). The constitutive gene for the peptide molecule of this enzyme was composed of 1163 bp including one 74 bp intron. A typical signal peptide region of 24 amino acids was followed by a 70 amino acid proprotein region, followed by the mature enzyme region of 269 amino acids. The post-translational cleavage point between proprotein regions and the mature enzyme region is between methionine and serine but the actual cleavage mechanism is unknown. Differential splicing of the intron gives rise to a shorter, inactive peptide of 13 amino acid residues as a result of a stop codon which is present in the intron.

The *R. niveus* lipase [192] has a prepro-sequence of 51 amino acid residues (the A-chain) followed by a catalytic region of 297 residues (B-chain). The B-chain itself is sufficient to show enzyme activity but the natural enzyme consists of a complex of the two chains.

The *R. delemar* lipase has the same amino acid sequence as that of the *R. niveus* protein [202]. However, its proprotein region contains only 31 amino acid residues. It is interesting to note that in mammalian lipase genes, the colipase pentapeptide of Val-Pro-Asp-Pro-Arg was also found in the same open reading frame with the lipase gene and conferred change on its specificity [213].

3.8.2 *Stereochemical studies*

The first lipase crystallization and X-ray analysis were studied with *A. niger* lipase [81] and *Geotrichum candicium* lipase [85]. From these studies the ellipsoidal structure of the lipase, which contains a large cleft, the helical content and the active centre containing a carboxyl, a histidyl and a hydroxy-amino acid residue, were determined [107–109]. *R. miehei* lipase has also been studied extensively [176, 177]. A triad structure with Ser^{144}, His^{257} and Asp^{203}, which is very similar to serine proteinases such as chymotrypsin or subtilisin, is formed at the catalytic centre. This resemblance is expected from their functional similarity to esterases. A lid structure of helical fragments from Ile^{85} to Asp^{91} was observed which covered over the catalytic triad with extensive hydrophobic and electrostatic interactions. The opening mechanism of the lid and the activation of the enzyme in the interface between the lipid and water was considered as follows. The helical lid of Ile^{85}-Arg-Asn-Trp-Ile-Ala-Asp^{92} is flanked by two hinge areas, Ser^{82}-Ser^{83}-Ser^{84} on one side and Leu^{92}-Thr^{93}-Phe^{94}-Val^{95} on the other. These hinges are flexible, but are very different in their hydrophilicity. When the enzyme is submerged in the lipid–water interface, the lid rolls back from the catalytic site. The hydrophilic side of the lid hinge becomes partly buried in a polar cavity previously filled by well-ordered water molecules. At the same time the hydrophobic side of the lid becomes completely exposed, significantly expanding the non-polar surface around the active site. Thus the non-polar surface is stabilized in the lipid environment and creates a stable and catalytically competent enzyme which can attack the triglycerides within the lipid phase.

A *Penicillium cyclopium* strain M1 lipase, which specifically hydrolysed only triacylglycerol, has been purified and crystallized. This lipase has two identical subunits each with a molecular weight of 54 000 kDa [186].

3.8.3 *Production of fungal lipases in heterologous hosts*

Although several lipase genes have been heterologously introduced into the bacterium *Escherichia coli* and the yeast *S. cerevisiae* for large-scale commercial production [205], success in the industrial production of *R. miehei* lipase, *H. lanuginosa* lipase and *Candida arctica* lipase using *A. oryzae* as a heterologous host has been attempted [175, 182, 184 and chapters 5 and 6]. These recombinant products are used commercially for washing detergent additives, as discussed before. For this purpose, characteristics, such as tolerance to detergents, alkaline pH, cold and/or warm temperatures and proteinase activity, are an advantage in cleaning situations. *A. oryzae* was used as a host for the cDNA copies from *R. miehei*, *H. lanuginosa* or *C. arctica* [185]. Prepro-lipases were cloned in plasmids pBoel-777 or pTAKA-17 behind the *A. oryzae* α-amylase promoter and its upstream region. The transcriptional terminator and polyadenylation site from the *A. niger* glucoamylase gene were placed between the lipase in a similar way to that taken for high level production of *R. miehei* aspartic proteinase [181]. The α-amylase promoter cloned from a high yielding mutant strain of *A. oryzae* (an industrial strain used to produce Taka-amylase) gave much higher levels of expression than the wild-type products. The recombinant plasmid was cotransformed into *A. oryzae* protoplasts together with the *amdS* plasmid (p35R2) for acetamidase [187 and chapter 1]. The guests were shown to integrate into the host chromosome with multiple gene copies. As a rule, the signal region of the produced preproprotein was cleaved. However, the internal cleaving point between methionine (−1) and N-terminal serine, in the case of *R. miehei* lipase, is not always perfect and the mechanisms of cleavage are as yet unknown [193]. Protein engineering work is presently being carried out to improve certain characteristics of recombinant lipases [194].

3.9 Fungal enzymes used in starch processing

Enzymes such as α-amylase and glucoamylase play a very important role in starch solubilization and saccharification in the brewing and food processing industries [110]. In Japan, for example, their commercial usage is estimated to be approximately 7000 tonnes per year, of which amylases make up about 80% [111]. Commercial amylases have been produced by many microorganisms including filamentous fungi of which *Aspergillus* and *Rhizopus* are the most important. In addition to industrial enzyme production, *Aspergillus* and *Rhizopus* strains have been employed for Koji production and in the manufacture of oriental alcoholic beverages such as saké, shochu and huangjiu [112, 113]. Accordingly, these fungal strains have been the subject of extensive gene manipulation to increase their enzyme yield. Fungal strain improvement has been attempted by classical genetic methods and has proved successful. Recently, fungal amylase genes have been isolated, sequenced and expressed in heterologous hosts, such as yeast and other filamentous fungi, and these have also produced higher yields. Furthermore, analyses of the mechanism regulating amylase gene expression have also been carried out by recombinant DNA techniques.

3.9.1 *α-Amylases*

In 1833, Payen and Persoz first discovered α-amylase activity in malt extract. Since then a number of α-amylase proteins have been isolated and purified from a variety of organisms and filamentous fungi. α-Amylase attacks the internal α-1,4-glucoside bonds of starch to yield maltodextrins and thus plays a crucial role in starch solubilization.

Among the fungal α-amylases, *A. oryzae* α-amylase (Taka-amylase A (TAA)) is the best known having been extensively studied and characterized. The enzyme was purified and crystallized from a commercial diastatic enzyme preparation [114] (sold under the trade name 'Takadiastase', named after Dr J. Takamine, who patented the process in 1894). This enzyme is a glycoprotein consisting of a single polypeptide chain of 478 amino acid residues [115]. Its carbohydrate chain structure [116] and X-ray crystallographic structure [117] have been well characterized. In contrast to the biochemical and biophysical analysis, there was until very recently little information regarding the organization and regulation of the gene(s) encoding α-amylase (*amy*).

A. oryzae is capable of secreting copious amounts of TAA, the synthesis of which is induced by oligosaccharides, such as maltose and starch [118, 119].

These observations indicate that the strong expression promoter of the *amy* gene from *A. oryzae* can be utilized for production of heterologous proteins in this organism when it is used as the host. Furthermore, the mechanism of regulation of *amy* gene expression is also of great interest and has been analysed at the molecular level.

Recently, the genes encoding α-amylase have been cloned and sequenced independently by four research groups [120–124]. Genomic or cDNA libraries were screened using oligonucleotide probes deduced according to the amino acid sequence of the α-amylase protein previously sequenced. Consequently, isolated DNAs were sequenced and shown to encode the entire α-amylase protein. Interestingly, genomic Southern blot analysis with a cloned DNA as a probe suggested the existence of multiple *amy* genes in *A. oryzae*. Such multiple gene families have nearly identical restriction maps in the coding regions and 5′ and 3′ flanking regions (Figure 3.15). Furthermore, their nucleotide sequences have a very high degree of homology, showing significant divergences only in the 3′ flanking region. Figure 3.16 shows the nucleotide sequences and deduced amino acid sequences of the three *A. oryzae amy* genes [117]. Unfortunately there are several designations currently in use: the gene *amy1* corresponds to *amyB* [121] and *AmyI* [123]; *amy2* corresponds to *amyC* [121] and *Taa-G2* [124]; *amy3* corresponds to *amyA* [121], *AmyII* [123] and *Taa-G1* [124]. The nucleotide sequences of the two genes *amy1* and *amy2* are completely identical throughout 617 bp of the 5′ non-coding region and the coding region, but are significantly different 69 bp downstream of the stop codon in the 3′ flanking region. A further gene (*amy3*) has almost the same sequence as the other two *amy* genes, having two nucleotide substitutions resulting in the corresponding two amino acid changes (^{35}Gln to ^{35}Arg and ^{151}Phe to ^{151}Leu) in the coding region as well as one nucleotide change in the

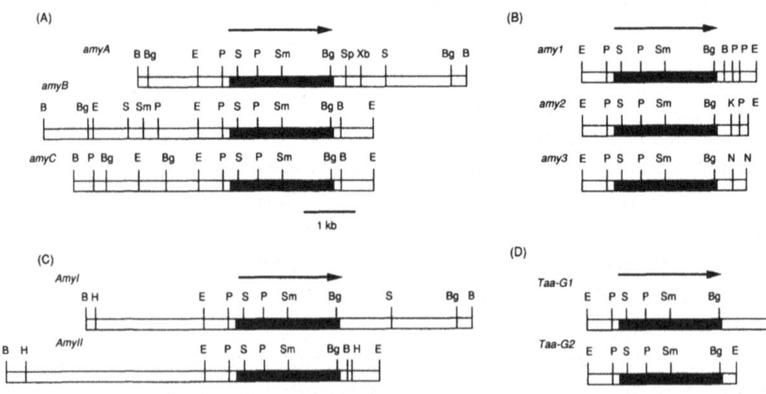

Figure 3.15 Restriction maps of multiple *amy* genes independently reported by four groups. (A) Tada *et al.* [120] and Gomi *et al.* [121]; (B) Wirsel *et al.* [122]; (C) Gines *et al.* [123]; (D) Tsukagoshi *et al.* [124]. The TAA coding region is indicated by a solid box. Key: B, *Bam*HI; Bg, *Bgl*II; E, *Eco*RI; H, *Hin*dIII; K, *Kpn*I; N, *Nru*I; P, *Pst*I, S, *Sal*I, Sm, *Sma*I, Sp, *Sph*I; Xb, *Xba*I.

```
                                                                                        -617         -601
                                                                                        aattcatggtgtttg
-600                                               -550                                              -501
atcatttaaatttttatatggcgggtggtgggcaactcgcttgcgcgggcaactcgcttaccgattacgttagggctgatatttacgtaaaatcgtca
-500                                               -450                                              -401
agggatgcaagaccaaagtagtaaaccccggagtcaacagcatccaagcccaagtccttcacggagaaaccccagcgtccacatcacgagcgaaggacc
-400                                               -350                                              -301
acctctaggcatcggacgcaccatccaaatagaagcagcaaagcgaaacagcccaagaaaaaggtcggcccgtcggccttttctgcaacgctgatcacgg
                              (t)
-300                                               -250                                              -201
gcagcgatccaaccaacaccctccagagtgactaggggcggaaatttaaagggattaatttccactcaaccacaaatcacagtcgtccccggtattgtcc
-200                                               -150                                              -101
tgcagaatgcaatttaaactcttctgcgaatcgcttggattccccgcccctggccgtagagcttaaagtatgtcccttgtcgatgcgatgtatcacaaca
-100                                               -50                                                 -1
tataaatactagcaagggatgccatgcttggaggatagcaaccgacaacatcacatcaagctctcccttctctgaacaataaaccccacagaaggcattt
1                                                  50                                                100
ATGATGGTCGCGTGGTGGTCTCTATTTCTGTACGGCCTTCAGGTCGCGGCACCTGCTTTGGCTGCAACGCCTGCGGACTGGCGATCGCAATCCATTTATT
[M  M  V  A  W  W  S  L  F  L  Y  G  L  Q  V  A  A  P  A  L  A] A  T  P  A  D  W  R  S  Q  S  I  Y  F
                                                   150                                               200
TCCTTCTCACGGATCGATTTGCAAGGACGGATGGGTCGACGACTGCGACTTGTAATACTGCGGATCAGgtgtgttgttacctactagctttcagaaagag
                                                                  (G)                        INTRON I
  L  L  T  D  R  F  A  R  T  D  G  S  T  T  A  T  C  N  T  A  D  Q(R)
                                                   250                                               300
gaatgtaaactgacttgatatagAAATACTGTGGTGGAACATGGCAGGGCATCATCGACAAGgtaaattgcccctttatcaaaaaaaaagaaggaaaagc
                         K  Y  C  G  G  T  W  Q  G  I  I  D  K                     INTRON II
                                                   350                                               400
agaagaaaaataaaataaaaagaactctagtcctaaccatcacatagTTGGACTATATCCAGGGAATGGGCTTCACAGCCATCTGGATCACCCCCGTTAC
                                                 L  D  Y  I  Q  G  M  G  F  T  A  I  W  I  T  P  V  T
                                                   450                                               500
AGCCCAGCTGCCCCAGACCACCGCATATGGAGATGCCTACCATGGCTACTGGCAGCAGGATATgtaagtcgatttctttaaatatctacctgtcatcttt
 A  Q  L  P  Q  T  T  A  Y  G  D  A  Y  H  G  Y  W  Q  Q  D  I                           INTRON III
                                                   550                                               600
tacatcaatatgaactaacttgatggtttagATACTCTCTGAACGAAAACTACGGCACTGCAGATGACTTGAAGGCGCTCTCTTCGGCCCTTCATGAGA
                                  Y  S  L  N  E  N  Y  G  T  A  D  D  L  K  A  L  S  S  A  L  H  E  R
                                                   650                                               700
GGGGGATGTATCTTATGGTCGATGTGGTTGCTAACCATATGgttcgtggtcctttgcaactgacttcgcggatatggttcatttcagtactgacaatgag
  G  M  Y  L  M  V  D  V  V  S  N  H  M                                INTRON IV
                                                   750                                               800
taatatcagGGCTATGATGGAGCGGGTAGCTCAGTCGATTACAGTGTGTTTAAACCGTTCAGTTCCCAAGACTACTTCCACCCGTTCTGTTTCATTCAAA
                                                                                             (C)
            G  Y  D  G  A  G  S  S  V  D  Y  S  V  F  K  P  F  S  S  Q  D  Y  F  H  P  F  C  F(L)I  Q  N
                                                   850                                               900
ACTATGAAGATCAGACTCAGGTTGAGGATTGCTGGCTAGGAGATAACACTGTCTCCTTGCCTGATCTCGATACCACCAAGGATGTGGTCAAGAATGAATG
  Y  E  D  Q  T  Q  V  E  D  C  W  L  G  D  N  T  V  S  L  P  D  L  D  T  T  K  D  V  V  K  N  E  W
                                                   950                                              1000
GTACGACTGGGTGGATCATTGGTATCGAACTACTCCAgtaagatatttctccctcattctacaacttggctgatcgatgatacttacgaaatcagTTGA
 Y  D  W  V  G  S  L  V  S  N  Y  S  I                         INTRON V                             D
                                                   1050                                             1100
CGGCCTCCGTATCGACACAGTAAAACACGTCCAGAAGGACTTCTGGCCCGGGTACAACAAAGCCGCAGGCGTGTACTGTATCGGCGAGGTGCTCGACGGT
 G  L  R  I  D  T  V  K  H  V  Q  K  D  F  W  P  G  Y  N  K  A  A  G  V  Y  C  I  G  E  V  L  D  G
                                                   1150                                             1200
GATCCGGCCTACACTTGTCCCTACCAGAACGTCATGGACGGCGTACTGAACTATCCCATgtatggttcctccaaccatgagccttcttgcaagtctcatc
D  P  A  Y  T  C  P  Y  Q  N  V  M  D  G  V  L  N  Y  P  I                           INTRON VI
                                                   1250                                             1300
tcctaacgaaacggctaaaaccagTTACTATCCACTCCTCAACGCCTTCAAGTCAACCTCCGGCAGCATGGACGACCTCTACAACATGATCAACACCGTC
                          Y  Y  P  L  L  N  A  F  K  S  T  S  G  S  M  D  D  L  Y  N  M  I  N  T  V
                                                   1350                                             1400
AAATCCGACTGTCCAGACTCAACACTCCTGGGCACATTCGTCGAGAACCACGACAACCCACGGTTCGCTTCgtaagtcttccctttatttccgttccc
K  S  D  C  P  D  S  T  L  L  G  T  F  V  E  N  H  D  N  P  R  F  A  S                     INTRON VII
                                                   1450                                             1500
aatttccacacagaaccccacctaacaagagcaaagTTACACCAACGACATAGCCCTCGCCAAGAACGTCGCAGCATTCATCATCCTCAACGACGGAATC
                                      Y  T  N  D  I  A  L  A  K  N  V  A  A  F  I  I  L  N  D  G  I
                                                   1550                                             1600
CCCATCATCTACGCCGGCCAAGAACAGCACTACGCCGGCGGAAACGACCCCGCGAACCGCGAAGCAACCTGGCTCTCGGGCTACCCGACCGACAGCGAGC
P  I  I  Y  A  G  Q  E  Q  H  Y  A  G  G  N  D  P  A  N  R  E  A  T  W  L  S  G  Y  P  T  D  S  E  L
                                                   1650                                             1700
TGTACAAGTTAATTGCCTCCGCGAACGCAATCCGGAACTATGCCATTAGCAAAGATACAGGATTCGTGACCTACAAGgtaagcacaacctctaagcatac
  Y  K  L  I  A  S  A  N  A  I  R  N  Y  A  I  S  K  D  T  G  F  V  T  Y  K
                                                   1750                                             1800
cctaatggcctatcttcagagtatctgacacaagagactaatcactggcaatacagAACTGGCCCATCTACAAAGACGACACAACGATCGCCATGCGCAA
      INTRON VIII                                          N  W  P  I  Y  K  D  D  T  T  I  A  M  R  K
                                                   1850                                             1900
GGGCACAGATGGGTCGCAGATCGTGACTATCTTGTCCAACAAGGGTGCTTCGGGTGATTCGTATACCCTCTCCTTGAGTGGTGCGGGTTACACAGCCGGC
  G  T  D  G  S  Q  I  V  T  I  L  S  N  K  G  A  S  G  D  S  Y  T  L  S  L  S  G  A  G  Y  T  A  G
                                                   1950                                             2000
CAGCAATTGACGGAGGTCATTGGCTGCACGACCGTGACGGTTGGTTCGGATGGAAATGTGCCTGTTCCTATGGCAGGTGGGCTACCTAGGGTATTGTATC
Q  Q  L  T  E  V  I  G  C  T  T  V  T  V  G  S  D  G  N  V  P  V  P  M  A  G  G  L  P  R  V  L  Y  P
                                                   2050                                             2100
CGACTGAGAAGTTGGCAGGTAGCAAGATCTGTAGTAGCTCGTGAagggtggagagtatatgatggtactgctattcaatctggcattggacagtgagttt
  T  E  K  L  A  G  S  K  I  C  S  S  S  Stop
                                                   2150                                             2200
gagtttgatgtacagttggagtcgttactgctgtcatccccttatactcttcgattgtttttcgaaccctaatgccaagcacgctagtctattataggaa   amy1
            tgtataaatctagtgtactttgcacccaccacgcaatgaaacggcaccgggccccgtctgagagcccgtctcgaatccctgttggtca   amy2
            acttgtctattctatgatgtatggtctttttgttctatagttggaaatcggaatgatctcaaatcttgaataaatataaaaggataa   amy3
                                                   2250                                             2300
aggatccggattaatgtgttttcataacgcggtactgtatggtacttctgtattatatcaccgaagctcatgtatcttacatgtatatattatacagaca   amy1
tcttccatcgcttcgtcctccagaggcgaggatgcaccccagccctgcgagctgtgagttgttaaggtgtaacgcttggtgggtcggtggcttctcggca   amy2
tactcacatccatcacaaccttacaaggttaattccgagctatattccaccgacacaaataggcagtatcttctctcgccagagaatcgcgatattattg   amy3
```

Figure 3.16 Nucleotide and deduced amino acid sequence of α-amylase genes of *A. oryzae* (Wirsel *et al.* [122]). Putative TATA box and CAAT boxes are overlined. The GT–AG sequence related to internal consensus sequence, PuCTPuAC, in introns and a putative polyadenylation signal in the 3′ flanking region are underlined. The signal peptide sequence is boxed. Different nucleotides and amino acids of *amy3* from those of *amy1* and *amy2* are shown in parentheses. Significant divergences observed in nucleotide sequences of three genes beyond 69 bp downstream of the stop codon are shown.

617 bp 5′ flanking region and considerable changes in the 3′ flanking region.

The nucleotide sequence of the genes indicates that the α-amylase coding region encodes a predicted polypeptide of 499 amino acids. The genes are interrupted by eight introns ranging in length from 55 to 86 bp (Figure 3.16). There are no differences in the positions of introns or in their sequences. All eight introns start with a GT sequence and terminate with an AG sequence, a general feature of introns [125]. In addition, a sequence homologous to the internal consensus sequence, PuCTPuAC, reported for fungal introns [125], is observed 8 to 18 bp from the 3′ splicing sites. An interesting finding is that the region of the ninth exon seems to correspond to a C-terminal domain by X-ray analysis [117] but the function of this still remains to be elucidated. Compared with the amino acid sequence of mature TAA [115], an additional 21 amino acids which have hydrophobic character are observed at the N-terminus of the cloned *amy* genes this is thought to be a signal sequence. Furthermore, one insertion and one deletion, as well as 10 amino acid substitutions, are found throughout the whole sequence of the two identical genes.

In the 5′ and 3′ flanking regions of fungal genes, there are several 'core promoter' consensus sequences [125]. A TATA sequence is found 100 bp upstream from the translation start codon and a CAAT box is present at −192 bp. However, in the 3′ flanking region, a putative polyadenylation signal, AATAAA, is found only in *amy3* about 140 bp downstream from the translational stop codon.

Since *A. oryzae* strains used by these groups are perhaps genealogically unrelated, the existence of multiple α-amylase genes in the genome may be a general characteristic for *A. oryzae*. *A. oryzae* is very closely related to *A. sojae*, which is utilized in soy sauce manufacture, as well as *A. flavus* and *A. parasiticus* [126]. Among these four species, *A. oryzae* is best known as an α-amylase hyperproducer and has been used extensively for industrial production [110]. Genomic Southern blot analysis of a number of strains of the four species has revealed that all strains of *A. oryzae* investigated have multiple *amy* genes, whereas *A. sojae* has a single *amy* gene (K. Gomi, unpublished). This finding of multiple *amy* genes is an intriguingly distinctive feature of *A. oryzae*. This suggests that the ability of the organism to make rapid use of available carbon sources by high α-amylase activity could have been selected preferentially during long-term cultivation on cereal grains such as rice which is used in saké manufacture.

The observation that there exist multiple *amy* genes in *A. oryzae* could explain the reason why this organism is a high producer of α-amylase. However, it has not been formally determined if some of the *amy* genes are non-functional. Gomi and co-workers [121] have demonstrated that the introduction of each of the three genes into *A. oryzae* and *A. nidulans* by transformation allowed the host to produce significantly higher α-amylase activities suggesting that all of these *amy* genes are indeed functional. Gines and co-workers [123] isolated cDNA clones of two *amy* genes (*AmyI* and *AmyII*), indicating that these two genes are expressed. Wirsel and colleagues [122] reported studies which showed that one of the three genes (*amy2*) is not expressed. Notwithstanding this, *amy2* has an identical nucleotide sequence to *amy1*, which was shown to be transcribed. Furthermore, Tsukagoshi and colleagues [124] showed that many *amy2* cDNAs could be isolated and that this gene is indeed actively transcribed.

Reports that isolated phage or cosmid DNA clones of *A. oryzae* contained only a single *amy* gene indicated that these gene families are not clustered. In order to elucidate the location of *amy* genes on individual chromosomes, physical mapping of the genes has been performed using orthogonal-field-alternation gel electrophoresis (OFAGE). Under appropriate conditions, at least six chromosome bands were resolved from two genealogically unrelated *A. oryzae* strains, RIB 40 and M-2-3 [127]. These chromosome bands were numbered consecutively from the smallest (I) to the largest (VI). At least two bands, I and VI, are probably doublets, as judged by the relative intensity of ethidium bromide staining. Consequently, there may be eight chromosomes present in *A. oryzae*, as has been found for *A. nidulans* [128] and *A. niger* [129]. The *amy* probe hybridized to three chromosome bands, III, V and VI, (Figure 3.17) indicating that the three *amy* genes are located on different chromosomes. Further analysis must be carried out to determine whether multiple *amy* genes are dispersed on different chromosomes in other strains.

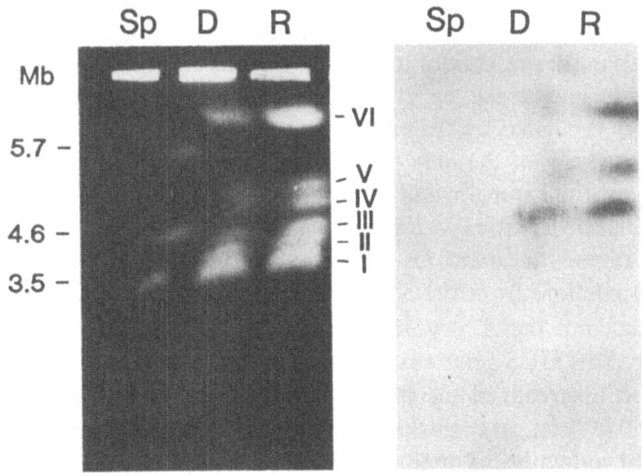

Figure 3.17 Chromosomal location of the three α-amylase genes in *A. oryzae*. (Gomi *et al.* [121]). Left panel: EtdBr staining of agarose gel. Lanes: Sp, *Schizosaccharomyces pombe* as molecular size markers, which are given in the panel; D, *A. oryzae* RIB 40; R, *A. oryzae* M-2-3. Six chromosome bands resolved are consecutively numbered from the smallest (I) to the largest (VI). Right panel: Autoradiogram of the blot following hybridization to the 3 kbp *Eco*RI-*Bam*HI fragment containing *amyB*.

It is known that the black aspergillus, *A. niger*, produces two forms of α-amylase; one which is acid stable, retaining most of its activity even at pH 2 and the other which is acid unstable, being sensitive at low pH, similar to *A. oryzae* TAA [130]. In addition, *A. kawachii*, used for shochu production and which also belongs to the *A. niger* group, produces two types of acid-stable α-amylase (MW 104 000 and MW 66 000) [131]. Recently, the nucleotide sequence and inferred amino acid sequence of an acid-stable α-amylase encoding gene of *A. niger* has been reported. The protein shows approximately 70% homology with the TAA protein [132]. Two genes encoding TAA-type α-amylase have been isolated from *A. awamori*. Their nucleotide sequence showed a very high degree of homology (more than 98%) to that of the *amy* gene of *A. oryzae* [133].

The production of α-amylase is induced in the presence of starch or oligosaccharides [118, 119]. This, of course, makes the *amy* gene an attractive model system for the analysis of regulational control of gene expression although the molecular mechanism of induction of α-amylase has not yet been well characterized. Northern blot analysis of the *amy* gene was performed to analyse the regulational control of the gene expression. Total RNA was extracted from a glucose- or starch-grown culture of *A. oryzae* and separated by formaldehyde-agarose gel electrophoresis. Hybridization with the *amy* gene as a probe revealed a discrete RNA species which was formed in abundance with starch as the sole carbon source [134]. This result confirmed that the inducible expression of *amy* is controlled at the transcriptional level. To analyse the molecular mechanism of *amy* induction, 617 bp of the *amy* gene promoter were fused to *E. coli uidA* gene (encoding β-glucuronidase, GUS) and transformed into *A. oryzae* [134]. Transformants express the fusion gene at high levels in the presence of starch and various malto-oligosaccharides, but at low levels in glucose or glycerin cultures, indicating that GUS gene expression is regulated by the *amy* promoter. In addition, it was shown that nucleotide (nt) 617 of the *amy* 5′ flanking region is sufficient to confer *amy* regulation on the heterologous reporter gene (Figure 3.18). Using this gene fusion, a series of deletions within the *amy* promoter has been made in order to locate the regions required for regulation and high-level expression [135]. Sequential deletions from the 5′ upstream region of the *amy* promoter were reintroduced into the *A. oryzae* recipient strain. Regulation and high-level expression of the GUS gene was observed in the transformants with deletions up to 377 nt upstream of the translation initiation point. The transformant with the 290 nt deletion construct showed significantly reduced levels of GUS activity and induction by maltose was abolished (Figure 3.18). Such results indicated that distinct DNA sequence elements from nts −377 to −290 appear to be responsible for the high-level expression and induction of the *amy* gene. Further experiments should permit identification of *cis*-acting elements (UAS- or enhancer or receptor sequences) within this 87 bp region and detection of DNA binding proteins interacting with those elements.

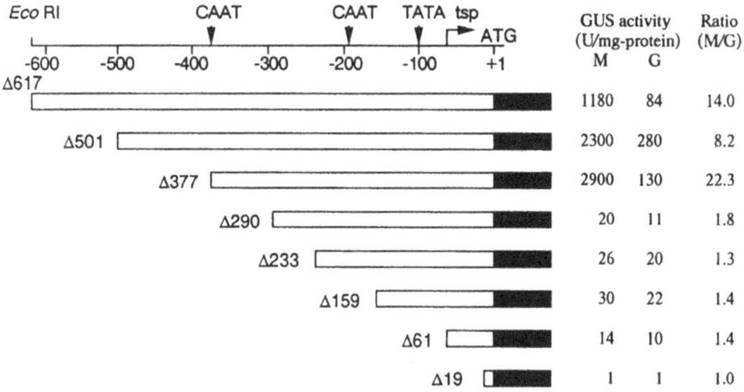

Figure 3.18 Effect of deletions of the *amy* promoter sequence on GUS gene expression [135]. Numbers of the deletions denote the distance in base pairs from the translation start point at +1 to the deletion endpoint. A TATA box exists at −100 bp and putative CAAT boxes exist at −192 bp and −377 bp. Major transcription start points (tsp) are located at −66, −64 and −61 bp. The *amy* promoter region is indicated by an open box and the GUS structural gene is indicated by a solid box.

Since the *amy* promoter is considered to be a strong expression promoter, it has been used to direct the expression of foreign genes at high levels using *A. oryzae* as a host (chapters 5 and 6). In this regard, an aspartic protease of *M. miehei*, which is known as microbial rennet, a substitute of calf chymosin in cheese making, has been secreted at levels in excess of 3 g/l when cDNA coding for the protease was fused downstream of *amy* promoter and expressed in *A. oryzae*.

3.9.2 *Glucoamylases*

Glucoamylase catalyses the release of glucose from the non-reducing termini of starch or related maltodextrins. Glucoamylases of *Aspergillus* and *Rhizopus* have been extensively characterized at the biochemical and molecular levels (Table 3.6). It is known that glucoamylase of *Aspergillus* (*A. niger* and *A. awamori*) exists as two major forms; a larger form (designated G1, MW 70 000) can absorb and degrade raw starch, whereas the smaller form (designated G2, MW 60 000) cannot absorb raw starch and can only degrade soluble starch or maltodextrins. These two forms of glucoamylase appear equal in activity toward soluble starch and oligosaccharides.

The G1 and G2 forms are very similar on the basis of partial amino acid sequence and immunological reactivity. However, although they share a common N-terminal amino acid sequence [136], the G2 form lacks the C-terminal region present in G1. The G2 form is detected in the culture filtrates of *A. niger* and is probably formed by limited proteolysis. *In vitro* treatment of the G1 form of glucoamylase from *A. awamori* var. *kawachi* with fungal acid protease or subtilisin, generated the smaller forms similar to those

Table 3.6 Characteristics of various glucoamylases.

Source	Optimum pH	Molecular weight	No. of isozymes	Reference
Aspergillus awamori	4.5	125 000	1	[161]
A. awamori var. *kawachi*	3.8	83 000, 90 000	3	[137]
A. niger	5.0	95 000, 112 000	2	[162]
A. oryzae	4.5	38 000, 76 000	3	[163]
A. phoenicis	4.6	63 600	2	[164]
A. saitoi	4.5	90 000	2	[165]
Cladosporium resinae	3.5–4.0	70 000, 82 000	2	[166]
Lentinus edodes	4.5	55 000	1	[167]
Mucor rouxianus	4.6 5.0	59 000 49 000	2	[168]
Paecilomyces varioti	4.5	69 000	1	[169]
Penicillium oxalicum	4.5 5.0	86 000 84 000	2	[170]
Rhizopus niveus	4.5	60 000(D)	5	[171]
Rhizopus sp.	4.5–5.0	59 000–74 000	3	[157]
Saccharomyces diastaticus	5.0	66 000–80 000	3	[172]

isolated from culture filtrates [137, 138]. Such smaller glycoamylases fail to bind and digest raw starch. There is a controversy regarding whether or not post-transcriptional or post-translational processing occurs in the formation of two forms of *A. niger* glucoamylase.

In order to elucidate the molecular relationships between the Gl and G2 forms, cDNA and genomic DNA encoding glucoamylase were isolated from *A. niger* [139]. Form G1 is synthesized as a precursor of 640 amino acid residues including a putative signal peptide of 18 amino acids with a short propeptide of 6 residues, which are probably processed by trypsin- or yeast Kex2-like protease (Figure 3.21). The sequence of the mature glucoamylase proteins of 616 amino acid residues is in good agreement with the results reported by Svensson and colleagues [140]. At the 3′ end of the cDNA the putative polyadenylation nucleotide motif AATAAA, is missing.

Total poly(A)$^+$ mRNA was translated *in vitro* using rabbit reticulocytes followed by immunoprecipitation with glucoamylase-specific antisera. Two proteins with molecular sizes of 71 and 61 kDa were observed by SDS-PAGE [139]. The sizes of the two proteins were consistent with those of the G1 and G2 forms of glucoamylase, indicating that these two forms may be synthesized as the primary translation products. The nucleotide sequences of several glucoamylase-specific cDNA recombinants showed that several clones lack a 169 bp intervening sequence that may be spliced out from mRNA coding for a full length G1 form of glucoamylase. The deletion of 169 bp nucleotides is flanked by the consensus sequences, GT and AG respectively at the 5′ and 3′ splicing sites. Moreover, deletion of this intervening sequence from the G1 mRNA results in a frame shift which gives a smaller protein with 8 additional

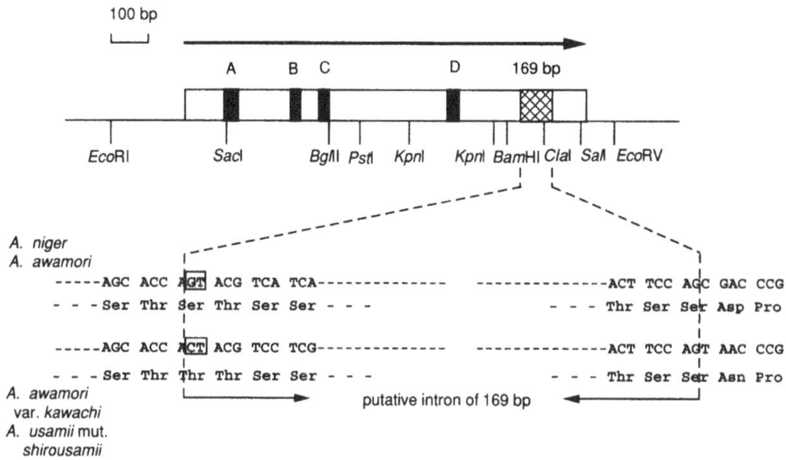

Figure 3.19 The structure of the glucoamylase genes of the *A. niger* group. The upper part shows the structure and restriction map of the glucoamylase gene from *A. niger*. Open boxes are exons and solid boxes are introns. A hatched box indicates a putative intron consisting of 169 bp. The lower part shows the nucleotide sequences around the putative 169 bp intron junctions of the glucoamylase genes from four species of *Aspergillus*. Shaded boxes indicate the differences in the exon/intron junction between *A. niger* (*A. awamori*) and *A. awamori* var. *kawachi* (*A. usamii* mut. *shirousamii*) genes.

amino acid residues at the C-terminus, ^{502}Thr. This mRNA form would code for a protein of 510 residues with a molecular weight of about 53 700, consistent with the size of the G2 form of glucoamylase. Boel *et al.* [139] have proposed that the G2 form may be synthesized from a differentially spliced mRNA, which is transcribed from a single gene which originally encodes the G1 form.

An *A. niger* genomic library was screened with a glucoamylase-specific cDNA as the probe. A DNA fragment containing a 2.5 kbp fragment was isolated and sequenced. Compared with the nucleotide sequence of cDNA and genomic DNA, the existence of four intervening sequences in addition to one of 169 bp was found. The four intervening sequences range in length from 55 to 75 bp, similar to *amy* gene of *A. oryzae*. All of the introns, including the 169 bp sequence, have typical features of fungal intron; the exon/intron junctions follow the GT–AG rule and an internal consensus sequence concerning a lariat formation in the splicing, PuCTPuAC, also is found in each intron. In the 5′ flanking regions, a TATA box and a CAAT box exist in the −35 bp and −100 bp respectively, relative to the transcription start site (+1) [141].

Nunberg and colleagues [142] also isolated glucoamylase cDNA and genomic DNA clones from *A. awamori*, a fungus which is very closely related to *A. niger*. The deduced amino acid sequence of the *A. awamori* glucoamylase was completely identical to that of *A. niger* [139] and the structures of the two genes were almost the same (Figure 3.20). Four intervening sequences were

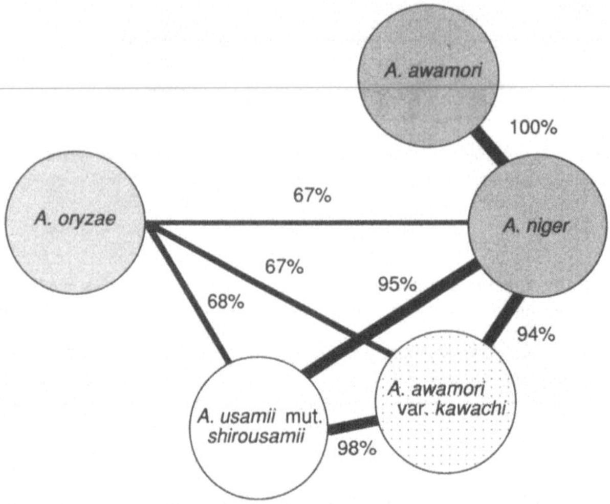

Figure 3.20 Diagrammatic representation of the level of the homology at amino acid sequences among *Aspergillus* glucoamylases.

identified at identical positions. In addition, a putative fifth intervening sequence of 169 bp was also found in the *A. awamori* glucoamylase gene. Boel and colleagues [141] and Nunberg and coworkers [142] have carried out Southern hybridization analysis of the DNA from both species, showing that glucoamylase exists as a single copy. However, Nunberg *et al.* failed to isolate a cDNA clone derived from the presumptive spliced G2 mRNA, although in their *in vitro* translation experiments a minor immunoprecipitable protein, probably corresponding to G2 form, was observed.

Glucoamylase G3 purified from a commercial preparation from *A. niger* has been characterized with respect to its amino acid sequence [143]. G2 consists of molecular species identical to residues ^{1}Ala-^{512}Pro and ^{1}Ala-^{514}Ala corresponding to the G1 polypeptic chain, which consists of 616 amino acid residues. Such proteins are almost similar in size to the presumptive product encoded by a differentially spliced mRNA derived from G1 mRNA [139]. The latter protein would consist of 510 amino acid residues and have a C-terminal peptide of 8 residues, i.e. Thr-Thr-Arg-Ser-Gly-Met-Ser-Leu. However, sequencing of the G2 protein failed to reveal Met, Arg or Leu in the C-terminal peptide fragment [143]. Therefore it is likely that G2 is generated by limited proteolysis in the C-terminal region of G1 rather than being synthesized from a spliced mRNA, although it is not ruled out that post-transcriptional processing may occur under certain conditions of growth.

Glucoamylase encoding genes have been cloned and sequenced from two further aspergilli belonging to the *A. niger* group, namely *A. awamori* var. *kawachi* [144] and *A. usamii* mut. *shirousamii* [145]. The deduced amino acid sequences of both these glucoamylases were nearly identical. Compared with

A. niger G1, both glucoamylases had 30 amino acid replacements and a deletion of one amino acid giving a total of 615 amino acid residues. In both genes the nucleotide change of GT to CT occurred at the 5′ splicing site of the putative 169 bp intervening sequence, which is a deviation from the GT-AG consensus (Figure 3.19). Furthermore, *in vitro* translation of total mRNA prepared from *A. awamori* var. *kawachi* mycelium resulted in the synthesis of only one protein product, which was detected by SDS-PAGE after immunoprecipitation with glucoamylase-specific antisera [146]. This protein was found to have a molecular weight of 69 kDa. Therefore, it is unlikely that the smaller forms of glucoamylase are generated through splicing out the putative 169 bp fifth intron of the mRNA encoding the larger form at least in *A. awamori* var. *kawachi* and *A. usamii* mut. *shirousamii.*

The amino acid sequences of *A. niger* G1 and G2 suggest that the G1 form has three domains (Figure 3.21): (i) a large catalytic domain (^{1}Ala-^{440}Thr) with a high degree of homology to glycoamylases from *Rhizopus* (and yeast); (ii) a starch-binding domain at the C-terminus (^{513}Thr-^{616}Arg), which confers the ability to absorb and degrade raw starch; and (iii) a highly *O*-glycosylated, Ser/Thr-rich region (^{441}Ser- ^{512}Pro) which links catalytic and starch-binding domains. To determine the extent of the *O*-glycosylated, Ser/Thr-rich region necessary for activity and stability of glucoamylase, a series of C-terminally truncated forms of G1 were made by insertion of a stop codon into appropriate restriction sites located within the gene [147] and expressed in yeast for analysis. Truncated forms with the C-terminus at position 513, which is comparable to the G2 form, 496 or 482 appeared to be fully active, as judged by bioassays on starch-containing plates. Moreover, these mutant enzyme forms showed similar thermal stability to the G1 form. In contrast, mutant forms at position 460 or 416 cause decrease or even loss of enzyme activity as well as loss of detectable protein in the culture supernatants. This may result from proteolytic degradation or failure of these forms to be secreted. Such results indicate that the *O*-glycosylated region (441–482) is essential for G2 to retain catalytic activity, structural integrity and secretion.

Limited proteolysis of G1 of *A. awamori* var. *kawachi* with subtilisin resulted in the liberation of a glycopeptide (Gp-I) of 45 residues corresponding to ^{471}Ala-^{515}Val, with regard to the *A. niger* G1 form [148]. Hayashida *et al* [148] proposed that this highly *O*-glycosylated region may function as the G1 affinity site for binding to raw starch. This region, however, is almost completely included in G2, and thus is assumed to be less important for starch binding. Neverthless, since Gp-I showed characteristic effects on the rate of raw starch digestion with G1 [149], the peptide region may act as a supplementary binding site.

To confer starch binding ability to fusion proteins by utilizing a starch binding domain of glucoamylase as a polypeptide 'tail', the 133 C-terminal amino acid residues (484–616) of G1 were fused to the C-terminus of the *E. coli* *β*-galactosidase [150]. The resultant fusion protein was able to bind strongly to raw starch granules and thus absorbed protein could be eluted at a high level of purity. Such results show (i) the existence of a strong starch-binding activity in the G1 C-terminal region and (ii) that this region is functional.

Glucoamylase produced by *A. oryzae* is assumed to be important in saké brewing because fermentation rates are dependent on glucoamylase activity. Furthermore, in contrast to the glucoamylase synthesized by the *A. niger* group, *A. oryzae* glucoamylases show a low level of absorbability and digestability of raw starch [151]. Hata *et al.* [152, 153] isolated glucoamylase (*glaA*) cDNA and genomic DNA clones from *A. oryzae* and determined their nucleotide sequence. Glucoamylase is synthesized as a precursor of 612 amino acids including a signal peptide, which so far has not been determined. Thus the *A. oryzae* glucoamylase is smaller than those of the *A. niger* group. The level of homology at the amino acid level between *A. oryzae* and *A. niger* glucoamylases is approximately 70%. However, both glucoamylases share significantly higher similarities (78–94%) in the five conserved regions from

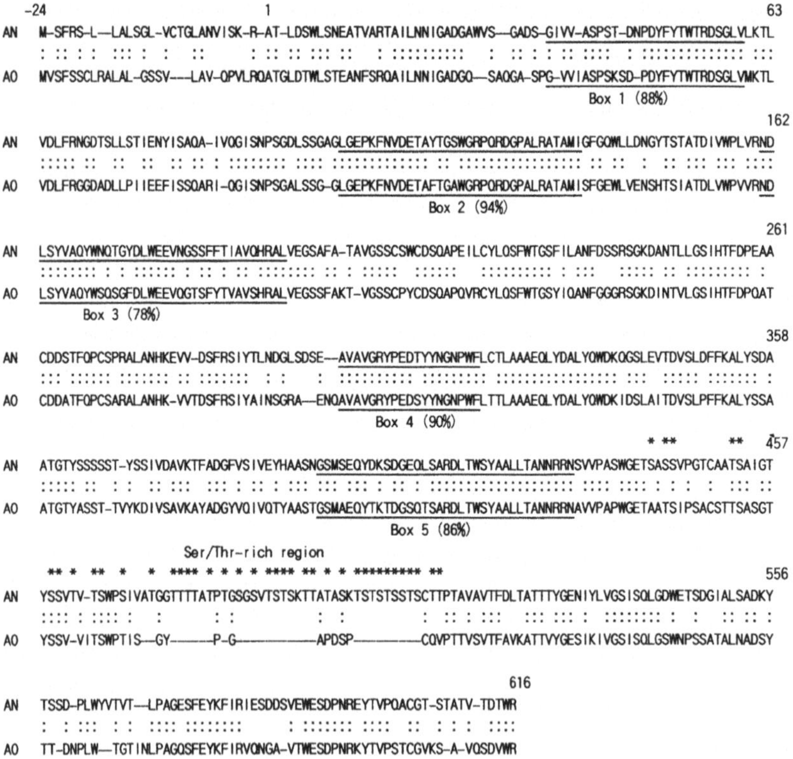

Figure 3.21 Sequence homology of glucoamylases from *A. oryzae* and *A. niger* [152]: AN, *A. niger* glucoamylase; AO, A. oryzae glucoamylase. Identical amino acid residues are indicated by a colon (:). Five highly conserved segments (boxes 1 to 5) [154] are shown in bold letters. Numbers in parentheses indicate the degree of homologies in each segment. Asterisks indicate the *O*-glycosylate residues of Thr and Ser proposed by Svensson *et al.* [160] in *A. niger* glucoamylase.

several fungal glucoamylases [154] (Figure 3.21). A distinctive feature of *A. oryzae* glucoamylase is that no peptide region abundant in Ser or Thr exists, although Ser/Thr-rich regions are thought to be involved in starch-binding ability [148]. Ser/Thr-rich regions may, however, be of minor importance for starch-binding and it will be interesting to discover whether the restricted activity towards raw starch and the deletion of the Ser/Thr-rich region will support this notion. It cannot be ruled out that the C-terminal region of glycoamylase is processed by limited proteolysis during processing through the secretory pathway. The positions of the four introns are conserved between *A. niger* and *A. oryzae* [153].

The expression of the *A. niger* glucoamylase gene is regulated by the carbon sources and assumed to occur at the transcriptional level [155]. Deletion analysis has been performed to determine more precisely the nucleotides responsible for the control of glucoamylase gene expression [156]. A series of deletion constructs within 1966 bp of the 5′ flanking region of *glaA* was introduced into an *A. niger* recipient (the resident glucoamylase gene had been disrupted by a gene replacement) and glucoamylase activity in transformants determined by immunoassays. When grown in media containing differing carbon sources, transformant 562 (transformant number is relative to the translation start point) showed wild-type regulation and high-level expression. In contrast, high-level expression was not observed in transformants 318 and 224. Therefore, between positions 318 and 562 bp upstream of the glucoamylase gene sequences are necessary for high-level gene expression [156]. The promoter of *A. niger* glucoamylase gene has been widely utilized for the efficient expression of a number of foreign genes, including mammalian genes (see chapter 5).

Rhizopus glucoamylase also exists as multiple forms, designated *Gluc1*, *Gluc2* and *Gluc3*, which differ in molecular size and possess different N-terminal regions [157, 158]. *Gluc2* and *Gluc3* are thought to arise from *Gluc1* by limited proteolysis. Clones from both *R. oryzae* genomic and cDNA libraries were isolated using synthetic oligonucleotides as probes [159]. Nucleotide sequencing analysis of the cDNA and genomic DNA encoding glucoamylase showed that the gene encodes a precursor protein of 604 amino acid residues, including a putative signal peptide at 25 residues. Southern blot analysis revealed the presence of a single copy of the glucoamylase gene in the *Rhizopus* chromosome, suggesting that multiple forms of the enzyme are formed from a single gene. The N-terminal amino acids of *Gluc1*, *Gluc2* and *Gluc3* are Ala, Glu and Lys, respectively. As judged from both carbohydrate contents and amino acid compositions of *Gluc2* and *Gluc3*, the N-termini could be assigned to ^{134}Glu for *Gluc2* and ^{85}Lys for *Gluc3*. Although all three forms are equally active towards soluble starch, only the largest form, i.e. *Gluc1*, has the ability to absorb and degrade raw starch. It is likely, therefore, that the N-terminal region of the *Rhizopus* glucoamylase is involved in the binding and digestion of raw starch.

3.10 Summary

Filamentous fungi are of great importance for traditional oriental food and beverage industries. It is interesting to note that the two major fungi used traditionally in China and in Japan, namely *Rhizopus* and *Aspergillus* respectively, have been the most important fungi for modern fermentation industries for enzyme production. Perhaps it is their ability to produce useful enzymes such as amylases, proteinases and lipases that allow such fungi to grow vigorously and abundantly. Industries based on the fungal enzymes in the oriental world do not require pressure or very high temperatures and the ability to produce enzymes is greater in fungi than any other class of microbes. Glucoamylase and other amylases produced by *Rhizopus* or *Aspergillus* reach over 20 g/l in liquid culture medium, and 30 g/kg in the form of the traditional solid Koji culture, the latter being still the best method to produce taka-amylase, using *Aspergillus oryzae*. Among bacteria, *Bacillus brevis* (Udaka's strain) is perhaps the best with 12 g of its outer-membrane protein per litre of culture secreted out of the cell although for other bacteria 1 g/l is more usual.

References

1. Sakaguchi, K. and Mori, H. (1969) *J. Gen. Appl. Microbiol.* **15**: 159–167.
2. Sakaguchi, K. (1959) *Bull. Agric. Chem. Soc. Japan* **23**: 100, 438, 443.
3. Onishi, H. (1957) *Bull. Agric. Chem. Soc. Japan* **21**: 137, 143, 151.
4. Mizunuma, T. (1988) In: *Jozo no Jiten* (Nojiro, K. *et al.*, eds), Asakura Press, Tokyo, p. 398.
5. Koizumi, T. (1988) Alcoholic beverages in China. In: *Jozo no Jiten* (Nojiro, K. *et al.*, eds), Asakura Press, Tokyo, pp. 384–397.
6. Su, Y.C. (1975) *J. Ferment. Soc.* **33**: 28.
7. Su, Y.C. *et al.* (1976) *Rep. Taiwan Natl Univ. Dept. Agric.* **16**: 93.
8. Fukumoto, J., Tsujisaka, Y. and Araki, M. (1960) *Chem. Abstr.* **55**: 24, 917.
9. Veda, S. and Saha, B.C. (1983) *Enzyme Microb. Technol.* **5**: 196.
10. Yamashita, I., Itoh, T. and Fukui, S. (1985) *Agric. Biol. Chem.* **49**: 3089.
11. Yamashita, I., Itoh, T. and Fukui, S. (1985) *Appl. Microbiol. Biotechnol.* **23**: 130.
12. Itoh, T., Ohtsuki, I., Yamashita, I. and Fukui, S. (1987) *J. Bacteriol.* **169**: 4171.
13. Itoh, T., Ohtsuki, I., Yamashita, I. and Fukui, S. (1987) *Agric. Biol. Chem.* **49**: 3089.
14. Hesseltine, C.W. (1965) *Mycologia* **57**: 1.
15. Kozaki, M. (1988) *Abstracts of 22nd Meeting*, Japan Food Industries Society, p. 12.
16. Shurtleff, W. and Aoyagi, A. (1979) *The Book of Tempeh*, Harper & Row.
17. Gandjar, J. (1981) *Adv. Biotechnol.* **2**: 531.
18. Madsen, G.B., Norman, B.E. and Slott, S. (1973) *Stärke* **25**: 304.
19. Sakano, Y. (1989) In: *Experimental Methods on Enzymes Related to Starch and Other Sugars* (Nakamura, M. and Kainuma, K., eds), Gakkai Shuppan Center, Tokyo, p. 164.
20. Yamanaka, K. (1963) *Agric. Biol. Chem.* **27**: 265.
21. Tsumura, N. and Sato, T. (1965) *Agric. Biol. Chem.* **29**: 1129.
22. Dekker, K., Yamagata, H., Sakaguchi, K. and Udaka, S. (1991) *J. Bacteriol.* **173**: 3078.
23. Hanai, S. (1988) In: *Jozo no Jiten* (Nojiro, K. *et al.* eds), Asakura Press, Tokyo, p. 352.
24. Yanai, K., Horiuchi, H., Takagi, M. and Yano, K. (1990) *Agric. Biol. Chem.* **54**: 2689.
25. Ito, H., Fukuta, Y., Murata, K. and Kimura, A. (1983) *J. Bacteriol.* **153**: 163.
26. Yanai, K., Horiuchi, H., Takagi, M. and Yano, K. (1991) *Curr. Genet.* **19**: 221.
27. Kamakura, T., Kobayashi, K., Tanaka, T., Yamaguchi, I. and Endo, T. (1987) *Agric. Biol. Chem.* **51**: 3165.

28. Kamakura, T., Yoneyama, K. and Yamaguchi, I. (1990) *Mol. Gen. Genet.* **223**: 332.
29. Ashikari, T., Kiuchi-Goto, N., Tanaka, Y., Shibano, Y., Amachi, T. and Yoshizumi, H. (1989) *Appl. Microbiol. Biotechnol.* **30**: 515.
30. Roncero, M.I.G., Jepsen, L.P., Strøman, P. and van Heeswijck, R. (1989) *Gene* **84**: 335.
31. Arnau, J., Murillo, F.J. and Torres-Martinez, S. (1988) *Mol. Gen. Genet.* **212**: 375.
32. Revuelta, J.L. and Jayaram, M. (1986) *Proc. Natl. Acad. Sci. USA* **83**: 7344.
33. Wöstemeyer, J., Burmester, A. and Weigle, C. (1987) *Curr. Genet.* **12**: 625.
34. Burmester, A., Wöstemeyer, A. and Wöstemeyer, J. (1990) *Curr. Genet.* **17**; 155.
35. van Heeswijck, R. and Roncero, M.I.G. (1984) *Carlsberg Res. Commun.* **19**: 691.
36. Horiuchi, H., Yanai, K., Okazaki, T., Takagi, M. and Yano, K. (1988) *J. Bacteriol.* **170**: 272.
37. Takahashi, K. (1987) *J. Biol. Chem.* **262**: 1468.
38. James, M.N.G. and Sielecki, R. (1983) *J. Mol. Biol.* **163**: 299.
39. Sogawa, K., Fujii-Kuriyama, Y., Mizukami, Y., Ichihara, Y. and Takahashi, K. (1983) *J. Biol. Chem.* **258**: 5306.
40. Horiuchi, H., Ashikari, T., Amachi, T., Yoshizumi, H., Takagi, M. and Yano, K. (1990) *Agric. Biol. Chem.* **54**: 1771.
41. Tonouchi, N., Shoun, H., Uozumi, T. and Beppu, T. (1986) *Nuc. Acids Res.* **14**: 7557.
42. Gray, G.L., Hayenga, K., Cullen, D., Wilson, L.J. and Norton, S. (1986) *Gene* **48**: 41.
43. Yamashita, T., Tonouchi, N., Uozumi, T. and Beppu, T. (1987) *Mol. Gen. Genet.* **210**: 462.
44. Hiramatsu, R., Aikawa, J., Horinouchi, S. and Beppu, T. (1989) *J. Biol. Chem.* **264**: 16862.
45. Aikawa, J., Yamashita, T., Nishiyama, M., Horinouchi, S. and Beppu, T. (1990) *J. Biol. Chem.* **265**: 13955.
46. Hiramatsu, R., Yamashita, T., Aikawa, J., Horinouchi, S. and Beppu, T. (1990) *Appl. Environ. Microbiol.* **56**: 2125.
47. Hiramatsu, R., Horinouchi, S. and Beppu, T. (1991) *Gene* **99**: 235.
48. Christensen, T., Woeldike, H., Boel, E., Mortensen, S.B., Hjortshoej, K., Thim, L. and Hansen, M.T. (1988) *Bio/Technology* **6**: 1419.
49. Dickinson, L., Harboe, M., van Heeswijck, R., Strøman, P. and Jepsen, L. P. (1987) *Carlsberg Res. Commun.* **52**: 243.
50. Suguna, K., Bott, R.R., Padlan, E.A., Subramanian, E., Sheriff, S., Cohen, G.H. and Davies, D.R. (1987) *J. Mol. Biol.* **196**: 877.
51. Delaney, R., Wong, R.N.S., Meng, G., Wu, N. and Tang, J. (1987) *J. Biol. Chem.* **262**: 1461.
52. Berka, R.M., Ward, M., Wilson, L.J., Hayenga, K.J., Kodama, K.H., Carlomagno, L.P. and Thompson, S.A. (1990) *Gene* **86**: 153.
53. Korman, D.R., Bayliss, F.T., Barnett, C.C., Carmona, C.L., Kodama, K.H., Royer, T.J., Thompson, S.A., Ward, M., Wilson, L.J. and Berka, R.M. (1990) *Curr. Genet.* **17**: 203.
54. Tatsumi, H., Ogawa, Y., Murakami, S., Ishida, Y., Murakami, K., Masaki, A., Kawabe, H., Arimura, H., Nakano, E. and Motai, H. (1989) *Mol. Gen. Genet.* **219**: 33.
55. Ogawa, Y., Tatsumi, H., Murakami, S., Ishida, Y., Murakami, K., Masaki, A., Kawabe, H., Arimura, H., Nakano, E., Motai, H. and Toh-e, A. (1990) *Agric. Biol. Chem.* **54**: 2521.
56. Tatsumi, H., Murakami, S., Tsuji, R.F., Ishida, Y., Murakami, K., Masaki, A., Kawabe, H., Arimura, H., Nakano, E. and Motai, H. (1991) *Mol. Gen. Genet.* **228**: 97.
57. Gysler, C., Harmsen, J.A.M., Kester, H.C.M., Visser, J. and Heim, J. (1989) *Gene* **89**: 101.
58. Harmsen, J.A.M., Kusters-van Someren, M.A. and Visser, J. (1990) *Curr. Genet.* **18**: 161.
59. Kusters-van Someren, M.A., Harmsen, J.A.M., Kester, H.C.M. and Visser, J. (1991) *Curr. Genet.* **20**: 293.
60. Bussink, H.J.D., Kester, H.C.M. and Visser, J. (1990) *FEBS Lett.* **273**: 127.
61. Ruttkowski, E., Labitzke, R., Khanh, N.Q., Löffler, F., Gottschalk, M. and Jany, K.-D. (1990) *Biochem. Biophys. Acta.* **1087**: 104.
62. Bussink, H.J.D., Brouwer, K.B., de Graaff, L.H., Kester, H.C.M. and Visser, J. (1991) *Curr. Genet.* **20**: 301.
63. Khanh, N.Q., Albrecht, H., Ruttkowski, E., Löffler, F., Gottschalk, M. and Jany, K-D. (1990) *Nuc. Acids Res.* **18**: 14.
64. Dean, R.A. and Timberlake, W.E. (1989) *Plant Cell* **1**: 275.
65. Takahashi, K. (1985) *J. Biochem.* (*Tokyo*) **98**: 815.
66. Watanabe, H., Ohgi, K. and Irie, M. (1982) *J. Biochem.* (*Tokyo*) **91**: 1495.
67. Hill, C., Dodson, G., Heinemann, U., Saenger, W., Mitsui, Y., Nakamura, K., Borisov, S., Tischenko, G., Polyakov, K. and Pavlovsky, S. (1983) *TIBS* October, 364.

68. Hoffman, E. and Ruterjans, H. (1988) *Eur. J. Biochem.* **177**: 539.
69. Takahashi, K. and Moore, S. (1982) In: *The Enzymes*, 3rd ed. (Boyer, P.D., ed.), Vol. XV, Part B, p. 435.
70. Ikehara, M., Ohtsuka, E., Tokunaga, T., Nishikawa, S., Uesugi, S., Tanaka, T., Aoyama, Y., Kikyodani, S., Fujimoto, K., Yamase, K., Fuchimura, K. and Morioka, H. (1986) *Proc. Natl. Acad. Sci. USA* **83**: 4695.
71. Nishikawa, S., Morioka, H., Fuchimura, K., Tanaka, T., Uesugi, S., Ohtsuka, E. and Ikehara, M. (1986) *Biochem. Biophys. Res. Commun.* **138**: 789.
72. Hakoshima, T., Toda, S., Sugio, S., Tomita, K., Nishikawa, S., Morioka, H., Fuchimura, K., Uesugi, S., Ohtsuka, E. and Ikehara, M. (1988) *Protein Eng.* **2**: 55.
73. Horiuchi, H., Yanai, K., Takagi, M., Yano, K., Wakabayashi, E., Sanda, A., Mine, S., Ohgi, K. and Irie, M. (1988) *J. Biochem.* (*Tokyo*) **103**: 408.
74. Kawata, Y., Sakiyama, F. and Tamaoki, H. (1988) *Eur. J. Biochem.* **176**: 683.
75. Watanabe, H., Naitoh, A., Suyama, Y., Inokuchi, N., Shimada, H., Koyama, T., Ohgi, K. and Irie, M. (1990) *J. Biochem.* (*Tokyo*) **108**: 303.
76. Irie, M., Watanabe, H., Ohgi, K. and Harada, M. (1986) *J. Biochem.* (*Tokyo*) **99**: 627.
77. Ohgi, K., Horiuchi, H., Watanabe, H., Takagi, M., Yano, K. and Irie, M. (1991) *J. Biochem.* (*Tokyo*) **109**: 776.
78. Ozeki, K., Kitamoto, K., Gomi, K., Kumagai, C., Tamura, G. and Hara, S. (1991) *Curr. Genet.* **19**: 367.
79. Iwai, M. (1991) *Lipase*, Saiwai Pub., Tokyo.
80. Iwai, M. (1988) *Yushi* **42**(2): 88, (5): 86, (6): 84, (8): 96, (9): 102, (10): 42.
81. Fukumoto, J., Iwai, M. and Tsujisaka, Y. (1963) *J. Gen. Appl. Microbiol.* **9**: 353.
82. Tomizuka, N., Ota, Y. and Tamada, K. (1966) *Agric. Biol. Chem.* **30**: 576.
83. Isobe, M. and Sugiura, M. (1977) *Chem. Pharm. Bull.* **25**: 1980.
84. Sugiura, M., Isobe, M., Moroya, N. and Yamaguchi, Y. (1974) *Agric. Biol. Chem.* **38**: 947.
85. Nagaoka, K. and Yamada, Y. (1973) *Agric. Biol. Chem.* **37**: 2791.
86. Liu, W-H., Beppu, T. and Arima, K. (1973) *Agric. Biol. Chem.* **37**: 157.
87. Omar, I.C., Hayashi, M. and Nagai, S. (1987) *Agric. Biol. Chem.* **51**: 37.
88. Iwai, M., Okumura, S. and Tsujisaka, Y. (1975) *Agric. Biol. Chem.* **39**: 1063.
89. Nakamura, M., Maejima, K. and Tomoda, K. (1976) *J. Takeda, Res. Lab.* (*Osaka*) **35**: 1.
90. Sugiura, M. and Oikawa, T. (1977) *Biochim Biophys. Acta* **489**: 262.
91. Nishio, T., Chikano, T. and Kamimura, M. (1987) *Agric. Biol. Chem.*, **51**: 181.
92. Japanese Patent S-59-156282; US Patent 4,665,029.
93. Iwai, M. and Tsujisaka, Y. (1974) *Agric. Biol. Chem.* **38**: 1241.
94. Aisaka, K. and Terada, O. (1981) *J. Biochem.* **89**: 817.
95. Yoshida, F., Motai, H. and Ichishima, E. (1968) *Biochim. Biophys. Acta* **154**: 586.
96. Shimada, Y., Tominaga, Y., Iwai, M. and Tsujisaka, Y. (1983) *J. Biochem.* **93**: 1655.
97. Borgström, B. (1954) *Biochim Biophys. Acta* **13**: 149.
98. Fritz, P.J. and Melius, P. (1963) *Can. J. Biochem. Biol.* **41**: 719.
99. Iwai, M., Tsujisaka, Y. and Fukumoto, J. (1970) *J. Gen. Appl. Microbiol.* **16**: 81.
100. Tsujisaka, Y., Iwai, M. and Tominaga, Y. (1972) *Proc. Fourth Internal Fermentation Technology Today.* p. 315.
101. Iwai, M., Tsujisaka, Y. and Fukumoto, J. (1979) *Proc. 15th Symposium of Enzyme Chemistry* (*Osaka*) p. 117.
102. Iwai, M. (1978) *Kagaku to Kogyo* **52**: 93.
103. Macrae, A.R. (1983) *J. Am. Oil. Chem. Soc.* **60**: (243A): 291.
104. Fukumoto, J., Iwai, M. and Tsujisaka, Y. (1962) *Proc. Symposium on Enzyme Chemistry* **18**: 53.
105. Fukumoto, J., Iwai, M. and Tsujisaka, Y. (1963) *J. Gen. Appl. Microbiol.* **9**: 353.
106. Tsujisaka, Y., Iwai, M. and Tominaga, Y. (1973) *Agric. Biol. Chem.* **37**: 1457.
107. Hata, Y., Matsuura, Y., Tanaka, N., Kakudo, K., Sugihara, A., Iwai, M. and Tsujisaka, Y. (1979) *J. Biochem.* **86** 1821.
108. Hata, Y., Matsuura, Y., Tanaka, N., Kakudo, K., Sugihara, A., Iwai, M. and Tsujisaka, Y. (1981) *Acta Cryst.* **A37**: C38.
109. Sugihara, A., Iwai, M. and Tsujisaka, Y. (1982) *J. Biochem.* **91**: 507.
110. Fogarty, M.W. (1983) In *Microbial Enzymes and Biotechnology* (Fogarty, M.W. ed.), Applied Science Publishers, London and New York, pp. 1–92.

111. Ohnishi, M., Sakano, Y. and Taniguchi, H. (1986) *Amylases*, Japanese Scientific Societies Press, Tokyo.
112. Hesseltine, C.W. (1983) *Ann. Rev. Microbiol.* **37**: 575–601.
113. Nunokawa, Y. (1972) In: *Rice: Chemistry and Technology* (Houston, D.F. ed.), American Association of Cereal Chemists Inc., St Paul, Minnesota, pp. 449–487.
114. Akabori, S., Hagihara, B. and Ikenaka, T. (1951) *Proc. Jpn. Acad.* **27**: 350–356.
115. Toda, H., Kondo, K. and Narita, K. (1982) *Proc. Jpn. Acad.* **58B**: 208–212.
116. Yamaguchi, H., Ikenaka, T. and Narita, K. (1971) *J. Biochem.* **70**: 587–594.
117. Matsuura, Y., Kusunoki, M., Harada, W. and Kakudo, K. (1984) *J. Biochem.* **95**: 697–702.
118. Tonomura, K., Suzuki, H., Nakamura, N., Kuraya, K. and Tanabe, O. (1961) *Agric. Biol. Chem.* **25**: 1–6.
119. Yabuki, M., Ono, N., Hoshino, K. and Fukui, S. (1977) *Appl. Environ. Microbiol.* **34**: 1–6.
120. Tada, S., Iimura, Y., Gomi, K., Takahashi, K., Hara, S. and Yoshizawa, K. (1989) *Agric. Biol. Chem.* **53**: 593–599.
121. Gomi, K., Tada, S., Kitamoto, K. and Takahashi, K. (1989) In: *Abstracts*, Annual Meeting of the Society of Fermentation Technology, Japan, Nagoya, October, p. 156.
122. Wirsel, S., Lachmund, A., Wildhardt, G. and Ruttkowski, E. (1989) *Mol. Microbiol.* **3**: 3–14.
123. Gines, M.J., Dove, M.J. and Seligy, V.L. (1989) *Gene* **79**: 107–117.
124. Tsukagoshi, N., Furukawa, M., Nagaba, H., Kirita, N., Tsuboi, A. and Udaka, S. (1989) *Gene* **84**: 319–327.
125. Gurr, S.J., Unkles, S.E. and Kinghorn, J.R. In: *Gene Structure in Eukaryotic Microbes* (Kinghorn, J.R. ed.) SGM Special Publications, Vol. 22, IRL Press, Oxford (1987) 93–139.
126. Raper, K.B. and Fennell, D.I. (1965) *The Genus Aspergillus*. The William and Wilkins Co., Baltimore.
127. Takata, Y., Goto, K., Hara, S. and Tamura, G. (1992) *Agric. Biol. Chem.* **55**: (in press).
128. Brody, H. and Carbon, J. (1989) *Proc. Natl Acad. Sci. USA* **86**: 6260–6263.
129. Debets, A.J.M., Holub, E.F., Swart, K., van den Broek, H.W.J. and Bos, C.J. (1990) *Mol. Gen. Genet.* **224**: 264–268.
130. Minoda, Y., Arai, M., Torigoe, Y. and Yamada, K. (1968) *Agric. Biol. Chem.* **32**: 110–113.
131. Mikami, S., Iwano, K., Shiinoki, S. and Shimada, T. (1987) *Agric. Biol. Chem.* **51**: 2495–2501.
132. Boel, E., Brady, L., Brzozowski, A.M., Derewenda, Z., Dodson, G.G., Jensen, V.J., Petersen, S.B., Swift, H., Thim, L. and Woeldike, H.F. (1990) *Biochemistry* **29**: 6244–6249.
133. Korman, D.R., Bayliss, F.T., Barnett, C.C., Carmona, C.L., Kodama, K.H., Royer, T.J., Thompson, S.A., Ward, M., Wilson, L.J. and Berka, R.M. (1990) *Curr. Genet.* **17**: 203–212.
134. Tada, S., Gomi, K., Kitamoto, K., Takahashi, K., Tamura, G. and Hara, S. (1991) *Mol. Gen. Genet.* **229**: 301.
135. Tada, S., Gomi, K., Kitamoto, K., Kumagai, C., Tamura, G. and Hara, S. (1991) *Agric. Biol. Chem.* **55**: 1939–1941.
136. Svensson, B.T., Pedersen, G., Svendsen, I., Sakai, T. and Ottesen, M. (1982) *Carlsberg Res. Commun.* **47**: 55–69.
137. Hayashida, S. and Yoshino, E. (1978) *Agric. Biol. Chem.* **42**: 927–933.
138. Yoshino, E. and Hayashida, S. (1978) *J. Ferment Technol.* **56**: 289–295.
139. Boel, E., Hjort, I., Svensson, B., Norris, F. and Fiil, N.P. (1984) *The EMBO J.* **3**: 1097–1102.
140. Svensson, B.T., Larsen, K. and Svendsen, I. (1983) *Carlsberg Res. Commun.* **48**: 517–527.
141. Boel, E., Hansen, M.T., Hoegh, I. and Fiil, N.P. (1984) *The EMBO J.* **3**: 1581–1585.
142. Nunberg, J.H., Meade, J.H., Cole, G., Lawyer, F.C., McCabe, P., Schweckart, V.S., Tal, R., Wittman, V.P., Flatgaard, J.E. and Innis, M.A. (1984) *Mol. Cell. Biol.* **4**: 2306–2315.
143. Sevensson, B.T., Larsen, K. and Gunnarsson, A. (1986) *Eur. J. Biochem.* **154**: 497–502.
144. Hayashida, S., Kuroda, K., Ohta, K., Kuhara, S., Fukuda, K. and Sakaki, Y. (1989) *Agric. Biol. Chem.* **53**: 923–929.
145. Shibuya, I., Gomi, K., Iimura, Y., Takahashi, K., Tamura, G. and Hara, S. (1990) *Agric. Biol. Chem.* **54**: 1905–1914.
146. Hayashida, S., Nakahara, K., Kuroda, K., Kamachi, T., Ohta, K., Iwanaga, S., Miyata, T. and Sakaki, Y. (1989) *Agric. Biol. Chem.* **52**: 273–275.
147. Evans, R., Ford, C., Sierks, M., Nikolov, Z. and Svensson, B. (1990) *Gene* **91**: 131–134.
148. Hayashida, S., Nakahara, K., Kuroda, K., Miyata, T. and Iwanaga, S. (1989) *Agric. Biol. Chem.* **53**: 135–141.

149. Hayashida, S., Nakahara, K., Kanlayakrit, W., Hata, T. and Teramoto, Y. (1989) *Agric. Biol. Chem.* **53**: 143–149.
150. Chen, L., Ford, C. and Nikolov, Z. (1991) *Gene* **99**: 121–126.
151. Saha, B.C., Mitsue, T. and Ueda, S. (1979) *Stärke* **31**: 307–312.
152. Hata, Y., Kitamoto, K., Gomi, K., Kumagai, C., Tamura, G. and Hara, S. (1991) *Agric. Biol. Chem.* **55**: 941–949.
153. Hata, Y., Tsuchiya, K., Kitamoto, K., Gomi, K., Kumagai, C., Tamura, G. and Hara, S. *Gene* **108**: 145.
154. Itoh, T., Ohtsuki, I., Yamashita, I. and Fukui, S. (1987) *J. Bacteriol.* **169**: 4171–4176.
155. Barton, L.L., Georgi, C.E. and Lineback, D.R. (1972) *J. Bacteriol.* **111**: 771–777.
156. Fowler, T., Berka, R.M. and Ward, M. (1990) *Curr. Genet.* **18**: 537–545.
157. Takahashi, T., Tsuchida, Y. and Irie, M. (1978) *J. Biochem.* **84**: 1183–1194.
158. Takahashi, T., Tsuchida, Y. and Irie, M. (1982) *J. Biochem.* **92**: 1623–1633.
159. Ashikari, T., Nakamura, N., Tanaka, Y., Kiuchi, N., Shibano, Y., Tanaka, T., Amachi, T. and Yoshizumi, H. (1986) *Agric. Biol. Chem.* **50**: 957–964.
160. Svensson, B., Clarke, A.J., Svensson, I. and Moller, H. (1990) *Eur. J. Biochem.* **154**: 29–38.
161. Yamashki, Y. *et al.* (1977) *Agric. Biol. Chem.* **41**: 2149.
162. Pazur, J.H. *et al.* (1971) *Carbohydr. Res.* **21**: 83.
163. Miah, M.N.N. and Ueda, S. (1977) *Stärke* **29**: 235.
164. Lineback, D.R. and Bauman, W.E. (1970) *Carbohydr. Res.* **14**: 341.
165. Takahashi, T. *et al* (1981) *J. Biochem* **89**: 125.
166. McCleavy, B.V. and Anderson, M.A. (1980) *Carbohydr. Res.* **80**: 77.
167. Yamasaki, Y. and Suzuki, Y. (1978) *Agric. Biol. Chem.* **42**: 971.
168. Tsuboi, A. *et al.* (1974) *Agric. Biol. Chem.* **38**: 543.
169. Takeda, Y. *et al.* (1985) *Agric. Biol. Chem.* **49**: 1633.
170. Yamasaki, Y. *et al.* (1977) *Agric. Biol. Chem.* **41**: 755.
171. Tanaka, A. *et al.* (1983) *Agric. Biol. Chem.* **47**: 573.
172. Yamashita, I. *et al.* (1984) *Agric. Biol. Chem.* **48**: 1611.
173. Antonian, E. (1988) *Lipids* **23**: 1101.
174. Ben-Avram, C.M., Ben-Zeev, O., Lee, T.D. and Schotz, M.C. (1986) *Proc. Natl Acad. Sci. USA* **84**: 4369.
175. Boel, E., Huge-Jensen, B., Christensen, M., Thim, L. and Fiil, N.P. (1988) *Lipids* **23**: 701.
176. Brady, L., Brzozowski, A.M., Derewenda, Z.S., Dodson, E., Dodson, G., Tolley, S., Turkenberg, J.P., Christensen, L., Huge-Jensen, B., Norskov, L., Thim, L. and Menge, V. (1990) *Nature* **343**: 767.
177. Brzozowski, A.M., Derewenda, U., Derewenda, Z.S., Dodson, G.G., Lawson, D.M., Tirkenberg, J.P., Bjorkling, F., Huge-Jensen, B., Patkar, S.A. and Thim, L. (1988) *Nature* **351**: 491.
178. Christensen, T., Woeldike, H., Boel, E., Mortensen, S.B., Hjortshoej, K., Thim, L., and Hansen, M.T. (1988) *Bio/Technology* **6**: 1419.
179. Datta, S., Luo, C., Liu, S. and Chan, L. (1988) *J. Biol. Chem.* **263**: 1107.
180. De Caro, J., Boudouard, M., Bonicel, J., Guidoni, A., Desnuelle, P. and Rovery, M. (1981) *Biochim. Biophys. Acta* **671**: 129.
181. Docherty, A.J.P., Bodmer, M.W., Augal, S., Verger, R., Riviere, C., Lowe, P.A., Lyons, A., Emtage, J.S. and Harris, T.J.R. (1985) *Nuc. Acids Res.* **13**: 1891.
182. Esper, B. and Huge-Jensen, I.B. (1988) European Patent Application 0305216A1.
183. Gotz, F., Popp, F., Korn, E. and Schleifer, K.H. (1985) *Nuc. Acids Res.* **13**: 5395.
184. Huge-Jensen, B., Andreasen, F., Christensen, T., Christensen, M., Thim, L. and Boel, E. (1989) *Lipids* **24**: 781.
185. Ishii, M. (1988) Novo Industri A/S, International Application Published under the Patent Cooperation Treaty (PCT) WO 88/02775 PCT/DK87/00127.
186. Isobe, K., Akiba, T. and Yamaguchi, S. (1988) *Agric. Biol. Chem.* **23**: 1101.
187. Kelly, J.M. and Hynes, M. (1985) *The EMBO Journal* **4**: 475.
188. Kerfelec, B., Laforge, K.S., Puigserver, A. and Scheele, G. (1986) *Pancreas* **1**: 430.
189. Kirchgessner, T.G., Srenson, K.L., Lusis, A.J. and Schotz, M.C. (1987) *J. Biol. Chem.* **262**: 8463.
190. Komaromy, M.C. and Shotz, M.C. (1987) *Proc. Natl Acad. Sci. USA* **84**: 1526.

191. Kugimiya, W., Otani, Y., Hashimoto, Y. and Takagi, Y. (1986) *Biochem. Biophys. Res. Commun.* **141**: 185.
192. Kugimiya, W., Otani, Y. and Hashimoto, Y. (1989) Japan Patent S-64-80290.
193. McLean, J., Fielding, C., Drayna, D., Dieplinger, H., Baer, B., Kohr, W., Henzel, W. and Lawn, R. (1986) *Proc. Natl Acad. Sci. USA* **83**: 2335.
194. Nielsen, T.B. (1991) *Symposium on Fermentation and Metabolism, Tokyo, Kanda*, June 18th.
195. Richardson, G.H. (1975) *Enzymes in Food Processing*, 2nd ed. (Reed, G. ed.), Academic Press, New York, p. 377.
196. Saha, B.C. and Ueda, S. (1983) *Biotechnol. Bioeng.* **25**: 1181.
197. Schormiller, J. (1968) *Adv. Food Res.* **16**: 231.
198. Semeriva, M. and Desnuelle, P. (1979) *Advances in Enzymology*, **48**: 360.
199. Senda, M., Oka, K., Brown, W.V., Quasaba, P. and Furuichi, Y. (1987) *Proc. Natl Acad. Sci. USA* **84**: 4369.
200. Sternby, B., Engström, Aa., Hellman, U., Vikert, A.M., Sternby, N.H. and Borgström, B. (1984) *Biochim Biophys. Acta* **784**: 75.
201. Tanimura, W., Sanchez, P.C. and Kozaki, M. (1977) *J. Agric. Sci. Tokyo Univ. Agric.* **22**: 118.
202. Tsuchiya, M. and Matsui, H. (1991) Japan Patent Application H3-8715.
203. Tyski, S., Hriniewicz, W. and Jeliazewicz, J. (1983) *Biochim Biophys. Acta* **749**: 312.
204. Uchimura, Y., Kozaki, M. *et al.* (1991) *J. Brew. Soc. Japan* **86**: 62, 818, 881.
205. Vandamme, E., Schank-Brodück, K.H., Colson, C. and Hantotier, J.D.V. (1986) *European Patent Application* 0243 338 A2.
206. Veda, S., Saha, B.C. and Koba, Y. (1984) *Microbiol. Sciences* **1**: 21.
207. Wion, K.L., Kirchgessner, T.G., Lusis, A.J., Schotz, M.C. and Lawn, R.M. (1987) *Science* **235**: 1638.
208. Yamaguchi, S. and Mase, T. (1991) *Appl. Microbiol. Biotechnol.* **34**: 720.
209. Yamaguchi, S. and Mase, T. (1991) *J. Ferment. Bioeng.* **72**: 162–167.
210. Yamaguchi, S., Mase, T. and Takeuchi, K. (1991) *Gene***103**: 61–67.
211. Yoritomi, K. (1977) In: *Handbook of Starch Sciences* (Nikuni, J., Nakamura, M. and Suzuki, S. eds), Asakura Press, Tokyo, p. 441.
212. Yoshii, H. (1988) In: *Jozo no Jiten* (Nojiro, K. *et al.* eds), Asadura Press, Tokyo, p. 441.
213. Yamiaguchi, S. Mase, T. and Takeuchi, K. (1992) *Biosci. Biotech. Biochem.* **56**: 315–319.

4 Fungal enzymes for lignocellulose degradation

D. CULLEN and P. KERSTEN

4.1 Introduction

Microbial degradation of lignocellulose has been intensively studied for decades. Interest in lignocellulose as a renewable resource prompted these investigations, although in recent years fungal enzymes have attracted much attention in pulp and paper manufacturing. Irrespective of considerable research effort, fundamental questions remain with regard to mechanisms of lignocellulose degradation. Progress has been hampered by numerous factors including substrate complexity and the multiplicity of enzymes.

Not surprisingly, this intense interest has resulted in voluminous literature. In this chapter literature relevant to lignocellulose degradation by filamentous fungi is outlined. The format of the text is organized around substrates of increasing complexity (cellulose, hemicellulose, and lignin) and recent developments are highlighted. Areas where knowledge is sketchy are identified. Except as points of reference, prokaryotic systems are not discussed, and the reader is referred to numerous review articles for additional information.

4.2 Cellulose degradation

Cellulose is the most abundant renewable organic resource. It comprises approximately 45% of dry wood weight. Crystalline cellulose is composed of long β-1,4-glucan chains (*c.* 40) that are hydrogen bonded into water-insoluble microfibrils. The crystalline cellulose coexists with less ordered amorphous regions. The fibers are embedded in a matrix of hemicellulose and lignin, but the precise organization is poorly understood [1]. Cellulose is often characterized by the degree of polymerization (DP), which is the average number of glucan residues per molecule. The average DP of native cellulose is 3500 [2].

Bacterial and fungal degradation of cellulose has been extensively studied and reviewed [3–10]. The soft-rot fungi, *Trichoderma viride* and *Trichoderma reesei* are, by far, the most extensively studied. During the past few years, these investigations have made significant progress toward elucidating the enzymology of cellulose degradation. Among prokaryotes, the Corynebacterium *Cellulomonas fimi* and the anaerobe *Clostridium thermocellum* are the best known examples.

All cellulase systems share common features. However, the mechanism of fungal brown-rot decay is clearly different from other systems, and the multicomponent enzyme complexes of some bacteria such as *C. thermocellum* are also distinct. Oxidative enzymes of fungi may also participate in cellulose degradation. This chapter focuses on recent advances in the enzymology of hydrolytic cellulases of fungal origin, in the molecular genetics of these fungi, and in the potential biotechnological applications.

4.2.1 *Cellulase biochemistry*

In submerged culture, most cellulolytic fungi secrete a complex array of degradative enzymes. Three classes of hydrolytic cellulases are recognized on the basis of substrate specificity:

(i) Endo-1,4-β-glucanases (EG) cleave randomly at 1,4-β-linkages within the cellulose chain. The endoglucanases are commonly assayed by viscosity reductions in carboxymethyl cellulose (CMC) solutions. Crystalline cellulose is not degraded by endoglucanases. Multiple chromatographically distinct endoglucanases have been identified in culture filtrates of *T. reesei* [8] and *Phanerochaete chrysosporium* [11].

(ii) Exo-1,4-β-glucanase (exo-1,4-β-D-glucan cellobiohydrolases, CBH) of *P. chrysosporium*, releases both glucose and cellobiose from the non-reducing ends of cellulose chains [12]. In contrast, the two cellobiohydrolases of *T. reesei* split off cellobiose only. Cellobiohydrolases have virtually no activity on CMC and slowly degrade crystalline cellulose. Both endoglucanases and cellobiohydrolases are active on amorphous (non-crystalline) cellulose.

(iii) 1,4-β-glucosidases hydrolyse cellobiose to glucose, and cellobionic acid to glucose and gluconolactone. Two *P. chrysosporium* enzymes, one cell-bound and one extracellular, have been identified [13]. Relative to endoglucanases and cellobiohydrolases, low levels of the *T. reesei* β-glucosidase are secreted in submerged culture. For certain applications, supplementing *T. reesei* cellulase preparations with additional β-glucosidase has proved beneficial [14, 15].

Fungal cellulases have frequently been reported to act synergistically in the degradation of crystalline cellulose. A commonly held model for the synergism between endoglucanases and cellobiohydrolases suggests that hydrolysis is initiated by endoglucanases that cleave random β-1,4-linkages within the cellulose chain. This activity is followed by cellobiohydrolase action that releases cellobiose from the non-reducing ends of the cellulose chain (reviewed in [3, 16]). Studies with *T. reesei* suggest a more complicated interaction [17] in that the extent of synergism may be related to the substrate used and the ratio of cellobiohydrolase to endoglucanase.

The discovery of synergism between two immunologically distinct cellobiohydrolases (CBHI and CBHII) from both *T. reesei* [18] and *Penicillium pinophilum* [19] indicates that the 'endo–exo' model is incomplete. A model for 'exo–exo' synergism, which is based upon substrate stereospecificity, has been advanced by Wood and McCrae [19]. Recent evidence demonstrates different stereochemical mechanisms for *T. reesei* CBHI and CBHII [20, 21].

Attempts to formulate a model of cellulose degradation have been stymied by uncertainties concerning the exact roles and specificities of *P. chrysosporium* and *T. reesei* cellulases (reviewed in [4]). For example, highly purified *T. reesei* endoglucanase has been reported to be active only on soluble β-1,4-glycans [22]. In addition, studies of purified *T. reesei* CBHI have demonstrated that the enzyme has endoglucanase-like properties including the ability to digest barley β-glucan [17] and to bind along the length of cellulose microfibrils [23]. To further complicate the issue, *T. reesei* cellobiohydrolase has been reported to be capable of degrading crystalline cellulose extensively, without assistance from any endoglucanase activity [24].

The wealth of conflicting results in the field of cellulose degradation [4] clearly calls into question the purity of many enzyme preparations that have been studied. Wood and co-workers [25] have demonstrated endoglucanase/cellobiohydrolase complexes that were resolved only after affinity chromatography [26]. When these highly purified enzymes were used to degrade microcrystalline cellulose in the form of cotton fiber, three enzymes (CBHI, CBHII, and trace endoglucanase) were necessary for rapid hydrolysis. Clearly, enzyme purity must be carefully established for studies of enzyme synergism.

Recombinant enzymes produced in heterologous hosts provide a powerful tool for resolving the uncertainties of cellulase substrate specificity. The finding that recombinant *T. reesei* CBHI lacked any β-glucanase activity [27] demonstrates the important role cellulase expression systems can play in answering such questions.

The mechanism of cellulose degradation by brown-rot fungi is poorly understood, but some progress has been made in recent years. Wood decay by brown-rot fungi such as *Poria placenta* is characterized by very rapid depolymerization of cellulose (reviewed in [3]). A large molecular weight multifunctional endoglucanase has been identified in culture filtrates but no exoglucanase has been found [28]. Formation of the enzyme is not repressed by glucose [29]. Perhaps the most striking feature of the brown-rot systems is that cellulose depolymerization appears to involve a non-enzymatic, diffusible agent. An oxidative mechanism is suspected although the exact agent is unknown [30].

4.2.2 *Molecular genetics and structure of fungal cellulases*

The molecular genetics of cellulolytic fungi, especially *T. reesei*, has advanced considerably over the past five years. Research has revealed extremely interesting structural features of the cellulase gene family that have important implications with regard to their mechanism(s) of action and evolution. From *T. reesei*, sequences have been reported for both genomic and cDNA clones encoding *cbh1* [31, 32], *cbh2* [33, 34], *egl* [35], *eg3* [36] and β-glucosidase [37]. Other sequenced fungal clones include *P. chrysosporium chb1* [38] and *Humicola grisea* var. *thermodea cbh1* [39]. An *Aspergillus niger* gene encoding β-glucosidase has been cloned [40].

Several lines of evidence indicate that cellulases are made of three structural regions: the catalytic 'core' which hydrolyses β-1,4-glycosidic linkages, the extended 'hinge' which is glycosylated, and the cellulose-binding 'tail' which has also been hypothesized to help solubilize the cellulose chain prior to hydrolysis (reviewed in [8, 41]). Comparisons of amino acid sequences of *T. reesei cbh1*, *cbh2*, *egl*, and *eg2* have shown the presence of two short consecutive blocks that are conserved in all four genes (Figure 4.1). Block A is approximately 30 amino acids long and is stabilized by 2 or 3 disulfide bridges. Sequence conservation is high within block A (*c.* 70% identical). Block B is approximately 40 amino acids long and is heavily glycosylated with >40% serine, threonine, or proline. Two copies of block B are present in *cbh2*.

Another line of evidence for functional domains is through extensive analyses of core and terminal domains prepared by limited proteolysis with papain [42]. The catalytic activity [43, 44], crystalline cellulose-binding capacity [42, 44, 45], and structure [46–48] of these cleavage products support the general model. *T. reesei* EGIII also requires the conserved blocks for efficient binding and hydrolysis of crystalline cellulose [49]. The functional domains can also be separated by native proteases in cellulolytic bacteria [50] and fungi [49, 51].

Based upon small angle X-ray scattering analysis [47, 48] and NMR spectroscopy [46], a 'tadpole-like' tertiary structure has been proposed for CBHI and CBHII (Figure 4.2).

A two-functional domain is a common design of fungal and bacterial cellulases. For example, Gilkes and co-workers [50] have separated and

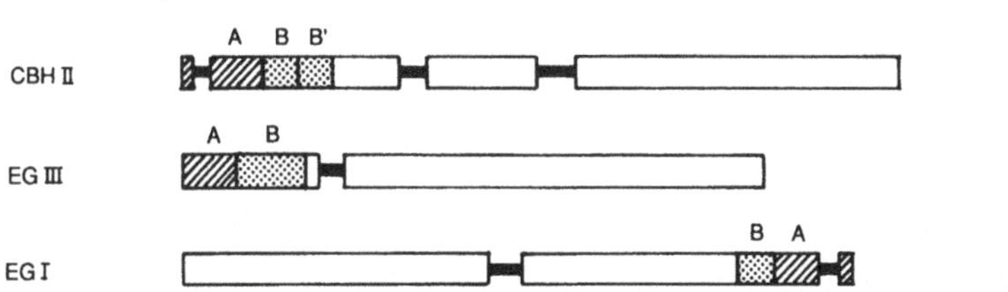

Figure 4.1 Representation of *T. reesei* cellulases. Microcrystalline binding regions and glycosylated hinge shown as blocks A and B, respectively. (Adapted from Knowles and colleagues [27].)

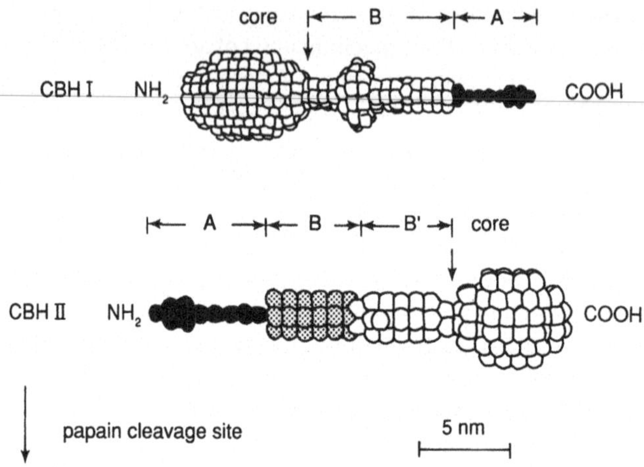

Figure 4.2 Model tertiary structure of *T. reesei* cellobiohydrolase. As in Figure 4.1, the binding domains and hinge region are designated A and B, respectively. (Adapted from Abuja and colleagues [47].)

characterized the catalytic and binding domains of *C. fimi* endoglucanase and exoglucanase. The domain structure has also been experimentally demonstrated in other bacteria including *C. thermocellum, Thermononospora fusca* and *Bacteroides succinogens* (reviewed in [4, 5]) as well as the fungi *P. chrysosporium* [52] and *P. pinophilum* [41].

Stahlberg and co-workers [49] have studied the kinetics of adsorption to crystalline cellulose by the isolated domains and by intact CBHI [43]. They propose a model in which the enzyme binds at two points (core and binding domains) and laterally diffuses on the crystalline surface.

Recent information has been gained concerning the active sites of cellulases. X-ray crystallography of the CBHII catalytic core suggests a tunnel-shaped active site [53, 54]. Consistent with studies of cellobiohydrolase degradation of cello-oligosaccharides [41, 55], four glucosyl binding sites have been identified within the tunnel. Cleavage of the cellulose chain is likely to occur within the tunnel between aspartic acid residues 175 and 221. Comparisons of amino acid sequences of CBHII and endoglucanases show deletions in the endoglucanases that would result in a more open active site or 'groove'. Rouvinen and co-workers [54] propose that these structural features may account for substrate specificities. The exoglucanases hydrolyse the ends of chains threaded within the tunnel, whereas the open grooves of endoglucanases can hydrolyse internal glycosidic bonds [54].

Hydrophobic cluster analysis has been used to classify 21 β-glucanases and to identify potential catalytic residues [56]. The six fungal cellulases included in the analysis were categorized into three of the six families. Only family A contained sufficient numbers of sequences to allow tentative identification of catalytic residues. These correspond to His-174 and Glu-218 of mature *T. reesei* EGIII.

Catalytic sites have also been tentatively identified by chemical modifications of cellulase. Tomme and Claeyssens [57] have implicated a Glu-126 in CBHI catalysis. Lending support to this is hydrophobic cluster analysis of EGI and CBHI, which shows considerable sequence conservation in the Glu-126 region [56]. Chemical modifications of *Schizophyllum commune* EGI have shown the importance of carboxyl groups in catalysis [58]. This, plus some weak sequence homology between T4 bacteriophage lysozyme and certain fungal cellulases, have suggested similar catalytic mechanisms [51, 59]. However, the putative catalytic *S. commune* EGI residues, Glu-33 and Asp-50, are not conserved in the *T. reesei* EGIII [36]. The tertiary structure of CBHII is also inconsistent with the lysozyme model [54].

In contrast to studies of cellulase structure and mechanisms, relatively little work has been done with regard to their regulation. The cellulases of *T. reesei* and *P. chrysosporium* are coordinately induced in media containing cellulose or derivatives (e.g., sophorose, cellobiose) as carbon source. The effectiveness of the inducers varies substantially with the particular disaccharide and the species. Thus, sophorose is a relatively poor inducer in *P. chrysosporium* [60], but very efficient in *T. reesei*. Enzyme synthesis is end-product inhibited by glucose, although low levels of *T. reesei* CBHII have been detected in glucose medium [61]. In the case of *T. reesei* and *P. chrysosporium*, induction has been shown to occur at the transcriptional level [34, 38].

The precise mechanism(s) of cellulase induction is unknown. It has been suggested [62] that low level, constitutive expression of cellulases might mediate the induction by producing low molecular weight, soluble compounds that could enter the cell and cause induction. Direct experimental evidence for this mechanism in *T. reesei* has been demonstrated by using cellulase-specific antibodies to block CBHI induction by cellulose [63]. The repression by antibodies could be overcome by sophorose induction [63]. The identity of the 'natural' inducer is still uncertain, although β-linked disaccharides are likely. Frischer and co-workers [64] emphasize the importance of β-glucosidase in the induction process.

In addition to the hydrolytic enzymes described thus far, several oxidative enzymes have been implicated in cellulose degradation by *P. chrysosporium*. These include cellobiose oxidase [65] and cellobiose : quinone oxidoreductase (CBQase) [66]. Cellobiose oxidase may stimulate cellulolytic activity by reducing cellobiose concentration, which appears to be a competitive inhibitor of cellulases, and by oxidizing reducing ends formed by endoglucanases thus preventing reformation of β-1,4-glycosidic bonds [65]. CBQase may be an important link to lignin degradation by reducing quinones while oxidizing cellobiose. However, recent work has shown that CBQase does not affect phenoxy radicals produced by lignin peroxidase or phenol polymerization catalysed by lignin peroxidases *in vitro* [67]. Lactonases may be important in cleaving the saccharide- lactone products of CBQase [68]. Both enzymes have been shown to bind microcrystalline cellulose while retaining enzymatic activity [69]. Recently, Henriksson and co-workers [70] have demonstrated papain cleavage of cellobiose oxidase into two domains: one containing flavin and the other containing a heme group. The flavin-containing domain binds crystalline cellulose and is functionally similar to CBQase leading Henriksson and co-workers to suggest that CBQase forms may be derived from cellobiose oxidase through proteolysis. Two acidic proteases have been identified in *P. chrysosporium* culture filtrates [71], although their specificities are unknown.

Relative to well-defined genetic systems, such as *Saccharomyces cerevisiae*, *Neurospora crassa*, and *Aspergillus nidulans*, genetic analysis of cellulolytic fungi has been extremely limited. *Trichoderma*, *Penicillium*, and *Humicola* species lack any meiotic or mitotic (parasexual) recombination system. Methods of genetic analysis are well established for *S. commune*, but the cellulases have not been studied. The mating behaviour of *P. chrysosporium* is somewhat controversial [72, 73], although it is clear that sexual crosses and meiotic recombination are possible [73, 74]. Importantly, there appears to be general agreement that the basidiospores of *P. chrysosporium*, although binucleate, are homokaryotic. Analyses of single basidiospore cultures have been used to determine allelic relationships [75, 76] and to generate a genetic map using restriction fragment length polymorphisms (RFLP) [77]. Linkage between a *cbh1* gene [38] and a lignin peroxidase gene has been established using the RFLP map [77, 78]. Recently developed transformation systems for *T. reesei* [79], *P. chrysosporium* [80, 81], and *S. commune* [82] have opened the way for more detailed genetic analyses.

4.2.3 *Applications*

Enzymatic saccharification of lignocellulose and subsequent fermentation to ethanol have been studied for many years (reviewed in [3]). A major objective of these studies has been to express the cellulase genes in heterologous systems. The *T. reesei* cDNAs encoding CBHI [83], CBHII [84], EGI [83, 85] and EGIII [27] have been expressed in *S. cerevisiae*. In each case, highly expressed yeast promoters were used, and 10–100 mg of heavily glycosylated product were obtained per liter of fermentation broth. The approximate native molecular weight was restored after treatment with glycosidase endo H [85, 86], indicating that glycosylation was N-linked. In one case, glycosylation appeared to confer thermostability [85]. Expression of *T. reesei eg2* in brewers' yeasts shows promise for reducing barley β-glucans [86]. Bacterial expression systems have not proven useful for *T. reesei* cellulases. Beguin [5] has reviewed expression of bacterial cellulase genes in heterologous hosts.

Cellulase expression has been altered in *T. reesei*. Strains lacking CBHI and others with increased EGII yields have been engineered [87]. Recently, Barnett and co-workers [37] have improved *T. reesei* cellulose saccharification rates by increasing β-glucosidase expression. The β-glucosidase gene was cloned, sequenced, and multiple copies re-introduced to *T. reesei.* One such transformant exhibited a five-fold increased rate of cellobiose hydrolysis.

Protein fusions involving cellulase-binding domains may provide an inexpensive and widely applicable method for immobilizing and purifying enzymes. In one case, the *Agrobacterium* gene encoding β-glucosidase was fused in frame to the cellulose-binding domain of an exocellobiohydrolase from *C. fimi.* The translational fusion was expressed in *E. coli.* When immobilized onto cellulose, 42% of the β-glucosidase activity was retained [88]. The immobilized enzyme is stable for long periods but can be desorbed at low ionic strengths or at alkaline pH [89]. Similarly, the cellulose-binding domain of the *C. fimi* endoglucanase was fused to alkaline phosphatase [90]. Although the approach to date has been limited to bacterial cellulases and *E. coli* expression systems, it is clear that the binding domains of fungal cellulases and fungal expression systems could be similarly exploited. Given the efficiency of heterologous expression in filamentous fungi such as *T. reesei* and *Aspergillus* spp. (chapters 5 and 6), relatively higher yields of fusion proteins might be expected.

An important application involving lignocellulolytic fungi is biomechanical pulping, which is defined as the use of fungi as a pretreatment for wood chips prior to mechanical pulping. Initial research on biomechanical pulping has been limited, but studies in several laboratories have shown that the fungal pretreatment could significantly decrease refining energy and enhance pulp strength properties [91–94]. Two studies have examined the effects of fungal treatment on crude mechanical pulps instead of on wood before pulping [95, 96]. The process has been refined in more recent studies [97–100].

A fundamental requirement for efficient biopulping performance is the elimination or reduction of cellulase activity. Towards this goal mutants of *P. chrysosporium* exhibiting altered cellulase secretion have been isolated [101, 102]. An ultraviolet-generated Cel$^-$ mutant, cel 44, has been intensively studied [103]. Treatment of wood chips with cel 44, prior to refiner pulping, reduced pulping energy requirements and increased strength properties of the resultant paper. However, such mutants are genetically undefined and pleiotropic effects such as lowered phenol oxidase production were noted [103]. This result is not surprising considering the mutagenesis procedure and the multiplicity of cellulase genes in *P. chrysosporium* [104].

In the absence of efficient Cel$^-$ mutants, glucose impregnation of chips before fungal inoculation has been necessary. The relative importance of glucose as a repressor of cellulases rather than as a source of easily assimilated carbon is unknown. However, it is clear that pretreatment with wild-type strains, in the absence of glucose repression, will lead to excessive cellulose degradation.

In summary, the exact roles and specificities of fungal cellulases are not clearly understood. This is particularly true in complex substrates such as wood chips. Further, the relationship between lignin and cellulose degradation is also poorly understood, although it is generally believed that cellulolytic and hemi-cellulolytic activity, as opposed to ligninolytic, provides most of the assimilated carbon for growth. This issue, the role and interactions of individual cellulases, is of paramount importance not only to understanding the mechanism of wood degradation but also to the successful development of processes such as biomechanical pulping.

4.3 Hemicellulose degradation

In contrast to cellulose, which contains only 1,4-β-glycosidic linkages, hemicellulose is structurally more complex. Accordingly, the enzyme systems giving complete degradation to monomeric sugars are similarly complex. Relative to cellulases, little is known concerning the molecular genetics of hemicellulases or their structure/activity relationships. However, research efforts have intensified in recent years because the hemicellulases have shown promise in pulp and paper production.

Hemicellulose is second only to cellulose in abundance, with dry wood concentrations ranging from 20 to 30%. Hemicellulose has three basic forms: 1,4-β-D-xylans, 1,3- and 1,4-β-D-galactans, and 1,4-β-D-mannans. The relatively short chains (average DP < 200) are frequently branched and the sugar residues are often acetylated or methylated. The chemistry and structure of hemicellulose have been reviewed [105–109]. Its biosynthesis is distinct from cellulases [110].

Hemicellulose is found in both secondary and primary walls and the composition varies substantially among species. Galactoglucomannans and arabinoglucuronoxylans are major components of softwoods. Glucuronoxylans, and to a lesser extent glucomannans, dominate in hardwood species. The arabinoglucuronoxylans and the glucuronoxylans, collectively referred to as xylans, represent approximately 7–12% and 15–30% of softwoods and hardwoods, respectively [2].

An important issue with respect to biotechnical exploitation of hemicellulases is the nature of lignin–hemicellulose bonds. Ester and ether linkages occur between hemicellulose sugar hydroxyls and the α-carbonyl of phenyl propane subunits of lignin (reviewed in [111]). Evidence suggests that difficulties in lignin removal during pulping processes are due to these bonds [112].

4.3.1 *Enzymes*

The complete hydrolysis of hemicellulose into monosaccharides requires the concerted action of several enzymes. These include β-D-xylanases, β-D-galactanases, β-D-mannanases, as well as glycosidases β-D-xylosidase, α- and β-D-galactosidase, and β-D-mannosidase. Esterases also participate by hydrolysis of acetylated carbohydrates.

A large number of filamentous fungi, yeasts, and bacteria are reported to produce 1,4-β-D-xylanases [3, 113]. These enzymes have been subdivided by their ability to release L-arabinose from arabinoxylans and from arabinoglucuronoxylans [114]. Arabinose-releasing xylanases are 'debranching' as opposed to the more common 'non-debranching' enzymes that do not liberate arabinose from the same substrates. Both types are capable of attacking unsubstituted 1,4-β-D-xylans and glucuronoxylans. Debranching and non-debranching 1,4-β-D-xylanases have been identified in cultures of *A. niger* [115] and *Ceratocystis paradoxa* [116, 117]. The xylanases can be further subdivided by a variety of physical characteristics (molecular weight (MW), p*I*, pH optima) which have been tabulated [3, 114].

Regulation of xylanase production has been studied in several fungi, virtually all of which also produce cellulases. In *Trichoderma* spp., sophorose and lactose induce both enzyme systems [118, 119]. However, separate control of cellulases and xylanases is observed with other substrates such as xylan and xylobiose. Like the cellulases, glucose repression of xylanase is common [119, 120].

The multiplicity of xylanases, especially those from *Aspergillus niger*, *Trichoderma* spp., *Bacillus* spp., and *Streptomyces* spp., has raised fundamental questions concerning their origin and function. Wong and co-workers [121] note a conserved relationship between molecular weight and p*I* when xylanases from the different species are compared. That is, low molecular weight xylanases tend to be basic whereas high molecular weight is associated with more acidic p*I*. A notable exception to this relationship is the *S. commune* xylanase (MW, 22000; p*I* 4) [122, 123], but the relationship is conserved even in yeast-like fungi that lack multiple xylanases, e.g., *Cryptococcus* [124–126]. Conservation of this relationship and observed differences in the activities of xylanases suggest distinct functions for the multiple xylanases.

Uncertainties remain concerning the roles and interactions of 1,4-β-D-xylanases in hemicellulose hydrolysis. Progress has been hampered by the complexity of natural substrates, the difficulties and expense associated with synthesis of model oligosaccharides, the problems inherent in preparing highly purified enzymes, and the lack of standardized assays. These issues have been reviewed by Erikkson *et al.* [3] and Wong *et al.* [121].

Recent progress has been made towards production of recombinant xylanases. The xylanase gene of *Clostridium thermocellum* (*xynZ*) has been cloned, sequenced, and expressed in *E. coli* [127, 128]. Interestingly, the 5′, non-catalytic region of the gene shares significant homology with endoglucanases from *C. thermocellum*, suggesting the possibility of conserved binding domains. The amino acid sequence within the catalytic region is 43% identical to the exoglucanase of *C. fimi*, which is reported to have xylanase activity [129]. The xylanase gene of *Streptomyces lividans* has also been cloned [130] and several *Bacillus* xylanase genes have been cloned and expressed in *E. coli* [131, 132]. The xylanase clone from *Butyrivibrio fibisolvens* [33] shows considerable amino acid identity with another *Bacillus xylanase* [134].

To our knowledge there have been no reports of xylanase clones from filamentous fungi. This is somewhat surprising given the variety and quantity of xylanases produced by filamentous fungi and the apparent conservation of amino acid sequence [122, 123]. As the following describes, recent interest in the use of xylanases for pulp and paper processing will inevitably change this situation. Both cDNA and genomic clones encoding the xylanases of *Colletotrichum albidus*, a yeast-like fungus, have been sequenced [135].

The complete breakdown of xylans requires hydrolysis of xylo-oligosaccharides to xylose by xylosidases. Most β-D-xylosidases show little or no activity toward xylans and hydrolysis rates decrease with increasing length of xylo-oligosaccharides [136–138]. Additional activities such as glycosyltransferase have been ascribed to several β-D-xylosidases [138–141]. The question of xylosidase specificity has been reviewed [3, 142]. Among the many fungi surveyed [142], xylosidase production was inducible by methyl-β-D-xylopyranosidase. The enzymes are primarily intracellular.

Several additional enzymes participate in xylan breakdown by cleaving side chains. α-L-arabinofuranosidases hydrolyse the α-1,3-linked arabinofuranosyl side chains of arabinans, arabinoxylans, and arabinogalactans [143–145]. α-D-glucuronidases cleave the 4-*O*-methyl glucuronic substituents from xylan chains [146]. Biely and co-workers [147, 148] purified and characterized acetyl xylan esterase from several fungi. Evidence supports synergistic relationships between these enzymes and the 1,4-β-D-xylanases [148, 149]. At least one fungus, *T. reesei*, produces all of these enzymes [150].

Enzymatic degradation of mannan components of hemicellulose involves 1,4-β-D-mannanases, β-mannosidases, β-glucosidases, and α-galactosidases. Dekker [114] and Erikkson and co-workers [3] have reviewed the mannan-degrading enzymes.

4.3.2 *Applications*

The enzymatic conversion of hemicellulose hydrolysates to monomeric sugars and subsequent fermentation to ethanol have attracted attention for many years [151]. Although hexose derivatives are efficiently utilized by *S. cerevisiae*, xylose fermentations have been problematic. Several yeast-like fungi, notably *Candida tropicalis*, *Pachysolens tannophilus*, and *Pichia stipitis*, have been shown to ferment xylose, but considerably more work is needed to improve the efficiency of the fermentations [152]. *S. cerevisiae* cells transformed with xylose reductase and xylitol dehydrogenase genes of *P. stipitis* will utilize xylose as the sole carbon source [153]. However, these transformants utilize xylose oxidatively, and ethanol yields are very low.

Hemicellulases, particularly, 1,4-β-D-xylanases, are becoming increasingly important in the treatment of pulps. Many studies have focused on reducing or eliminating the oxidative chlorine extractions that are used to bleach kraft pulps. The impetus for these efforts has been largely due to environmental concerns over discharge of chlorinated organic compounds, especially dioxin. The industry's scale is worth noting. Approximately 48×10^6 tons of kraft pulp were produced in the US in 1988. This represents $>75\%$ of total pulp production of which more than half is bleached [2, 154]. Given the relatively low value of bleached kraft pulp, commercially viable processes will require massive quantities of inexpensive enzyme.

Recent trials with a variety of hemicellulases show substantial reductions in chlorine consumption, and paper properties (e.g. strength, brightness) are comparable to conventional chlorine bleaching [155–159]. The mechanism(s) involved is unclear, in part because chemical modifications of lignin and hemicellulose occurring during alkali cooking are poorly defined. However, it is clear that kraft pulps, particularly when derived from hardwoods, retain substantial concentrations of deacetylated xylan components. Further, a portion of the modified hemicellulose is redeposited onto cellulose fibers during the process [154]. It has been suggested that the hemicellulases cleave lignin-carbohydrate bonds or that swelling of the pulps facilitates extraction of residual lignin.

Future development of enzymatic bleaching processes will be dependent on availability of large quantities of inexpensive, highly specific enzymes. Even small amounts of contaminating cellulase activity drastically reduce paper strength, thus highly purified preparations have been sought [157–160]. Further improvements require consideration of the physical characteristics of the enzyme and their relationship to current kraft processes. In practical terms, this means that enzyme treatments will be at a high temperature, alkaline pH, and last only for a short duration. Pedersen *et al.* [158] have used alkaline xylanases for treating softwood kraft pulps. Several commercial enzyme preparations are currently available [161], and recombinant enzyme sources are under investigation [128, 131–133, 162].

Ultimately, increased understanding of the mechanism(s) of enzymatic bleaching will facilitate process development. Recent studies have examined accessibility of the substrate to xylanases [163], the adsorption of xylanase to lignin and cellulose [164], and the presence of inhibitory materials in pulp [164].

Hemicellulases also show promise for the production of dissolving chemical pulps [165] and for improving pulp beatability of unbleached pulps [156, 166]. Complete removal of hemicellulose is sought in dissolving pulps to allow more efficient synthesis of cellulose derivatives (e.g. rayon) [154]. This is in distinction to papermaking where exhaustive removal of xylans reduces strength characteristics.

Recently, a combination of xylanase and peroxidase treatments has been shown to be useful [167]. A multi-step process is described in which at least one xylanase treatment followed by one or more peroxidase treatments effectively delignifies pulps. Hardwood and softwood kraft pulps and either manganese peroxidase or lignin peroxidase could be used.

4.4 Lignin degradation

Unlike cellulose and hemicellulose, the lignin polymer is not principally linear nor does it have a repeating hydrolysable interunit bond. Instead, lignin is a complex, three-dimensional, non-stereoregular aromatic polymer composed of phenylpropanoid units linked through several major types of carbon–carbon and ether bonds (Figure 4.3). It is synthesized by higher plants, reaching levels of 20–30% of the dry weight in the tissue of woody plants. The bulk of the lignin is in the thick secondary cell walls, but the highest concentrations of lignin are in the middle lamellae (intercellular regions) where the lignin cements the plant cells together thereby providing rigidity and strength for the plant.

The highly irregular structure of lignin is the result of free radical polymerization of the precursors, *p*-coumaryl, coniferyl, and sinapyl alcohols, the relative proportions depending on the type of plant and tissue. Plant

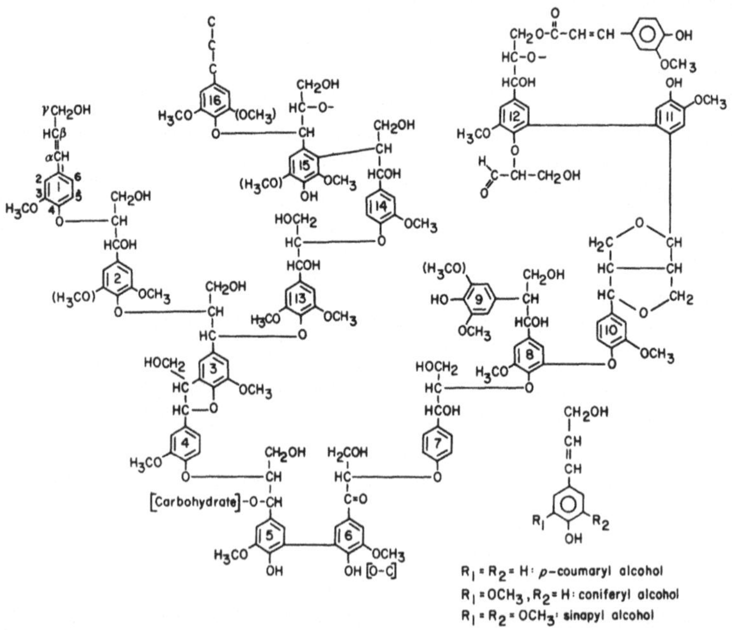

Figure 4.3 Structure of lignin and precursors. An example of a *β*-*O*-4 linkage is shown between residues 1 and 2, and a *β*-1 between residues 8 and 9. (From Kirk and Farrell [195].)

peroxidases catalyse the one-electron oxidation of these precursors to generate phenoxy radical intermediates that diffuse away from the enzyme and couple with each other and the growing lignin polymer. This random coupling generates oligomeric quinone methides susceptible to nucleophilic attack at the benzyl carbons by water, phenolic hydroxyls of other lignols, and also by the hydroxyls of hemicellulose to form lignin-carbohydrate complexes. The interunit bonds are characterized by the points of attachment. The standard nomenclature designates positions on the propyl side chains as α, β, γ, (α being proximal to the aromatic ring) and positions on the aromatic ring as 1–6 (indicating the point of attachment of the propyl side chain). Thus, the β-*O*-4 ether bond is quantitatively the most important interunit bond in spruce lignin. Details of lignin synthesis, structure and chemistry are reviewed by Adler [168], Harkin [169], Higuchi [170], and Sarkanen and Ludwig [171].

The white-rot basidiomycetes degrade lignin more extensively and rapidly than any other known group of organisms. These fungi are also well adapted for utilizing the other major plant components. Because the complex lignin polymer encrusts the cellulosic microfibrils of plants and is chemically bonded to the hemicelluloses, the fungi capable of lignin degradation in fully lignified tissue play a key role in the recycling of carbon, not only from the lignin polymer, but also from the plant polysaccharides.

The most extensively characterized white-rot fungus is *P. chrysosporium*, previously known as *Chrysosporium lignorum* and *Sporotrichum pulverulentum* [172, 173]. In addition to *P. chrysosporium* being one of the fastest lignin degraders so far characterized, it also grows rapidly on defined glucose media, has a high temperature optimum for growth, conidiates profusely, and produces basidiospores under established conditions, making it particularly suited for study in the laboratory.

Important culture parameters required by *Phanerochaete* for lignin degradation to carbon dioxide and water have been determined using ^{14}C-lignin as the substrate in defined media. Ligninolysis is triggered by nutrient limitation; cultures starved for carbon, nitrogen, or sulfur are able to oxidize lignin to carbon dioxide [174–177]. There is also great influence by oxygen partial pressure [176–178], agitation [176, 179], metal ion balance [174] and pH [176]. To induce the ligninolytic system, the fungus is typically grown in stationary liquid medium at pH 4.5, 39°C, high oxygen partial pressure, with glucose as the carbon source and ammonium as the limiting nitrogen source. Although lignin is not required to induce the ligninolytic system, *P. chrysosporium* synthesizes veratryl alcohol (3,4-dimethoxybenzyl alcohol), which has a lignin aromatic substituent pattern [180, 181].

A defined medium for lignin degradation has allowed detailed studies of the metabolism of lignin-model compounds (e.g., β-*O*-4 and β-1 lignin substructure models) without the complications associated with the complex heterogeneous lignin polymer. The oxidations of the lignin substructures (e.g. C_α–C_β

cleavage and C_α oxidation) [182–189] agree well with those deduced from the studies with lignin polymer [190–192].

The ability of many other organisms to degrade lignin is somewhat uncertain. This is partially due to the inability of microbes, including the white-rot basidiomycetes, to use lignin as the sole source of carbon or energy. Consequently, there is no easy selection for lignin degradation by conventional methods. Also, the characterizations of minor modification and depletion of lignin in native lignocellulose are often equivocal due to its intimate association with the other plant constituents. However, other fungi such as certain ascomycetes and fungi imperfecti typify the soft-rot decay of wood. This is characterized by softening under high humidity conditions with extensive polysaccharide and lignin loss, but the rates are not as high as they typically are for the white-rot fungi. Other basidiomycetes, even closely related taxonomically to the white-rot fungi, cause brown rot. This is characterized by extensive decay of the hemicellulose and cellulose of wood, leaving a brown modified lignin residue [193]. The microbiology of lignin degradation has been reviewed [3, 194–196].

4.4.1 *Enzymology*

For the purposes of this chapter, the enzymology of lignin biodegradation will be limited principally to those enzymes secreted by *P. chrysosporium*. Growing evidence indicates that the lignin peroxidase (or ligninase) plays a central role in the initial degradation of the complex aromatic polymer lignin with this organism. Lignin peroxidase was first discovered based on the H_2O_2-dependent C_α–C_β cleavage of lignin model compounds and subsequently shown to catalyse the partial depolymerization of methylated lignin *in vitro* [197–200]. Since the first reports of lignin peroxidase, isozymic forms have been detected. The principal methods for distinguishing these have been by their p*I* and their order of elution from a Mono Q anion exchange column [201–203]. Ten peroxidases are separated by Mono Q chromatography and are designated H1 through to H10. Six of these, H1 (p*I* 4.7), H2 (p*I* 4.4), H6 (p*I* 3.7), H7 (p*I* 3.6), H8 (p*I* 3.5), and H10 (p*I* 3.3), have veratryl alcohol oxidation activity characteristic of lignin peroxidase [204]. H3, H4, H5, and H9 are manganese peroxidases (discussed below). Analytical isoelectric focusing has resolved fifteen proteins with lignin peroxidase activity [202]. The number of isozymes observed depends on the growth conditions (e.g. N v. C starved), the means of purification and storage. The isozymes are glycoproteins of molecular weights estimated at 38–46 kDa and also appear to be structurally related in that polyclonal antibodies raised against H8 cross react with the various other lignin peroxidases. It is clear from *N*-terminal and genetic analyses that some of these lignin peroxidases are separate gene products and not solely due to post-translational modifications (see below). Although the various lignin peroxidases have similar substrate specificities towards aromatic substrates [204, 205], their kinetic behavior varies more clearly when *tert*-butyl hydroperoxide is substituted for hydrogen peroxide [205]. Crystallization of a lignin peroxidase should allow the determination of the three-dimensional structure and key structure/function relationships [206].

Lignin peroxidases are true peroxidases [207–209] and the kinetics of enzyme intermediates have been studied in detail [210–212]. Lignin peroxidase isozymes have absorption spectra characteristic of other hemeproteins, and pyridine hemochromogen complex formation indicates 1 mol of protoheme IX per mole of lignin peroxidase. Like other peroxidases, lignin peroxidase has a ping-pong mechanism, i.e. hydrogen peroxide oxidizes resting enzyme by two electrons to give Compound I enzyme intermediate. Compound I oxidizes aromatic substrates by one electron to give Compound II (a one-electron oxidized enzyme intermediate) that can again oxidize substrate to return the enzyme to resting state. This is depicted in the following, in which LiP represents the ferric state resting lignin peroxidase and S represents an aromatic substrate:

$$\mathrm{LiP} + H_2O_2 \rightarrow \text{Compound I} + H_2O$$

$$\text{Compound I} + \mathrm{S} \rightarrow \text{Compound II} + \mathrm{S}^{+\bullet}$$

$$\text{Compound II} + \mathrm{S} \rightarrow \mathrm{LiP} + \mathrm{S}^{+\bullet}$$

Lignin peroxidase catalyses a variety of oxidations, all of which are dependent on H_2O_2. These include C_α–C_β cleavage of the propyl side chains of lignin and lignin models, hydroxylation of benzylic methylene groups, oxidation of benzyl alcohols to the corresponding aldehydes or ketones, phenol oxidation, and even aromatic ring cleavage of non-phenolic lignin model compounds [200, 203, 213–215]. The underlying principle behind this array of reactions was initially perplexing but can now be explained. Simply stated, lignin peroxidase oxidizes the aromatic nuclei of substrates by one electron (Figure 4.4). The resulting aryl cation radicals degrade spontaneously via many reactions dependent on the structure of the substrate and on the presence of reactants (Figure 4.5). Evidence for the involvement of cation radical intermediates was provided using electron spin resonance [216], using

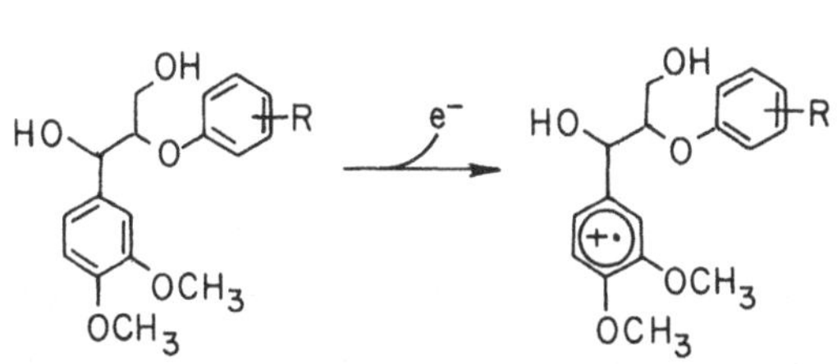

Figure 4.4 Formation of aryl cation radical intermediate by the oxidation of substrate by one electron.

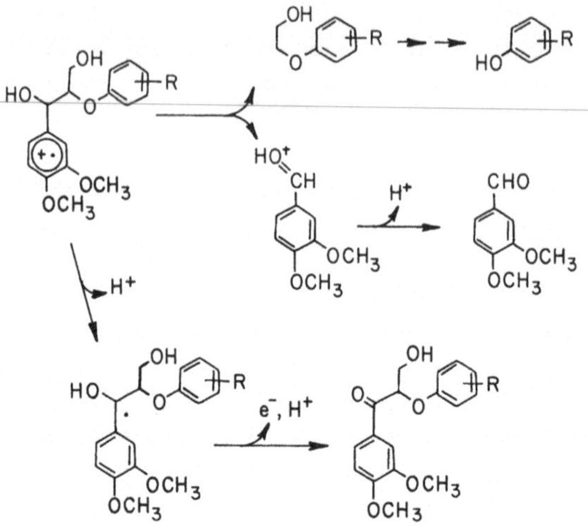

Figure 4.5 C_α–C_β cleavage and C_α oxidation as a result of spontaneous reactions subsequent to aryl cation radical formation. (Adapted from Kirk and colleagues [220].)

purified lignin peroxidase and methoxybenzenes as substrates. Using more lignin-related compounds, Hammel *et al.* [217] further proved the involvement of radical intermediates by identifying radical–dimer products as well as carbon-centered and peroxyl radical intermediates. Other researchers, using more indirect methods, also concluded that cation radical intermediates were involved [218, 219].

Although the reactions subsequent to cation radical formation are complex, they follow chemical principles that readily explain the predominant products. Thus, the C_α–C_β cleavage (Figure 4.5) that is predominant with model compounds most closely related to lignin [217, 220] can be understood based on the degree of substitution of the C_β carbon stabilizing a C-centered radical in purely chemical systems [221]. The cation radical mechanism also explains the oxygenase properties of lignin peroxidase; O_2 reacts with carbon-centered radical intermediates that result from C–C bond cleavage of cation radicals [222, 223]. Lignin peroxidase activity is also expected to break the C_α lignin–carbohydrate bond [220]. Detailed reviews on the radical chemistry of lignin peroxidase catalysed reactions are provided elsewhere [170, 224].

A distinguishing characteristic of lignin peroxidase is that its oxidized enzyme intermediates (Compound I and Compound II) must be sufficiently positive to remove one electron from its non-phenolic substrates. Hammel *et al.* [225] demonstrated that lignin peroxidase oxidizes polycyclic aromatics with ionizing potentials of approximately 7.5 eV or lower, whereas horseradish peroxidase will oxidize only those polycyclics with ionization potentials below 7.35 eV [226]. Similarly, the homologous series of methoxybenzenes (half-wave potentials range from 0.81 V to 1.76 V v. a saturated calomel electrode) has been useful in characterizing the relative oxidation potentials of oxidative enzymes. In the presence of H_2O_2, lignin peroxidase oxidizes the ten congeners with the lowest half-wave potentials, whereas horseradish peroxidase oxidizes the four lowest and laccase (with O_2 as electron acceptor) oxidizes only the lowest [227]. The higher E_{m7} values of the ferrous–ferric states with lignin peroxidases, when compared to other peroxidases, also suggest that the heme active site of lignin peroxidase is more electron deficient, thereby allowing the oxidized intermediates of the enzyme to oxidize substrates of high oxidation-reduction potential [228].

Consistent with the characterization of lignin peroxidase as a peroxidase are the polymerization reactions it catalyses with phenols. This is of special interest because phenolics are produced by the action of lignin peroxidase on non-phenolic lignin structures, and therefore explains why the treatment of lignin with the enzyme can cause not only the depolymerization of lignin [199] but also the repolymerization of lignin fragments *in vitro* [67, 229]. Recently, the partial depolymerization of a synthetic lignin with crude lignin peroxidase and H_2O_2 has been demonstrated *in vitro* [230]. Dilute lignin dispersions and low steady-state H_2O_2 concentrations are thought to be important in minimizing the bimolecular coupling of phenoxy radicals that would lead to polymerization *in vitro* [230]. It has also been proposed that glycosylation of lignin breakdown products may be important in favoring the depolymerizing reactions [231].

The importance of lignin peroxidase in depolymerization of lignin *in vivo* was convincingly demonstrated by Leisola *et al.* [232]. Addition of exogenous lignin peroxidase to carefully washed mycelial pellets greatly stimulated the conversion of ^{14}C-lignin to $^{14}CO_2$. In this case, the presence of mycelia may play an important role in favoring overall depolymerization by removing lignin fragments as they are released. Adding veratryl alcohol, in addition to the lignin peroxidase, to the washed mycelia had a further stimulatory effect. Conversely, horseradish peroxidase had no effect.

Another heme peroxidase found in the extracellular fluid of ligninolytic cultures of *P. chrysosporium* is manganese peroxidase [233, 234]. The principal function of the enzyme is to oxidize Mn^{2+} to Mn^{3+} using H_2O_2 as oxidant. Activity of the enzyme is stimulated by simple organic acids that stabilize the Mn^{3+} and allow it to oxidize organic compounds including phenolic lignin model compounds [235, 236]. As with lignin peroxidase, the prosthetic group is iron protoporphyrin IX and several isozymes can be detected [202, 237–239]. The 46 kDa glycoprotein does not share antigenic determinants with lignin peroxidase, and peptide patterns are different to those observed with lignin peroxidase [202].

Manganese peroxidase exhibits enzyme intermediates analagous to other peroxidases [239, 240]. Compound I is formed by the oxidation of manganese peroxidase by H_2O_2, and this enzyme intermediate can then be reduced by Mn^{2+} and phenols to generate Compound II. Compound II is then reduced back to resting state by Mn^{2+} but not by phenols [240]. Therefore, Mn^{2+} is necessary to complete the catalytic cycle. The enzyme also shows saturation kinetics with Mn^{2+}

Typical assays for monitoring manganese peroxidase activity use phenolics that are easily oxidized by Mn^{3+} and that give useful color changes for detection. However, there has been some question as to the role of the enzyme, in particular in relation to oxidation of lignin and its depolymerization. The biomimetic oxidation of lignin model compounds by Mn^{3+} suggests that it may play a role in oxidizing both phenolic and non-phenolic residues of lignin [241]. More recently, the *in vitro* partial depolymerization of synthetic lignin by manganese peroxidase has been demonstrated [242].

Clearly, an important component of the ligninolytic system of *P. chrysosporium*, in addition to the extracellular peroxidases, is the source of H_2O_2 required as oxidant. Several oxidases have been proposed to play a role in this regard. However, the only one that appears to be secreted in this system under standard ligninolytic conditions is glyoxal oxidase [195].

The temporal correlation of glyoxal oxidase, peroxidase, and oxidase substrate appearances in cultures suggests a close physiological connection between these components [243, 244] The oxidase is a glycoprotein of 68 kDa with two isozymic forms (p*I* 4.7 and 4.9). The active site of the enzyme has not been characterized but Cu^{2+} appears to be important in maintaining activity of purified enzyme. Glyoxal oxidase is produced in cultures when *P. chrysosporium* is grown on glucose or xylose, the major sugar components of lignocellulosics. However, the physiological substrates for glyoxal oxidase are not these growth-carbon compounds but apparently intermediary metabolites. Several simple aldehyde, α-hydroxycarbonyl, and α-dicarbonyl compounds are substrates. Interestingly, proposed products of ligninolysis, such as glycolaldehyde [200], are also substrates for glyoxal oxidase, suggesting that the oxidase–peroxidase coupled system can partially be perpetuated by the action of lignin peroxidase on lignin itself [244].

Perhaps the property of glyoxal oxidase of most interest and of considerable physiological significance is that, in the absence of a peroxidase system, the oxidase is reversibly inactivated [244]. However, the enzyme is reactivated by reconstituting the complete peroxidase system, including both lignin peroxidase and substrate. This suggests that the supply of H_2O_2 by glyoxal oxidase is responsive to the demand of the peroxidases, thereby providing an extracellular regulatory mechanism for control of the coupled enzyme systems. On a protein basis, the amount of glyoxal oxidase appears to be relatively minor when compared to the extracellular peroxidases. However, the specific activity of the oxidase appears to be considerably greater than that of lignin peroxidase. Interestingly, the relative amounts of glyoxal oxidase appear to be much greater when extracted from wood solid substrate compared to defined liquid medium [245].

Another oxidase secreted by *P. chrysosporium* is cellobiose oxidase previously described. This enzyme is not normally detected in standard ligninolytic cultures in the laboratory, perhaps as a result of repression by glucose in cultures. However, the possibility that it may play a role in supplying H_2O_2 during lignin degradation on native lignocellulosic substrates has not been excluded. Also, an extracellular aryl-alcohol has been reported from the white-rot fungus *Bjerkandera adusta* as well as a lignin-peroxidase [246].

Unlike many white-rot fungi, *Phanerochaete* has no detectable laccase activity. However, recent evidence suggests that laccase, like lignin peroxidase, plays a role in lignin degradation by fungi [247–249]. The laccase of *Trametes versicolor* is a blue copper oxidase that catalyses the four-electron reduction of O_2 to H_2O during its oxidation of phenolics, aromatic amines, ascorbate, and metal cyanides [250]. The enzyme has four copper atoms with type-1 and type-3 copper sites of particularly high redox potential in comparison to other blue copper proteins [251]. As with lignin peroxidase, the substrate radicals formed subsequent to oxidation undergo further nonenzymatic reactions [252]. The chemistry of laccase oxidation of lignin-related compounds has recently been reviewed [253]. Also, the synergism between the laccase and manganese peroxidase of *Rigidoporus lignosus* is an example of the possible complex relationships that may be important in lignocellulose degradation [254].

4.4.2 *Molecular genetics of ligninolytic fungi*

Knowledge concerning the molecular genetics of the *P. chrysosporium* ligninolytic system has advanced considerably in the past few years. Standard procedures have been established for auxotroph production [255], recombination analysis [74], and rapid DNA and RNA purification [256, 257].

Several groups have developed transformation systems for *P. chrysosporium*. Randall and co-workers [258] reported transformation to G418 resistance. This system appears to involve extremely low copy numbers of autonomously replicating plasmid. An *ade1* strain of *P. chrysosporium* has been transformed with the corresponding wild-type gene from *Schizophyllum commune* [80, 81], and from *P. chrysosporium* [259]. These transformants involved stable integration of the complementing vectors into the genome, although the frequency of integration at the homologous *ade1* locus has not been determined. Homologous recombination frequencies are high in *Ustilago maydis* [260] and relatively low in *Coprinus cinereus* [261], the only basidiomycetes where this has been examined and reported. This issue is important because it dictates strategies required for gene disruptions/replacements.

Since Tien and Tu [262] first reported cloning of the cDNA encoding lignin peroxidase H8, much has been learned concerning the number, structure, and organization of the peroxidase genes of *P. chrysosporium*. It is now clear that lignin peroxidases are encoded by a family of at least six closely related genes and their allelic variants, and that many of these genes are linked. Regrettably,

Table 4.1 Nucleotide homologies based on optimized alignments of nucleotides. −100 bp of translational start codons to +100 bp beyond translational stops are shown in parentheses following coding sequence homology. Only cDNA sequence is available for CLG4 (266).

	Amino acid							
	H8	LiPA	O282	LPOB	LiPB	GLG5	GLG2	CLG4
H8		99.5	96.5	91.1	91.4	82.6	82.8	72.1
LiPA	99.0(97.8)		96.5	91.4	91.9	82.6	82.6	71.8
O282	88.1(79.8)	87.5(79.9)		89.5	89.8	81.2	81.2	70.5
LPOB	86.6(73.6)	86.2(76.6)	84.3(72.8)		99.5	81.0	81.0	69.7
LiPB	86.9(76.1)	86.7(76.3)	84.4(72.3)	98.5(96.3)		81.2	81.2	69.4
GLG5	80.8(70.5)	80.9(70.4)	78.4(70.2)	79.3(69.0)	79.9(71.4)		100	66.8
GLG2	80.5(70.5)	80.6(70.0)	78.1(69.2)	79.3(69.0)	79.9(70.4)	99.2(97.5)		66.8
CLG4	73.7	73.2	71.1	73.3	73.1	71.8	72.1	

Nucleotide

no uniform nomenclature has been adopted for clone/gene designations. Also, when some sequences were initially published, their allelic relationships had not been established. A list of BKM-1767 synonyms follows:

1. ML1, cDNA encoding isozyme H8 [262]
2. ML4, cDNA encoding allelic variant of ML1 [263]
3. LPOA, genomic clone corresponding to ML1 [264, 265]
4. LiPA, genomic clone corresponding to ML4 [76]
5. CLG4, a cDNA [266]
6. CLG5, a cDNA [266]
7. GLG5, genomic clone corresponding to CLG5 [76]
8. GLG2, genomic clone probably allelic to CLG5/GLG5 [267]
9. O282, genomic clone [75]
10. LiPB, genomic clone allelic to LPOB [76, 265]
11. V4, genomic clone partially sequenced [75].

Sequence, intron position, and intron length are highly conserved among lignin peroxidase genes. Nucleotide and amino acid sequence homology among clones derived from *P. chrysosporium* BKM-1767 are shown in Table 4.1. The lignin peroxidase genes contain eight relatively small introns, except GLG5 contains nine [76]. Intron position is conserved particularly around residues Arg-43, His-47, and His-176, which are believed to be essential for peroxidase activity [262] (Figure 4.6). Also, the last intron of all clones is adjacent to a short, proline-rich exon, and the first intron splits a putative signal sequence-propeptide junction in several genes [75].

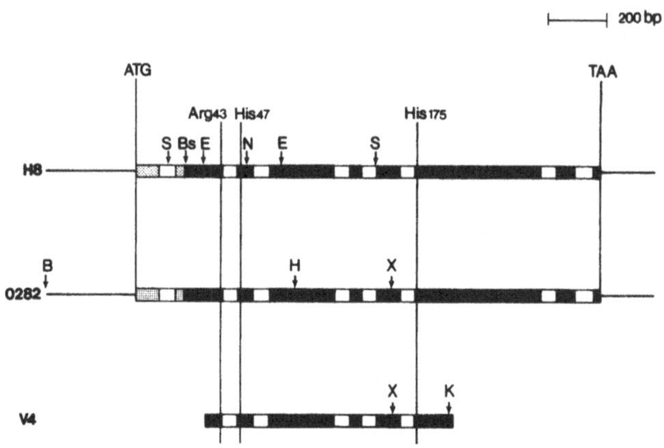

Figure 4.6 Representation of genomic clone showing alignments of signal sequences (stippled box), introns (open box), and coding regions for mature polypeptide (blackened box). The positions of essential amino acid residues are shown by cDNA coordinates [262]. For cosmid V4, only sequences surrounding essential amino acid residues were determined. Restriction site abbreviations: S, *Sph*I; Bs, *Bss*HII; E, *Eco*RV; N, *Nco*I; H, *Hin*dIII; X, *Xho*I; K, *Kpn*I.

A cDNA encoding manganese peroxidase [268] and its corresponding genomic clone [269] have been isolated from a derivative of strain ME446. A manganese peroxidase cDNA sequence derived from BKM-1767 has also been reported [270]. Amino sequence homology to lignin peroxidase H8 is 50–60%, and the residues essential for peroxidase activity are conserved.

The lignin peroxidase genes are linked. An RFLP-based genetic map localized lignin peroxidase genes of ME446 to two linkage groups [77]. Following chromosome separation by clamped homogeneous electrical field (CHEF) electrophoresis, five lignin peroxidase genes have been assigned to a single chromosome [76]. In agreement with the RFLP map, another lignin peroxidase clone (GLG4) has been assigned to the same chromosome as a *cbh1* cluster [104]. The number of manganese peroxidase genes and their chromosomal organization have not been reported, although preliminary (unpublished) results show that at least one manganese peroxidase gene resides on the same chromosome as five lignin peroxidase genes.

Both lignin peroxidase-containing chromosomes are dimorphic with respect to migration on CHEF gels [76]. Homologous chromosomes differing in electrophoretic mobility have also been observed in *Plasmodium falciparum* [271], *S. cerevisiae* [272], *Candida albicans* [273], *U. maydis* [274], and *C. cinereus* (M. Zolan, personal communication). The mechanism(s) giving rise to these chromosome-length polymorphisms may be related to variations in lignin peroxidase activity. In *P. falicurum*, polymorphisms are caused by homologous recombination in distal regions of chromosomes, and it has been suggested that recombination in these dynamic regions may be related to antigenic variation [271]. Extensive mapping of chromosome homologues will be required to determine if a similar mechanism is operative in *P. chrysosporium.*

Studies of the transcriptional regulation of peroxidases have been hampered by difficulties in distinguishing closely related genes. Although it is clear that lignin peroxidase genes are transcriptionally regulated [262, 263, 275] and that expression of manganese peroxidase genes is Mn-dependent [276], the specificity of the transcripts observed on Northerns is questionable. S1 mapped transcriptional start points are equally tenuous [275, 277].

In addition to confusion concerning transcript specificity, considerable uncertainty remains concerning the precise translational products of various clones. Identification by matching deduced amino acid sequence with short regions of experimentally determined sequence is particularly risky [266]. This is also complicated by the presence of allelic forms that encode slightly different isozymes [76]. Recently, Glumoff *et al.* [205] purified five lignin peroxidase isozymes from carbon-starved cultures and determined their N-terminal amino acid sequences. On this basis, three isozymes with p*I*s of 4.65, 3.85, and 3.70 can be tentatively identified as products of CLG4, LiPA/LPOA, and LiPB/LPOB, respectively.

Large quantities of purified lignin peroxidases are needed for basic biochemical investigations and for assessing various commercial applications. Unfortunately, expression of lignin peroxidase genes in heterologous systems has proven to be problematic. In *E. coli*, relatively low levels of aggregated apoprotein are produced in inclusion bodies. Attempts to recover and reconstitute the active enzyme have had limited success (reviewed in [278]). *S. cerevisiae* systems are also inadequate, yielding little or no intracellular levels of apoprotein [278]. Even when under the control of the highly expressed and inducible *cbhl* promoter, expression of peroxidases in *T. reesei* is limited to transcripts only, i.e., no protein is detected [279]. Recently, a baculovirus expression system has been used to produce active recombinant lignin peroxidase isozyme H8 [280]. Although not appropriate for large-scale production, such systems may be useful for experiments requiring limited quantities of recombinant protein, e.g. site specific mutagenesis.

Little is known concerning the molecular genetics of other lignin degrading fungi. Saloheimo and co-workers [279] cloned and sequenced a lignin peroxidase gene from *Phlebia radiata*. The deduced amino acid sequence of this clone is 62%, identical to *P. chrysosporium* isozyme H8 (ML1). On the basis of Southern blot hybridization to the H8 gene, multiple lignin peroxidases appear to be present in *Bjerkandera adjusta*, *Coriolus versicolor* and *Fomes lignosus* [265]. A phenol oxidase gene of *Colletotrichum hirsutus* has been cloned, sequenced and expressed in *S. cerevisiae* [281].

4.5 Applications

A potential application for the ligninolytic system of white-rot fungi is the biomechanical pulping process previously described. The mechanism of biopulping is not well understood but a working hypothesis is that the limited removal of lignin facilitates the separation of the cellulose microfibrils. Elucidation of the mechanism has been hampered by the complexity of the substrates, although lignin peroxidase, manganese peroxidase, and glyoxal oxidase have been identified in *P. chrysosporium* cultured on aspen pulp [245].

The non-specific oxidative ligninolytic system of white-rot fungi also has potential in solving certain environmental problems. The first alkaline extraction stage of Kraft pulp bleach plants contains oxidized and chlorinated lignin fragments and a complex mixture of low molecular aromatics that are a major source of water pollution. *P. chrysosporium* can decolorize the effluent, degrade the lignin residues, and remove the chloro-aromatics [282–284].

P. chrysosporium also is able to oxidize and, in some cases, mineralize to CO_2 and H_2O certain xenobiotics that are not lignin derived. Examples are polychlorinated biphenyls [285], benzo(a)pyrene, DDT, lindane [286], and other chloro-organics [284]. Recently, Lamar and co-workers have demonstrated depletion of the wood preservative pentachlorophenol in soils by *Phanerochaete* spp. [287]. The metabolism of these compounds appears to correlate with the ligninolytic system, and in some cases the direct involvement of lignin peroxidase in oxidizing these compounds has been shown [225, 288–291]. Although lignin peroxidase may catalyze the initial oxidations of such xenobiotics, the subsequent steps are not well understood and should be an important goal of future research.

References

1. Meier, H. (1985) *Biosynthesis and Biodegradation of Wood Components* (Higuchi, T. ed.). Academic Press, New York, pp. 43–50.
2. Smook, G. (1982) *Handbook for Pulp and Paper Technologists*, Joint Textbook Committee of the Paper Industry, TAPPI, Atlanta, Georgia.
3. Eriksson, K.-E.L., Blanchette, R.A. and Ander, P. (1990) *Microbial and Enzymatic Degradation of Wood and Wood Components*, Springer-Verlag, Berlin.
4. Wood, T.M. and Garcia-Campayo, V. (1990) *Biodegradation* **1**: 147–161.
5. Beguin, P. (1990) *Ann. Rev. Microbiol.* **44**: 219–248.
6. Eriksson, K.E. and Wood, T.M. (1985) *Biosynthesis and Biodegradation of Wood Components* (Higuchi, T. ed.) Academic Press, New York, 469–503.
7. Coughlan, M.P. and Ljugdahl, L.G. (1988) *Biochemistry and Genetics of Cellulose Degradation.* (Aubert, J.P., Beguin, P. and Miller, J. eds). Academic Press, London, 11–30.
8. Knowles, J., Lethovaara, P. and Teeri, T. (1987) *TIBTECH* **5**: 255–261.
9. Eriksson, K.-E. (1978) *Biotechnol. Bioeng.* **20**: 317–332.
10. Wood, T.M. (1989) *Enzyme Systems for Lignocellulose Degradation* (Coughlan, M.P. ed.). Elsevier Applied Science, New York, 17–35.
11. Eriksson, K.-E. and Pettersson, B. (1975) *Eur. J. Biochem.* **51**: 193–206.
12. Eriksson, K.-E. and Pettersson, B. (1975) *Eur. J. Biochem.* **51**: 213–218.
13. Despande, V., Eriksson, K.-E. and Pettersson, B. (1978) *Eur. J. Biochem.* **90**: 191–198.
14. Sternberg, D. (1976) *Appl. Environ. Microbiol.* **31**: 648–654.
15. Vallander, L. and Eriksson, K.-E. (1987) *Enzyme Microb. Technol* **9**: 714–720.
16. Coughlan, M.P. (1985) *Biotechnol. Gen. Eng. Rev.* **3**: 39–109.
17. Henrissat, B., Driquez, H., Viet, C. and Schulein, M. (1985) *Bio/Technology* **3**: 722–726.
18. Fagerstam, L.G. and Pettersson, L.G. (1980) *FEBS Lett.* **119**: 97–100.
19. Wood, T.M. and McCrae, S.I. (1986) *Biochem. J.* **234**: 93–99.
20. Knowles, J.K.C., Lehtovaara, P., Murray, M. and Sinnott, M. (1988) *J. Chem. Soc., Chem. Commun.* 1401–1402.
21. Claeyssens, M., Tomme, P., Boewer, C.F. and Hehre, E.J. (1990) *FEBS Lett.* **263**: 89–92.
22. Niku-Paavola, M.L., Lappalainene, A., Enari, T.M. and Nummi, M. (1985) *Biochem. J.* **231**: 75–81.
23. Chanzy, H., Henrissat, B. and Vuong, R. (1984) *FEBS Lett.* **172**: 193–197.
24. Enari, T.M. and Niku-Paavola, M.L. (1987) *CRC Crit. Revs. Biotechnol.* **5**: 67–87.
25. Wood, T.M., McCrae, S.I. and Bhat, K.M. (1989) *Biochem. J.* **260**: 37–43.
26. Van Tilbeurgh, H., Bhikhabhai, R., Pettersson, L.G. and Claeyssens, M. (1984) *FEBS Lett.* **169**: 215–218.
27. Knowles, J.K., Teeri, T.T., Lehtovaara, P., Penttila, M. and Saloheimo, M. (1988) *Biochemistry and Genetics of Cellulose Degradation*, (Aubert, J.P., Beguin, P. and Miller, J. eds). Academic Press, London. 153–169.
28. Highley, T. and Wolter, K.E. (1982) *Mater. Org.* **17**: 127–134.
29. Highley, T.L. (1973) *Wood Fiber*, **5**: 50–58.
30. Illman, B. and Highley, T. (1989) *Biodeterioration Research II*, (O'Rear, C.E. and Llewellyn, G.E. eds). Plenum Press, New York, 465–484.
31. Shoemaker, S., Schweickart, V., Ladner, M., Gelfand, D., Kwok, S., Myambo, K. and Innis, M. (1983) *Bio/Technology* **1**: 691–696.
32. Teeri, T., Salovuori, I. and Knowles, J.K. (1983) *Bio/Technology* **1**: 696–699.
33. Chen, C.M., Gritzali, M. and Stafford, D.W. (1987) *Bio/Technology* **5**: 274–278.

34. Teeri, T., Lehtovaara, P., Kauppinen, S., Salovuori, I. and Knowles, J. (1987) *Gene* **51**: 43–52.
35. Penttila, M., Lehtovaara, P., Nevalainen, H., Bhikabhai, R. and Knowles, J. (1986) *Gene* **45**: 253–263.
36. Saloheimo, M., Lehtovaara, P., Penttila, M., Teeri, T., Stahlberg, J., Johansson, G., Claeyssens, M., Tomme, P. and Knowles, J.K. (1988) *Gene* **63**: 11–21.
37. Barnett, C., Berka, R. and Fowler, T. (1991) *Bio/Technology* **9**: 562–567.
38. Sims, P., James, C. and Broda, P. (1988) *Gene* **74**: 411–422.
39. Azevedo, M., Felipe, M.S., Astolfi-Filho, S. and Radford, A. (1990) *J. Gen. Microbiol.* **136**: 2569–2576.
40. Penttila, M., Nevalainen, H., Raynal, A. and Knowles, J.K. (1984) *Mol. Gen. Genet.* **194**: 494–499.
41. Claeyssens, M. and Tomme, P. (1989) *Enzyme Systems for Lignocellulose Degradation*, (Coughlan, M.P. ed.). Elsevier Applied Science, New York 37–49.
42. Van Tilbeurgh, H., Tomme, P., Claeyssens, M., Bhikhabhai, R. and Pettersson, G. (1986) *FEBS Lett.* **204**: 223–227.
43. Stahlberg, J., Johansson, G. and Pettersson, G. (1991) *Bio/Technology* **9**: 286–290.
44. Tomme, P., van Tilbeurgh H., Pettersson, G., van Damme, J., Vandekerckhove, J., Knowles, J., Teeri, T. and Claeyssens, M. (1988) *Eur. J. Biochem.* **170**: 575–581.
45. Tomme, P., Heriban, V. and Claeyssens, M. (1990) *Biotechnol. Lett.* **12**: 525–530.
46. Kraulis, P.J., Clore, G.M., Nilges, M., Jones, T.A., Pettersson, G., Knowles, J.K. and Gronenborn, A.M. (1989) *Biochemistry* **28**: 7241–7257.
47. Abuja, P.M., Pilz, I., Claeyssens, M. and Tomme, P. (1988) *Biochem. Biophys. Res. Commun.* **156**: 180–185.
48. Abuja, P.M., Schmuck, M., Pilz, I., Tomme, P., Claeyssens, M. and Esterbauer, H. (1988) *Eur. Biophys. J.* **15**: 339–342.
49. Stahlberg, J., Johansson, G. and Pettersson, G. (1988) *Eur. J. Biochem.* **173**: 179–183.
50. Gilkes, N.R., Warren, R., Miller, R. and Kiburn, D. (1988) *J. Biol. Chem.* **263**: 10401–10407.
51. Paice, M.G., Desrochers, D., Rho, D., Jurasek, J., Roy, C., Rollin, C.F., Migel, E.D. and Myaguchi, M. (1984) *Bio/Technology* **2**: 535–539.
52. Johansson, G., Stahlberg, J., Lindeberg, G., Engstrom, A. and Pettersson, G. (1989) *FEBS Lett.* **243**: 389–393.
53. Bergfors, T., Rouvinen, J., Lethovaara, P., Caldentey, X., Tomme, P., Claeyssens, M., Pettersson, G., Teeri, T., Knowles, J. and Jones, T. (1989) *J. Mol. Biol.* **209**: 167–169.
54. Rouvinen, J., Bergfors, T., Teeri, T., Knowles, J.K.C. and Jones, T.A. (1990) *Science* **249**: 380–386.
55. Claeyssens, M., van Tilbeurgh, H., Tomme, P., Wood, T.M. and McCrae, S.I. (1989) *Biochem. J.* **261**: 819–825.
56. Henrissat, B., Claeyssens, M., Tomme, P., Lemesle, L. and Mornon, J.P. (1989) *Gene* **81**: 83–95.
57. Tomme, P. and Claeyssens, M. (1989) *FEBS Lett.* **243**: 239–343.
58. Clarke, A.J. and Yaguchi, M. (1985) *Eur. J. Biochem.* **149**: 233–238.
59. Yaguchi, M., Roy, C., Rollin, C.F., Paice, M. and Jurasek, L. (1983) *Biochem. Biophys. Res. Commun.* **116**: 408–411.
60. Eriksson, K.-E. and Hamp, S.G. (1978) *Eur. J. Biochem.* **90**: 183–191.
61. Messner, R. and Kubicek, C.P. (1991) *Appl. Environ. Microbiol.* **57**: 630–635.
62. Mandels, M. and Reese, E.T. (1960) *J. Bacteriol.* **79**: 816–826.
63. El-Gogary, S., Leite, A., Crivellaro, O., Eveleigh, D.E. and El-Dorry, H. (1989) *Proc. Natl. Acad. Sci. USA* **86**: 6138–6141.
64. Fritscher, C., Messner, R. and Kubicek, C.P. (1990) *Exp. Mycol.* **14**: 405–415.
65. Ayers, A.R., Ayers, S.B. and Eriksson, K. (1978) *Eur. J. Biochem.* **90**: 171–181.
66. Westermark, U. and Eriksson, K.-E. (1974) *Acta Chem. Scand.* **28**: 209–214.
67. Odier, E., Mozuch, M.D., Kalyanaraman, B. and Kirk, T.K. (1988) *Biochemie* **70**: 847–852.
68. Bruchmann, E., Schach, H. and Graf, H. (1987) *Biotechnol. Appl. Biochem.* **9**: 146–159.
69. Renganathan, V., Usha, S.N. and Lindenburg, F. (1990) *Appl. Microbiol. Biotechnol.* **32**: 609–613.
70. Henriksson, G., Pettersson, G., Johansson, G., Ruiz, A. and Uzcategui, E. (1991) *Eur.. J. Biochem.* **196**: 101–106.
71. Eriksson, K. and Pettersson, B. (1982) *Eur. J. Biochem.* **124**: 635–642.

72. Thompson, W. and Broda, P. (1987) *Tr. Br. Mycol. Soc.* **3**: 285–294.
73. Alic, M., Letzring, C. and Gold, M.H. (1987) *Appl. Environ. Microbiol.* **53**: 1464–1469.
74. Alic, M. and Gold, M.H. (1985) *Appl. Env. Microbiol.* **50**: 27–30.
75. Schalch, H., Gaskell, J., Smith, T.L. and Cullen, D. (1989) *Mol. Cell. Biol.* **9**: 2743–2747.
76. Gaskell, J., Dieperink, E. and Cullen, D. (1991) *Nuc. Acids Res.* **19**: 599–603.
77. Raeder, U., Thompson, W. and Broda, P. (1989) *Mol. Microbiol* **3**: 911–918.
78. Raeder, U., Thompson, W. and Broda, P. (1989) *Mol. Microbiol* **3**: 919–924.
79. Penttila, M., Nevalainen, H., Ratto, M., Salminen, E. and Knowles, J. (1987) *Gene* **61**: 155–164.
80. Alic, M., Kornegay, J., Pribnow, D. and Gold, M.H. (1989) *Appl. Environ. Microbiol.* **55**: 406–411.
81. Alic, M., Clark, E.K., Kornegay, J.R. and Gold, M.H. (1990) *Curr. Genet.* **17**: 305–312.
82. Munoz-Rivaz, A., Specht, C.A., Drummond, B.J., Froeliger, E., Novotony, C.P. and Ullrich, R.C. (1986) *Mol. Gen. Genet.* **205**: 103–106.
83. Gelfand, D., Innis, M., Kwok, S., Ladner, M. and Schweickart, V. (1985) Recombinant fungal cellobiohydrolases, European Patent Application No. 84110305.4.
84. Knowles, J., Penttila, M., Teeri, T., Nevalainen, H., Salovuori, I. and Lehtovaara-Helenius, P. (1985) Yeast strains producing cellulolytic enzymes and methods and means for constructing them, International Patent Publication No. WO 85/04672.
85. Van Arsdell, J., Kwok, S., Schweickart, V., Ladner, M., Gelfand, D. and Innis, M. (1987) *Bio/Technology* **5**: 60–64.
86. Penttila, M., Lehtovaara, P. and Knowles, J. (1989) *Yeast Genetic Engineering*, Barr, P., Brake, A. and Valenzuela, P. (eds.). Butterworth, Boston 247–267.
87. Nevalainen, H., Penttila, M., Harkki, A., Teeri, T. and Knowles, J. (1990) *Molecular Industrial Mycology: Systems and Applications*, (Leong, S.A. and Berka, R. eds). Marcel Dekker, New York.
88. Ong, E., Gilkes, N.R., Warren, R.A., Miller, R. and Kilburn, D.G. (1989) *Bio/Technology* **7**: 604–607.
89. Ong, E. (1991) *Enzyme Microb. Technol.* **13**: 59–65.
90. Greenwood, J.M., Gilkes, N.R., Kilburn, D.G., Miller, R.C. and Warren, R.A. (1989) *FEBS Lett.* **244**: 127–131.
91. Kirk, T.K. and Moore, W.E. (1972) *Wood and Fiber* **4**: 72–79.
92. Akamatsu, I., Yoshihara, H., Kamishima, H. and Fujii, T. (1984) *Mokuzai Gakk.* **30**: 697–702.
93. Samuelsson, L., Mjoberg, P.J., Hartler, L., Vallander, L. and Eriksson, K. (1980) *Svensk Papperstidn.* **83**: 221–225.
94. Eriksson, K. and Vallander, L. (1982) *Svensk Papperstidn.* **6**: 33–38.
95. Pilon, L., Varbe, M.C., Desrochers, M. and Jurasek, L. (1982) *Biotechnol. Bioeng.* **24**: 2063–2076.
96. Bar-Lev, S.S., Kirk, T.K. and Chang, H.M. (1982) *TAPPI J.* **65**: 111–113.
97. Leatham, G. and Myers, G. (1990) *TAPPI J.* **73**: 192–197.
98. Leatham, G., Myers, G. and Wegner, T. (1990) *TAPPI J.* **73**: 249–255.
99. Leatham, G., Myers, G. and Wegner, T. (1990) *TAPPI J.* **73**: 197–200.
100. Sachs, I., Leatham, G. and Myers, G. (1989) *Wood and Fiber Sci.* **21**: 331–342.
101. Eriksson, K., Jonsrud, S.C. and Vallander, L. (1983) *Arch. Microbiol.* **135**: 161–168.
102. Ruel, K., Joseleau, J., Jonsrud, S.C. and Eriksson, K. (1986) *Holzforschung* **40**: 5–9.
103. Eriksson, E.K. and Kirk, T.K. (1985) *Comprehensive Biotechnology Vol. III*, (Cooney, C.L. and Humphrey, A.E. eds). Pergamon Press, Oxford 271–294.
104. Covert, S.F. (1990) Ph.D. Thesis, University of Wisconsin, Madison (1990).
105. Aspinall, G.O. (1970) *Polysaccharides*, Pergamon Press, Oxford.
106. Timell, T.E. (1967) *Wood Sci. Technol.* **1**: 54–70.
107. Timell, T.E. (1982) *Wood Sci. Technol.* **16**: 83–122.
108. Wilkie, K.C.B. (1979) *Adv. Carbohydrate Chem. Biochem.* 215–264.
109. Sjostrom, E. (1981) *Wood Chemistry: Fundamentals and Applications*, Academic Press, New York.
110. Delmer, D.P. (1982) *CRC Handbook of Biosolar Resources*, (Zaborsky, O. ed.). 357–361.
111. Jeffries, T. (1990) *Biodegradation* **1**: 163–176.
112. Jiang, J., Cahng, H., Bhattahcharjee, S. and Kwoh, D. (1987) *J. Wood Chem. Technol.* **7**: 81–86.

113. Dekker, R.F.H. and Richards, G.N. (1976) *Adv. Carbohyd. Chem. Biochem.* **31**: 277–352.
114. Dekker, R.F.H. (1985) *Biosynthesis and Biodegradation of Wood Components*, (Higuchi, T. ed.). Academic Press, New York 505–533.
115. Takenishi, S. and Tsujisaka, Y. (1975) *Agric. Biol. Chem.* **39**: 2315–2323.
116. Dekker, R.F.H. and Richards, G.N. (1975) *Carbohyd. Res.* **39**: 97–114.
117. Dekker, R.F.H. and Richards, G.N. (1975) *Carbohyd. Res.* **42**: 107–123.
118. Hrmova, M., Biely, P. and Vrsanska, M. (1986) *Arch. Microbiol.* **144**: 307–311.
119. Royer, J.C. and Nakas, J.P. (1990) *Appl. Environ. Microbiol.* **56**: 2535–2539.
120. Leathers, T.D., Detroy, R.W. and Bothast, R.J (1986) *Biotechnol. Lett.* **8**: 867–872.
121. Wong, K.K.Y., Tan, L.U.L. and Saddler, J.N. (1988) *Microbiol. Rev.* **52**: 305–317.
122. Ujiie, M., Roy, C. and Yaguchi, M. (1991) *Appl. Environ. Microbiol.* **57**: 1860–1862.
123. Paice, M.G., Jurasek, L., Carpenter, M.R. and Smillie, L.B. (1978) *Appl. Environ. Microbiol.* **36**: 802–808.
124. Biely, P., Vrsanska, M. and Kratky, Z. (1980) *Eur. J. Biochem.* **112**: 313–321.
125. Morosoli, R., Roy, C. and Yaguchi, M. (1986) *Biochim Biophys Acta* **970**: 473–478.
126. Nakanishi, K., Arai, H. and Yasui, T. (1984) *J. Ferment. Technol.* **62**: 361–369.
127. Grepinet, O., Chebrou, M. and Beguin, P. (1988) *J. Bacteriol.* **170**: 4576–4581.
128. Grepinet, O., Chebrou, M. and Beguin, P. (1988) *J. Bacteriol.* **170**: 4582–4588.
129. Gilkes, N.R., Langsford, M.L., Kilburn, D.G., Miller, R.C. and Warren, R.A.J. (1984) *J. Biol. Chem.* **259**: 10455–10459.
130. Mondou, F., Shareck, F., Morosoli, R. and Kluepfel, D. (1986) *Gene* **49**: 323–329.
131. Yang, R.C., MacKenzie, C.R., Bilous, D., Seligy, V.L. and Narang, S.A. (1988) *Appl. Environ. Microbiol.* **54**: 1023–1029.
132. Yang, R.C., MacKenzie, C.R., Bilous, D. and Narang, S.A. (1989) *Appl. Environ. Microbiol.* **55**: 568–572.
133. Mannarelli, B.M., Evans, S. and Lee, D. (1990) *J. Bacteriol.* **172**: 4247–4254.
134. Hamamoto, T., Honda, H., Kudo, T. and Horikoshi, K. (1987) *Agric. Biol. Chem.* **51**: 953–955.
135. Boucher, F., Morosoli, R. and Durand, S. (1988) *Nuc. Acids Res.* **16**: 9874.
136. Van Doorslaer, V., Kersters-Hilderson, H. and DeBruyne, C.K. (1985) *Carbohyd. Res.* **140**: 342–346.
137. Matsuo, M. and Yasui, T. (1984) *Agric. Biol. Chem.* **48**: 1853–1860.
138. Rodionova, N.A., Tavobilov, I.M. and Bezborodov, A.M. (1983) *J. Appl. Biochem.* **5**: 300–312.
139. Despphande, V., Lachke, A., Mishra, C., Keskar, S. and Rao, M. (1986) *Biotechnol. Bioeng.* **28**: 1832–1837.
140. Matsuo, M. and Yasui, T. (1984) *Agric. Biol. Chem.* **48**: 1845–1852.
141. Deleyn, F., Claeyssens, M. and DeBruyne, C.K. (1982) *Meth. Enzymol.* **83**: 639–644.
142. Reese, E.T., Maguire, A. and Parrish, F.W. (1973) *Can. J. Microbiol.* **19**: 1063–1074.
143. Lee, S.F. and Forsberg, C.W. (1987) *Can. J. Microbiol.* **33**: 1011–1016.
144. Poutanen, K. (1988) *J. Biotechnol.* **7**: 271–282.
145. Kaji, A. (1984) *Adv. Carbohydr. Chem. Biochem.* **42**: 383–394.
146. Puls, J., Schmidt, O. and Granzow, C. (1987) *Enz. Microb. Technol.* **9**: 83–88.
147. Biely, P., Puls, J. and Schneider, H. (1985) *FEBS Lett.* **186**: 80–84.
148. Biely, P., MacKenzie, C.R., Puls, J. and Schneider, H. (1986) *Bio/Technology* **4**: 731–733.
149. Poutanen, K. and Sundberg, M. (1988) *Appl. Microb. Biotechnol.* **28**: 419–424.
150. Poutanen, K., Puls, J. and Linko, M. (1986) *Appl. Microb. Biotechnol.* **23**: 487–490.
151. Jeffries, T. (1985) *TIB* **3**: 208–212.
152. Jeffries, T. (1990) *Yeast Biotechnology and Biocatalysis* (Verachtert, H. and DeMot, R. eds). Marcel Dekker, New York, 349–394.
153. Kotter, P., Amore, R., Hollenberg, C.P. and Ciriacy, M. (1990) *Curr. Genet.* **18**: 493–500.
154. Grace, T.M., and Malcolm, E.W. (1989) *Pulp and Paper Manufacture, Vol. 5, Alkaline Pulping, Joint Textbook* Committee of the Paper Industry, TAPPI, Atlanta, Georgia.
155. Viikari, L., Kantelinen, A., Poutanen, K. and Ranua, M. (1990) *Biotechnology in Pulp and Paper Manufacture.* (Kirk, T.K. and Chang, H.M. eds), Butterworth-Heinemann Press, Boston 145–151.
156. Clark, T.A., McDonald, A.G., Senior, D.J. and Mayers, P.R. (1990) *Biotechnology in Pulp and Paper Manufacture*, (Kirk, T.K. and Chang, H.M. eds). Butterworth-Heinemann Press, Boston 153–167.

157. Skerker, P.S., Farrell, R.L. and Chang, H.M. (1991) Chlorine-Free Bleaching with Catazyme HS Treatment, *Proc. Int. Bleaching Conf., Swedish Assoc. Pulp and Paper Engineers*, Stockholm, June 11–14.
158. Pedersen, L.S., Nissen, A.M., Elm, D.D. and Choma, P.P. (1991) Bleach Boosting of Kraft Pulp Using Alkaline Hemicellulases, *Proc. Int. Bleaching Conf., Swedish Assoc. Pulp and Paper Engineers*, Stockholm, June 11–14.
159. duManoir, J.R., Hamilton, J., Senior, D.J., Bernier, R.L., Grant, J.E., Moser, L.E. and Dubelsten, P. (1991) Biobleaching of Kraft Pulps with Cellulase-Free Xyanases, *Proc. Int. Bleaching Conf., Swedish Assoc. Pulp and Paper Engineers*, Stockholm, June 11–14.
160. Tan, L.U.L., Yu, E.K.C.U., Loui-Seize, G.W. and Saddler, J.N. (1987) *Biotechnol. Bioeng.* **30**: 96–100.
161. Grant, R. (1991) *Pulp and Paper International* **33**: 61–63.
162. Paice, M.G., Bernier, R. and Jurasek, L. (1988) *Biotechnol. Bioeng.* **32**: 235–239.
163. Puls, J., Poutanen, K. and Lins, J.J. (1990) *Biotechnology in Pulp and Paper Manufacture.* (Kirk, T.K. and Chang, H.M. eds). Butterworth-Heinemann Press, Boston 183–190.
164. Senior, D.J., Mayers, P.R., Breuil, C. and Saddler, J.N. (1990) *Biotechnology in Pulp and Paper Manufacture* (Kirk, T.K. and Chang, H.M. eds). Butterworth-Heinemann Press, Boston 169–182.
165. Paice, M.G. and Lurasek, L. (1984) *J. Wood Chem. Technol.* **4**: 187–198.
166. Noe, P., Chevalier, J., Mora, F. and Comtat, J. (1986) *J. Wood Chem. Technol.* **6**: 167–184.
167. Olsen, W.L., Gallager, H.P., Burris, K.A., Bhattacharjee, S.S., Slocomb, J.P. and DeWitt, D. (1991) Enzymatic delignification of lignocellulosic materials, European Patent Application No. 90111620.2.
168. Adler, E. (1977) *Wood. Sci. Technol.* **11**: 169–218.
169. Harkin, J.M. (1967) Lignin–a natural polymeric product of phenol oxidation. In: *Oxidative Coupling of Phenols* (Taylor, W.I. and Battersby, A.R. eds). Marcel Dekker, New York, (1967) ch. 6.
170. Higuchi, T. (1990) *Wood Sci. Technol.* **24**: 23–63.
171. Sarkanen, K.V. and Ludwig, C.H. (eds). (1971) *Lignins. Occurrence, Formation, Structure and Reactions.* Wiley-Interscience, New York.
172. Burdsall, H.H. and Eslyn, W.E. (1974) *Mycotaxon* **1**: 123–133.
173. Raeder, U. and Broda, P. (1984) *Curr. Gen.* **8**: 499–506.
174. Jeffries, T.W., Choi, S. and Kirk, T.K. (1981) *Appl. Environ. Microbiol.* **42**: 290–296.
175. Keyser, P., Kirk, T.K. and Zeikus, J.G. (1978) *J. Bacteriol.* **135**: 790–296.
176. Kirk, T.K., Schultz, E., Conners, W.J., Lorenz, L.F. and Zeikus, J.G. (1978) *Arch. Microbiol.* **117**: 277–285.
177. Reid, I.D. (1979) *Can. J. Bot.* **57**: 2050–2058.
178. Bar-Lev, S.S. and Kirk, T.K. (1981) *Biochem. Biophys. Res. Comm.* **99**: 373–378.
179. Jäger, A., Croan, S. and Kirk, T.K. (1985) *Appl. Environ. Microbiol.* **50**: 1274–1278.
180. Lundquist, K. and Kirk, T.K. (1978) *Phytochemistry* **17**: 1676.
181. Shimada, M., Nakatsubo, F., Kirk, T.K. and Higuchi, T. (1981) *Arch. Microbiol.* **129**: 321–324.
182. Enoki, A. and Gold, M.H. (1982) *Arch. Microbiol.* **132**: 123–130.
183. Enoki, A., Goldsby, G.P. and Gold, M.H. (1980) *Arch. Microbiol.* **125**: 227–232.
184. Goldsby, G.P., Enoki, A. and Gold, M.H. (1980) *Arch. Microbiol.* **128**: 190–195.
185. Kamaya, Y. and Higuchi, T. (1984) *Wood Res.* **70**: 25–28.
186. Kirk, T.K. and Nakatsubo, F. (1983) *Biochim Biophys. Acta.* **756**: 376–384.
187. Nakatsubo, F., Reid, I.D. and Kirk, T.K. (1982) *Biochim Biophys. Acta.* **719**: 284–291.
188. Umezawa, T. and Higuchi, T. (1984) *Agric. Biol. Chem.* **48**: 1917–1921.
189. Umezawa, T. and Higuchi, T. (1985) *FEBS Lett.* **192**: 147–150.
190. Chen, C.-L., Chang, H.-m. and Kirk, T.K. (1982) *Holzforschung* **36**: 3–9.
191. Chen, C.-L., Chang, H.-m. and Kirk, T.K. (1983) *J. Wood Chem. Technol.* **3**: 35–57.
192. Chua, M.G.S., Chen, C.-L., Chang, H.-m. and Kirk, T.K. (1982) *Holzforschung* **36**: 165–172.
193. Kirk, T.K. and Adler, E. (1970) *Acta Chem. Scand.* **24**: 3379–3390.
194. Buswell, J.A. and Odier, E. (1987) *CRC Crit. Rev. Biotechnol.* **6**: 1–60.
195. Kirk, T.K. and Farrell, R.L. (1987) *Ann. Rev. Microbiol.* **41**: 465–505.
196. Kirk, T.K. and Shimada, M. (1985) *Biosynthesis and Biodegradation of Wood Components* (ed). Higuchi, T. Uni, Tokyo (1985) pp. 579–605.

197. Glenn, J.K., Morgan, M.A., Mayfield, M.B., Kuwahara, M. and Gold, M.H. (1983) *Biochem. Biophys. Res. Comm.* **114**: 1077–1083.
198. Gold, M.H., Kuwahara, M., Chiu, A.A. and Glenn, J.K. (1984) *Arch. Biochem. Biophys.* **234**: 353–362.
199. Tien, M. and Kirk, T.K. (1983) *Science (Washington DC)* **221**: 661–663.
200. Tien, M. and Kirk, T.K. (1984) *Proc. Natl Acad. Sci. USA* **81**: 2280–2284.
201. Kirk, T.K., Croan, S., Tien, M., Murtagh, K.E. and Farrell, R.L. (1986) *Enzyme Microb. Technol.* **8**: 27–32.
202. Leisola, M.S.A., Kozulic, B., Meusdoerffer, F. and Fiechter, A. (1987) *J. Biol. Chem.* **262**: 419–424.
203. Renganathan, V., Miki, K. and Gold, M.H. (1985) *Arch. Biochem. Biophys.* **241**: 304–314.
204. Farrell, R.L., Murtagh, K.E., Tien, M., Mozuch, M.D. and Kirk, T.K. (1989) *Enzyme Microb. Technol.* **11**: 322–328.
205. Glumoff, T., Harvey, P.J., Molinari, S., Goble, M. Frank, G., Palmer, J.M., Smit, J.D.G. and Leisola, M.S.A. (1990) *Eur. J. Biochem.* **187**: 515–520.
206. Troller, J., Smit, J.D.G., Leisola, M.S.A., Kallen, J., Winterhalter, K.H. and Fiechter, A. (1988) *Bio/Technology* **6**: 571–573.
207. Tien, M., Kirk, T.K., Bull, C. and Fee, J.A. (1986) *J. Biol. Chem.* **261**: 1687–1693.
208. Kuila, D., Tien, M., Fee, J.A. and Ondrias, M.R. (1985) *Biochemistry* **24**: 3394–3397.
209. Renganathan, V. and Gold, M.H. (1986) *Biochemistry* **25**: 1626–1631.
210. Andrawis, A., Johnson, K.A. and Tien, M. (1988) *J. Biol. Chem.* **263**: 1195–1198.
211. Marquez, L., Wariishi, H., Dunford, H.B. and Gold, M.H. (1988) *J. Biol. Chem.* **263**: 10549–10552.
212. Wariishi, H., Marquez, L., Dunford, H.B. and Gold, M.H. (1990) *J. Biol. Chem.* **265**: 11137–11142.
213. Leisola, M.S.A., Schmidt, B., Thanei-Wyss, U. and Fiechter, A. (1985) *FEBS Lett.* **189**: 267–270.
214. Umezawa, T., Shimada, M., Higuchi, T. and Kusai, K. (1986) *FEBS Lett.* **205**: 287–292.
215. Umezawa, T. and Higuchi, T. (1987) *FEBS Lett.* **218**: 255–260.
216. Kersten, P.J., Tien, M., Kalyanaraman, B. and Kirk, T.K. (1985) *J. Biol. Chem.* **260**: 2609–2612.
217. Hammel, K.E., Kalyanaraman, B. and Kirk, T.K. (1986) *Proc. Natl Acad. Sci. USA* **83**: 3708–3712.
218. Harvey, P.J., Schoemaker, H.E., Bowen, R.M. and Palmer, J.M. (1985) *FEBS Lett.* **183**: 13–16.
219. Schoemaker, H.E., Harvey, P.J., Bowen, R.M. and Palmer, J.M. (1985) *FEBS Lett.* **183**: 7–12.
220. Kirk, T.K., Tien, M., Kersten, P.J., Mozuch, M.D. and Kalyanaraman, B. (1986) *Biochem. J.* **236**: 279–287.
221. Snook, M.E. and Hamilton, G.A. (1974) *J. Am. Chem. Soc.* **96**: 860–869.
222. Hammel, K.E., Tien, M., Kalyanaraman, B. and Kirk, T.K. (1985) *J. Biol. Chem.* **260**: 8348–8353.
223. Renganathan, V., Miki, K. and Gold, M.H. (1986) *Arch. Biochem. Biophys.* **246**: 155–161.
224. Schoemaker, H.E. (1990) *Recl. Trav. Chim. Pays-Bas* **109**: 255–272.
225. Hammel, K.E., Kalyanaraman, B. and Kirk, T.K. (1986) *J. Biol. Chem.* **261**: 16948–16952.
226. Cavalieri, E. and Rogan, E. (1985) *Environ. Health Perspect.* **64**: 69–84.
227. Kersten, P.J., Kalyanaraman, B., Hammel, K.E., Reinhammar, B. and Kirk, T.K. (1990) *Biochem. J.* **268**: 475–480.
228. Millis, C.D., Cai, D., Stankovich, M.T. and Tien, M. (1989) *Biochemistry* **28**: 8484–8489.
229. Haemmerli, S.D., Leisola, M.S.A. and Fiechter, A. (1986) *FEMS Microbiol. Lett.* **35**: 33–36.
230. Hammel, K.E. and Moen, M.A. (1991) *Enzyme Microb. Technol.* **13**: 15–18.
231. Kondo, R., Limori, T., Imamura, H. and Nishida, T. (1990) *J. Biotechnol.* **13**: 181–188.
232. Leisola, M.S.A., Haemmerli, S.D., Waldner, R., Schoemaker, H.E., Schmidt, H.W.H. and Fiechter, A. (1988) *Cellulose Chem. Technol.* **22**: 267–277.
233. Kuwahara, M., Glenn, J.K., Morgan, M.A. and Gold, M.H. (1984) *FEBS Lett.* **169**: 247–250.
234. Paszczynski, P., Huynh, V.-B. and Crawford, R.L. (1985) *FEMS Microbiol. Lett.* **29**: 37–41.
235. Glenn, J.K. and Gold, M.H. (1985) *Arch. Biochem. Biophys.* **242**: 329–341.
236. Glenn, J.K., Akileswaran, L. and Gold M.H. (1986) *Arch. Biochem. Biophys.* **251**: 688–696.

237. Mino, Y., Wariishi, H., Blackburns, N.J., Loehr, T.M. and Gold, M.H. (1988) *J. Biol. Chem.* **263**: 7029–7036.
238. Paszczynski, A., Huynh, V.-B. and Crawford, R.L. (1986) *Arch. Biochem. Biophys.* **244**: 750–765.
239. Wariishi, H., Akileswaran, L. and Gold, M.H. (1988) *Biochemistry* **27**: 5365–5370.
240. Wariishi, H., Dunford, H.B., MacDonald, I.D. and Gold, M.H. (1989) *J. Biol. Chem.* **264**: 3335–3340.
241. Hammel, K.E., Tardone, P.J., Moen, M.A. and Price, L.A. (1989) *Arch. Biochem. Biophys.* **270**: 404–409.
242. Wariishi, H., Valli, K. and Gold, M.H. (1991) *Biochem. Biophys. Res. Comm.* **176**: 269–275.
243. Kersten, P.J. and Kirk, T.K. (1987) *J. Bacteriol.* **169**: 2195–2201.
244. Kersten, P.J. (1990) *Proc. Natl Acad. Sci. USA* **87**: 2936–2940.
245. Datta, A., Bettermann, A. and Kirk, T.K. (1991) *Appl. Environ. Microbiol.* **57**: 1453–1460.
246. Muheim, A., Leisola, M.S.A. and Schoemaker, H.E. (1990) *J. Biotechnol.* **13**: 159–167.
247. Kawai, S., Umezawa, T. and Higuchi, T. (1988) *Arch. Biochem. Biophys.* **262**: 99–110.
248. Kawai, S., Umezawa, T. and Higuchi, T. (1989) *Wood Res.* **76**: 10–16.
249. Morohoshi, N., Fujita, K., Wariishi, H. and Haraguchi, T. (1987) *Mokuzai Gakkaishi* **33**: 310–315.
250. Malmström, B.G., Andréasson, L.-E. and Reinhammar, B. (1975) Academic Press. New York.
251. Reinhammer, B. (1984) *Laccase*. CRC Press, Boca Raton, FL.
252. Andréasson, L.-E. and Reinhammar, B. (1976) *Biochim Biophys. Acta* **445**: 579–597.
253. Higuchi, T. (1989) In: *Plant Cell Wall Polymers Biogenesis and Biodegradation* (Lewis, N.G. and Paice, M.G. eds). ACS symp. series 399, American Chemical Society, Washington DC 482–502.
254. Galliano, H., Gas, G., Seris, J.L. and Boudet, A.M. (1991) *Enzyme Microb. Technol.* **13**: 478–482.
255. Gold, M.H., Cheng, T.M. and Mayfield, M.B. (1982) *Appl. Environ. Microbiol.* **44**: 996–1000.
256. Raeder, U. and Broda, P. (1985) *Lett. Appl. Microbiol.* **1**: 17–20.
257. Haylock, R., Liwicki, R. and Broda, P. (1985) *Appl. Environ. Microbiol.* **46**: 691–702.
258. Randall, T., Reddy, C.A. and Boominathan, K. (1991) *J. Bacteriol.* **173**: 776–782.
259. Alic, M., Mayfield, M.B., Akileswaran, L. and Gold, M. (1991) *Curr. Genet.* (in press).
260. Wang, J., Holden, D. and Leong, S. (1988) *Proc. Natl Acad. Sci. USA* **85**: 865–869.
261. Mellon, F., Litle, P. and Casselton, L. (1987) *Mol. Gen. Genet.* **210**: 352–357.
262. Tien, M. and Tu, C. (1987) *Nature* **326**: 520–523.
263. Andrawis, A., Pease, E., Kuan, I., Holzbaur, E. and Tien, M. (1989) *Biochem. Biophys. Res. Commun.* **162**: 673–680.
264. Smith, T., Schalch, H., Gaskell, J., Covert, S. and Cullen, D. (1988) *Nuc. Acids Res.* **3**: 1219.
265. Huoponen, K., Ollikka, P., Kalin, M., Walther, I., Mantsala, P. and Reiser, J. (1990) *Gene* **89**: 145–150.
266. deBoer, H.A., Zhang, Y., Collins, C. and Reddy, C.A. (1987) *Gene* **60**: 93–102.
267. Zhang, Y. (1987) Ph.D Thesis, Michigan State University.
268. Pribnow, D., Mayfield, M.B., Nipper, V.J., Brown, J.A. and Gold M.H. (1989) *J. Biol. Chem.* **264**: 5036–5040.
269. Godfrey, B.J., Mayfield, M.B., Brown, J.A. and Gold, M. (1990) *Gene* **93**: 119–124.
270. Pease, E.A., Andrawis, A. and Tien, M. (1989) *Biol. Chem.* **264**: 13531–13535.
271. Corcoran, L.M., Thompson, J.K., Walliker, D. and Kemp, D.J. (1988) *Cell.* **53**: 807–813.
272. Ono, B. and Ishino-Arao, Y. (1988) *Curr. Genet.* **14**: 413–418.
273. Rustehenko-Bulgac, E.P., Sherm, F. and Hicks, J.B. (1990) *J. Bacteriol.* **172**: 1276–1283.
274. Kinscherf, T. and Leong, S.A. (1988) *Chromosoma* **96**: 427–433.
275. Holzbaur, E. and Tien, M. (1988) *Biochem. Biophys. Res. Commun.* **155**: 626–633.
276. Brown, J., Alic, M. and Gold, M. (1991) *J. Bacteriol.* **173**: 4101–4106.
277. Walther, I., Kaelin, M., Reiser, J., Suter, F., Fritsche, B., Saloheimo, M., Leisola, M., Teeri, T., Knowles, J. and Fiechter, A. (1988) *Gene* **70**: 127–137.
278. Pease, E.A. and Tien, M. (1991) *Biocatalysts for Industry* (Dordick, J.S. ed.). Plenum Press, New York 115–135.
279. Saloheimo, M., Barajas, V., Niku-Paavola, M. and Knowles, J. (1989) *Gene* **85**: 343–351.

280. Johnson, T.M. and Li, J. (1991) Heterologous expression of an active lignin peroxidase from *Phanerochaete chrysosporium* using recombinant baculovirus. 75th Ann. Meeting FASEB, Atlanta, Georgia Abstract No. 6818.
281. Kojima, Y., Tsukuda, Y., Kawai, Y., Tsukamoto, A., Sugiura, J., Sakaino, M. and Kita, Y. (1990) *J. Biol. Chem.* **256**: 15224–15230.
282. Kirk, T.K. and Chang. H. (1981) *Enzyme Microb. Technol.* **3**: 189–196.
283. Eaton, D.C., Chang, H., Joyce, T.W., Jeffries, T.W. and Kirk, T.K. (1982) *TAPPI J.* **65**: 89–92.
284. Huynh, V.-B., Chang, H.-m. and Joyce, T.W. (1985) *TAPPI J.* **68**: 98–102.
285. Eaton, D.C. (1985) *Enzyme Microb. Technol.* **7**: 194–196.
286. Bumpus, J.A., Tien, M., Wright, D. and Aust, S.D. (1985) *Science* **228**: 1434–1436.
287. Lamar, R.T. and Dietrich, D.M. (1990) *Appl. Environ. Microbiol.* **56**: 3093–3100.
288. Haemmerli, S.D., Leisola, M.S.A., Sangland, D. and Fiechter, A. (1986) *J. Biol. Chem.* **261**: 6900–6903.
289. Hammel, K.E. and Tardone, P.J. (1988) *Biochemistry* **27**: 6563–6568.
290. Sanglard, D., Leisola, M.S.A. and Fiechter, A. (1986) *Enzyme Microb. Technol.* **8**: 209–212.
291. Valli, K. and Gold, M.H. (1991) *J. Bacteriol* **173**: 345–352.

5 Foreign proteins

D.I. GWYNNE

5.1 Introduction

The advent of recombinant DNA technology in the 1970s allowed the development of methods to clone or duplicate large quantities of DNA in prokaryotic microorganisms such as the human intestinal bacterium *Escherichia coli*. Provided that the DNA of interest was inserted into the appropriate bacterial plasmid, large quantities of a given sequence could be generated by taking advantage of the replicative abilities of the plasmid and the host cell. Those plasmids which were used for the production of DNA were termed cloning vectors because they allowed the user to shuttle a given piece of DNA into a heterologous host. A derivative of this technique was soon developed which allowed the insertion of a piece of DNA into a bacterial plasmid containing a bacterial transcriptional initiation region or promoter. This sort of construct, provided that the DNA was inserted in the appropriate orientation with respect to the promoter, allowed transcription of the heterologous DNA and subsequent translation of the accumulated messenger RNA into protein. These specialized plasmids were termed expression vectors. The wide range of *E. coli* hosts and expression vectors now available have provided indispensible research tools and have also allowed the development of an industry, based on the production of recombinant proteins, which has impacted on many fields.

While *E. coli* expression systems have become the workhorse of the recombinant protein industry, several groups have turned to other microorganisms for the development of expression systems. The fungi have offered alternative systems which have allowed the expression of many active, recombinant proteins. The budding yeast *Saccharomyces cerevisiae* in particular has been used for the expression and production of proteins which were either difficult to express in bacterial systems or were encumbered by process patents involving *E. coli* [1]. The filamentous fungi have also provided useful alternatives to bacterial systems.

Recombinant proteins can be expressed intracellularly or can be secreted. Several characteristics of filamentous fungal expression hosts give them potential advantages over *E. coli* and *S. cerevisiae*. Problematical production of inactive intracellular inclusion bodies and the lack of certain eukaryotic post-translational modifications in *E. coli* has precluded its use in certain situations. While *Saccharomyces* has solved many of the problems of 'eukaryotic' type modifications, it has been limited in its capability to secrete large amounts of protein. In light of these concerns surrounding the use of bacterial and yeast expression hosts, the filamentous fungi are attractive targets for the production of economically important recombinant peptides due to their efficient secretion of endogenous proteins. In addition, new fungal systems offer alternatives for the commercial production of those proteins already encumbered by process patent applications involving other organisms.

Tractable expression technology in filamentous fungi is dependent, as in any expression system, on the acquisition and development of efficient expression vectors and the development of methods for incorporating functional expression vector DNA into the host cell (DNA mediated transformation). Successful transformation is often host (species) dependent (see chapter 1) and useful expression vectors rely on the isolation of efficient promoter regions with the appropriate regulatory characteristics (see chapter 2). Since the first descriptions of the successful expression of a foreign gene in a filamentous fungus [2, 3], several species from this group of organisms have been used to express a wide variety of foreign proteins. The genus which has been most exploited for the purpose of expression of foreign proteins has been *Aspergillus*. This genus and other filamentous fungi have been used to express proteins of pharmaceutical interest as well as proteins which can be included in the industrial category.

5.2 Industrial enzymes and food products

Filamentous fungi, especially those of the genus *Aspergillus*, have traditionally been used for the manufacture of food products and industrial enzymes and were an obvious target when the expression of recombinant industrial proteins was being considered (see chapter 3). Since the production of Takadiastase in *Aspergillus oryzae*, industry has taken advantage of the ability of this group of organisms to secrete large quantities of homologous hydrolytic enzymes.

Many of the traditional products of the enzyme industry, as described above, have been derived from fungi. These include enzymes used in the manufacture of food, animal feed and in biodegradation. Recombinant DNA approaches give the potential for increased efficiency of production of these fungal proteins. In addition, the availability of cloned genes encoding non-fungal products has provided the opportunity for established fungal fermentation technology to be used for the production of other valuable products.

Scientists at Genencor have developed a successful production system in *Aspergillus niger* (variety *awamori*) for recombinant chymosin [3]. The aspartyl protease chymosin is found in the fourth stomach of unweaned calves and is an important enzyme used in the production of cheese. This enzyme is synthesized as a preproprotein containing a 16 amino acid signal peptide and a 42 amino acid pro region. The signal peptide is removed during the normal secretion process and the propeptide is subsequently removed in an autocatalytic manner. Natural chymosin, which is often in limited supply, has occasionally been replaced by acid proteases derived from microbial sources (such as the fungus *Mucor miehei*). These substitute proteases, however, have not been entirely adequate due to lack of specificity and the production of off-flavours in the cheese.

Ward *et al.* [4] first expressed chymosin in *Aspergillus nidulans* using the highly expressed, constitutive *oliC* gene as a source of a promoter for their expression vector. Cullen *et al.* [3] described the expression of chymosin in *A. nidulans* which was driven by a promoter derived from the *A. niger* glucoamylase (*glaA*) gene. A series of expression constructs were described in which the transcription, translation and secretory control elements of the *glaA* gene were coupled to a cDNA sequence encoding bovine prochymosin. These fusion constructs contained varying lengths of sequence derived from the *A. niger glaA* coding sequence. Portions of the *glaA* sequence ranged from the secretion signal only to a construct containing the signal, the *glaA* pro region and a small portion of the region encoding the mature *glaA* peptide. An additional construct contained the chymosin signal peptide region in place of the *glaA* signal. Active chymosin was secreted from the *A. nidulans* host using each of the constructs described above. A mixture of chymosin and prochymosin was detectable in the medium using Western blot analysis suggesting that some prochymosin was autocatalytically cleaved at low pH. More than 90% of the product was secreted into the fermentation broth. Transformants containing the construct with the *glaA* secretion signal sequence fused to the chymosin propeptide sequence gave the highest level of expression (average of 150 μg/g dry weight of mycelia).

In an attempt to increase secretion yields, the range of constructs described above were transformed into *A. niger* (var. *awamori*). *A. niger* strain UVK143 (derived from NRRL strain 3112) was selected because it had been shown to produce high levels of glucoamylase and it was thought that recombinant chymosin expression, driven by the *glaA* promoter, would be maximized in this strain [5]. The best level of expression was again achieved by the *glaA* signal prochymosin fusion and the levels were close to that achieved in *A. nidulans* (approximately 10 mg/l in a shake flask culture). In both *A. nidulans* and *A. niger* no correlation was observed between the copy number of the construct which was integrated into the genome and the level of expression which was observed. This differs from data derived from the expression of other recombinant proteins [2, 6] and suggests that the rate limiting step in the production of chymosin in recombinant *Aspergillus* may occur after transcription.

Recently Ward *et al.* [7] described a fusion between the full length *glaA* coding region and the prochymosin sequence. The active product seemed to be either autocatalytically released from the fusion protein or released by an endogenous protease. Significant increases in yield were obtained over the constructs described above. The increased yields were attributed to the fact that glucoamylase is secreted efficiently and that the fusion facilitated the secretion of prochymosin.

Despite the extent of the genetic manipulations described above the yields of chymosin were not in the range which would be commercially viable for this enzyme. Some manipulation of the *A. niger* host was performed to decrease the levels of endogenous secreted proteases which may potentially affect the yield of product. Berka *et al.* [8] cloned the gene encoding the native aspergillopepsin A and used the clone to delete the gene from their production strain by recombinant means. Again some progressive gains in expressed levels of secreted chymosin were observed. The most significant increases in yield and the method which allowed the development of a viable production process was achieved by using a traditional approach; mutagenesis, screening and selection. Using a recombinant parent strain containing copies of the preprochymosin cDNA (chymosin secretion signal sequence and driven by the glucoamylase promoter), Lamsa and Bloebaum [9] performed several rounds of mutagenesis and used an agar plate screen to select strains with increased secreted levels of chymosin.

The agar plate method was developed by incorporating a colony size restrictor 2,6-dichloro-4-nitroaniline and an acid protease inhibitor diazoacetyl-norleucine methyl ester to reduce the high background concentration of the native acid protease which interferes with the assay. This combination allowed the screening of 5 to 50 000 colonies per screen. Selected putative hyperproducers were tested in a further screen which involved growing mutants in liquid medium in a 24 well culture plate. A final test was performed in shake flasks and several selected strains were then evaluated in fermenters. Yields in excess of 100 mg/l were obtained in test fermenter runs.

The Genencor chymosin production process has been approved by the FDA and recombinant protein is currently being marketed.

Haarki *et al.* [10] have reported the expression and secretion of bovine chymosin in *Trichoderma reesei*. This host was selected due to its capacity to secrete large quantities of cellulolytic enzymes (greater than 30 mg/l) [11]. Prochymosin cDNA was inserted between the promoter and terminator regions of the cellobiohydrolase (*cbh1*) gene of *T. reesei*. Four variations included a direct promoter-preprochymosin fusion, a promoter-*cbh1* signal-prochymosin fusion, a promoter double signal-prochymosin fusion and a *cbh1* signal-prochymosin fusion which included the first 20 amino acids of the *cbh1* coding sequence. Initial results showed the presence of active enzyme in the growth medium from all constructs in the growth medium. The predominant form detectable on immunoblots was of the same size as mature chymosin and levels of the best transformants (containing the construct with the 20 amino acids of the *cbh1* coding region) approached 40 mg/l in fermenter tests. This finding, that fusion with a portion of the *cbh1* coding region gave the best results, differs from that reported for *Aspergillus* spp. where the partial glucoamylase fusion tended to decrease yields.

Another enzyme which can be used in cheesemaking is an aspartyl protease derived from *Mucor miehei*. Gray *et al.* have expressed this protein, using the native *Mucor* promotor, in *Aspergillus nidulans* [12]. The secreted product was biologically active although immunoblot analysis showed that the product was significantly larger than the native enzyme. This significant difference in molecular weight between the native and recombinant enzymes is probably attributable to aberrant glycosylation since removal of the N-linked carbohydrate (with endoglycosidase F) from both the native and recombinant enzymes caused them to migrate at similar molecular weights on a polyacrylamide gel.

The *M. miehei* aspartyl protease has also been produced in *Aspergillus oryzae*. Christensen *et al.* [13, 14] have reported the expression of this enzyme in a vector containing the *A. oryzae* α-amylase promoter. The promoter sequence was derived from a mutant of *A. oryzae* which produced α-amylase in high amounts. The cDNA encoding the preproenzyme form of the aspartyl protease was used. The construct also contained the *A. niger* glucoamylase terminator. Several transformants of the wild-type strain IFO 4177 were tested in a fermenter and one transformant produced 3.3 g/l of culture medium. The continuously fed biorector which was used allowed a biomass of 120 g/l dry weight to be obtained. As in the case of the enzyme produced by *A. nidulans*, the product was biologically active. The specific activity of the recombinant form was identical to the naturally produced *Mucor* enzyme, but the product was hyperglycosylated.

The zygomycete *Mucor circinelloides* has also been used for the expression of the *M. miehei* aspartyl protease. This particular expression system is unique in the sense that the system uses an expression vector which autonomously replicates in *M. circinelloides*. The entire *M. miehei* gene was expressed in the vector system and the recombinant enzyme was biologically active, immunodetectable and secreted at levels up to 12 mg/l of culture. The product did not seem to be hyperglycosylated [15].

Christensen *et al.* have also reported the expression of a thermophilic lipase in the *A. oryzae* expression system used for the *M. miehei* protease. The enzyme is intended for detergent use and the recombinant product is being produced at Novo Industri in Japan (T. Christensen, pers. commun.).

The cellulose-degrading enzyme complex of *Trichoderma reesei* and other cellulolytic organisms has been an attractive target for recombinant expression. While *T. reesei* produces large quantities of these enzymes, the ability to produce unique activities (which are also free of other contaminating enzymes), which are tailored to specific uses, is an attractive goal.

Barnett *et al.* [16, 17] have reported the expression of endoglucanase I (EGI) and cellobiohydrolase I (CBHI) genes in *A. nidulans*. Entire *T. reesei* genes were expressed in this situation indicating that *A. nidulans* has the ability to recognize and utilize *T. reesei* transcription and intron processing signals. Some aberrant glycosylation of the product may have occurred in the *Aspergillus* host but the products were demonstrated to be biologically active in substrate degradation assays. *A. nidulans* has also been used to secrete biologically active bacterial endoglucanase. In this case a genomic clone derived from the prokaryote *Cellulomonas fimi* was used as a source of the region encoding the mature protein. This sequence was fused to a consensus fungal secretion signal and expression was driven by the *A. nidulans alcA* promoter. Biologically active product was detectable in a plate assay when the recombinant host was grown under conditions which induced the *alcA* promoter. In a parallel series of experiments similar results were obtained using the *A. niger* glucoamylase promoter and secretion signal sequence [18].

Chen *et al.* [19] described the expression of *T. reesei* CBHII in *A. niger* (var. *awamori*). The intent here was to express and secrete CBHII into a low cellulolytic background. The expression of a genomic copy and an intronless copy of the CBHII sequence was driven by the *A. niger glaA* promoter. Enzymatic activity was measured by the ability of samples of the growth medium to produce reducing sugars from phosphoric acid swollen cellulose. Both intron-containing and intronless genes produced biologically active material indicating that *A. niger*, like *A. nidulans*, recognizes *T. reesei* splicing signals. The product also seemed to be hyperglycosylated as indicated by the heterogeneous pattern of the product on SDS polyacrylamide immunoblots of the protein [19].

Several groups have used the availability of cloned DNA encoding fungal enzymes to raise the expression level of these endogenous proteins. The group at Panlabs Research showed that the level of secreted glucoamylase in *A. niger* could be raised several fold by inserting multiple copies of the glucoamylase gene into the host. Expression level correlated with gene copy number [20]. Christensen *et al.* used a similar approach when they transformed a wild-type *A. oryzae* strain with copies of the cDNA encoding the entire α-amylase gene derived from an overproducing mutant. One transformant produced approximately 12 g/l of the enzyme in a 2 l fermenter as compared to the host production level of 1–2 g/l. It was unclear whether the increased yield was due to a gene dosage effect, as described by the Panlabs group, or whether the gene derived from the mutant strain contained a superior promoter. Barnett *et al.* [21] have cloned and characterized the β-glucosidase (*bgl1*) gene of *T. reesei*. This clone was inserted into *Trichoderma* and multiple copy hosts were selected. One transformant produced an extracellular cellulase with a significant increase in activity which was observable on several substrates.

5.3 Proteins of pharmaceutical interest

Recombinant filamentous fungal systems have been used for the expression of therapeutically important proteins in the same way in which *E. coli* and *Saccharomyces cerevisiae* have been used. The expression of these sorts of proteins, often derived from human sequences, has its own set of unique characteristics when compared to industrial enzymes. Quite often these proteins are complex and have evolved to be secreted into human biological fluids such as serum. These differences make them difficult to express and to assay routinely. The filamentous fungi can potentially offer production advantages, however, in the form of increased yields and proprietary process alternatives.

The Allelix Biopharmaceuticals group have focused on the development of an expression system in *A. nidulans* based on the *alcA* promoter (derived from the alcohol dehydrogenase I gene), a consensus fungal secretion signal sequence and the *A. niger glaA* transcription terminator sequence. The system offers good potential for the production of recombinant pharmaceutical proteins because the *alcA* promoter is repressed in the presence of glucose and is inducible in the absence of glucose by such agents as ethanol, various ketones and threonine. The expression of a cDNA clone encoding the processed, mature sequence of human α-interferon was used to test and to develop the system. Initial results showed that recombinant *A. nidulans* would secrete biologically active, immunodetectable product into the growth medium. Biological activity was detectable with a viral plaque reduction assay and the highest level of activity was observable in the presence of the *alcA* inducer threonine. Expression levels correlated with gene copy number and the maximal levels observed in shake flask, minimal media experiments were in the low microgrammes per litre range [18]. Expression was improved several orders of magnitude by manipulating the host strain, the expression vector and the culture conditions. The host strain proved to exert some constraints on the expression of multiple copies of the *alcA* promoter due to the lack of expression of sufficient *alcR* gene product. A trans acting factor, which is required for induction of *alcA*, is encoded by *alcR*. The titration of *alcR* gene product was overcome by overexpressing *alcR* as a multicopy chromosomal insert in the host [22]. The new host allowed increased expression of α-interferon. Further expression improvements were effected by changing the sequence environment of the translation start codon allowing it to conform to the consensus described by Kozak [23]. Media improvements included the development of rich media where expression of the recombinant product could be separated from the growth phase by using glucose as the carbon source. The appropriate inducer could be added when glucose concentrations dropped below the level which repressed *alcA* expression.

A process for the production of a human therapeutic enzyme was developed with the *alcA* expression system. This enzyme is a natural serum protein with anti-inflammatory activity. The recombinant enzyme can potentially reduce much of the tissue damage which occurs during inflammation and reperfusion injury following heart attack, stroke or organ transplant. The N-terminus of the enzyme is normally acetylated post-translationally to produce a fully active peptide. A synthetic DNA was prepared which represented the normal human enzyme sequence. This was inserted into the *alcA* expression vector between the *alcA* promoter-messenger RNA leader and the *A. niger glaA* transcription terminator. No secretion signal was included since the enzyme is an intracellular protein and the objective was to retain the recombinant product inside the fungal mycelia. Recombinant human enzyme was recoverable by homogenizing intact mycelia. The purified product was indistinguishable from the human blood cell protein in all respects including SDS gel profile, biological activity and N-terminal sequence. The N-terminal methionine was also acetylated by the *Aspergillus* host in an identical way to the authentic human protein. A two-step fermentation process was developed similar to that described above where mycelia were cultured in the presence of glucose and the production of the enzyme was induced when glucose was in sufficiently low concentration (Gwynne *et al.*, unpublished data).

Several other proteins of therapeutic interest have been expressed in the *alcA* system. These include human epidermal growth factor (hEGF), human growth hormone (hGH), human interleukin-6 (hIL-6) and two human therapeutic glycoproteins. All of these proteins were secreted into the extracellular medium and were biologically active. Several of these recombinant proteins have been sequenced and it has been determined that the secretion signals have been removed leaving the correct N-termini on each mature protein. Their expression level in shake flask cultures was in the 10 to 50 mg per litre range. Several of the proteins described contained disulphide bonds which are required for biological activity. This indicates that the host system allowed correct folding and bond formation. The recombinant human glycoprotein appears to be glycosylated to a similar extent as the natural human protein as observed on SDS polyacrylamide gel immunoblots (Gwynne, unpublished data).

Upshall and colleagues at Zymogenetics have reported the expression and secretion of human tissue plasminogen activator (htPA) in *A. nidulans* [24]. This protein is involved in blood clot removal and has recently been approved for the treatment of heart attacks. tPA is a 68 kDa secreted glycoprotein with many disulphide bonds. It functions as a serine protease in its activation of plasminogen and it is produced via proteolytic cleavage of a precursor molecule. The prepro form of the protein undergoes normal secretion signal removal and the pro sequence is subsequently removed to generate a protein with glycine as the *N*-terminal amino acid. Aminopeptidase activity next removes the four amino terminal residues to generate a protein with serine as the *N*-terminal residue. Both the glycine and the serine forms of the molecule have plasminogen activating activity but fibrinogen is required for this function. A subsequent processing step, which is mediated by plasmin, converts the single-chain form into an active two-chain form linked by disulphide bonds. This homodimer does not require fibrinogen for activity [25, 26]. A cDNA clone encoding the entire htPA sequence (including the secretion signal sequence) was inserted into an expression vector between the promoter and terminator sequences derived from the triosephosphate isomerase (*tpiA*) gene of *A. nidulans*. The final construct contained a fusion at the translation initiation codon between the *tpiA* mRNA leader and the htPA sequence. Transformants which tested positive on Southern blots for the presence of the htPA sequence were cultured and tested for their ability to secrete active htPA into the culture medium. The assay involved testing the ability of the samples to lyse fibrin in a plate assay. Samples were also tested for their ability to be inhibited by an antibody directed against natural human tPA since other proteolytic enzymes can lyse fibrin. The addition of these antibodies inhibited the ability of the samples to lyse fibrin. Samples of recombinant protein were semi-purified and were analysed on Western blots and only the single-chain form was detectable in the culture medium. The material ran as a doublet on the gels in a similar manner to the natural protein isolated from a mammalian source. The two mammalian species represent different glycoforms. These data suggest that *Aspergillus* glycosylates the protein in a quantitatively similar manner to the mammalian system. N-terminal analysis of the purified recombinant protein showed that the serine form was the predominant species with the glycine form existing as a less abundant species. The concentration of the two forms in the sample isolated from a mammalian source differed from the recombinant with respect to the relative concentrations of both forms. The fact that the glycine and serine forms are detectable in the *Aspergillus* culture medium suggests that the fungus is able to mimic both the correct processing of the proprotein and the subsequent aminopeptidase activity which removes the four N-terminal residues [24]. The mammalian aminopeptidase activity cleaves tPA after an Arg–Arg pair. This is reminiscent of an activity which is involved in the processing of glucoamylase in *A. niger* which also cleaves after a pair of basic amino acids [27, 28].

The level of product which was secreted into the culture medium was approximately 100 μg/l in the shake flask cultures as measured by the fibrinolytic assay. Higher levels (1 mg/l) were reported for strains containing expression constructs driven by the *A. nidulans alcC* promoter (derived from the alcohol dehydrogenase 3 gene) or by the *A. niger* alcohol dehydrogenase A promoter [24].

Usphall *et al.* have also reported the secretion of human α-1-antitrypsin and human granulocyte macrophage colony stimulating factor (hGMCSF) at levels of up to 1 mg/l in *A. nidulans* [24]. α-1-antitrypsin is an inhibitor of the serine protease elastase which causes tissue damage during inflammation and emphysema. GMCSF is a protein involved in the maturation of white blood cells. The *A. nidulans* transformants contained constructs driven by the same alcohol dehydrogenase promoters used to express htPA [24].

The oomycete water mould *Achlya ambisexualis* has been used to express proteins of therapeutic interest. Gamma interferon has been expressed and secreted in a transformant containing a construct driven by the simian virus 40 (SV 40) promoter. Secretion was mediated by the interferon secretion signal sequence [29–31]. Most of the interferon activity remained associated with the cell. The secreted material seemed to be N-glycosylated at both available sites. An interesting aspect of this work was the fact that a promoter which normally functions in a mammalian environment was able to operate in a fungal system.

5.4 Expression of other proteins

Recombinant expression systems are often used to provide sufficient levels of material for biochemical or structural analysis. High level expression systems are often the only way to provide sufficient amounts of material if the target protein is present at very low levels in its natural source or if large amounts are required for crystallography or nuclear magnetic resonance studies. In addition, the availability of cloned DNA and a system for expression allow the easy generation and analysis of sequence variants. These sorts of tools are invaluable for protein structure function studies. Filamentous fungi have been utilized in this regard because of their ability to secrete large amounts of material. Additionally since fungi are eukaryotic microbes the 'turnaround time' from gene to protein is generally faster than in mammalian systems.

Archer *et al.* used an *A. niger* system to produce sufficient amounts of correctly folded hen egg white lysozyme (HEWL) for their structure/function studies. HEWL has been extensively studied since it was the first enzyme to have its structure determined by X-ray crystallography [32]. The accumulated structural knowledge about this protein makes it an ideal candidate for structure/function studies. The *Aspergillus* system was considered when other heterologous systems produced either insufficient quantities of protein or product which was folded incorrectly. A cDNA clone encoding the complete HEWL coding region was inserted into two types of expression vectors: one driven by the *A. niger* (var. *awamori*) glucoamylase promoter (*glaA* terminator) and a second driven by the *A. nidulans* glyceraldehyde-3-phosphate (*gpdA*) promoter (*trpC* terminator). Levels of HEWL activity were determined in shake flask cultures of transformants containing either of the two constructs. The starch inducible *glaA* promoter produced significantly higher levels of protein than the constitutive *gpdA* promoter (12 mg/l and 1 mg/l respectively). Polyacrylamide gel analysis of the proteins in mycelial extracts and in culture media showed that approximately 50% of the product was retained inside the cell in strains containing the *glaA* driven construct. The *glaA* strains were used for further analysis of the protein. The recombinant protein was purified from culture medium filtrates by FPLC and SDS gel analysis showed that the recombinant protein co-migrated with the natural egg white protein. Some proteolysis of the protein may have occurred since anti-HEWL antibodies detected faint bands at molecular weights which were lower than those seen for the full length protein. This result was confirmed when N-terminal sequencing revealed the presence of an internal sequence. The protein was further purified by ion exchange chromatography to separate and further purify the intact protein. N-terminal analysis revealed the sequence KVFGRCELA which corresponds exactly to the sequence of the authentic egg white protein. This indicates that the fungal processing system recognizes the chicken HEWL secretion signal sequence. The specific activities of the recombinant and the egg white proteins were identical. The purified protein was further subjected to two-dimensional nuclear magnetic resonance spectroscopy. This powerful analytical procedure gives structural data by positioning specific atoms within the structure. The spectra generated for both proteins were essentially identical, confirming the activity and sequence data. In addition, this confirmed that the protein was correctly processed and folded, and that the appropriate disulphide bonds had formed.

5.5 Scale-up, production and regulatory aspects

The success of a recombinant process rests on its ability to produce high levels of active product from maximum biomass in a manner which does not compromise downstream processing and which is acceptable by the regulatory authorities. In this sense the development of a process for the production of a recombinant protein from a filamentous system is no different from development of other microbial production systems. Processes for the manufacture of therapeutic proteins and industrial proteins have many features in common. Significant differences, however, exist in the regulatory environment, the level of purity required, and the part of the process where maximum costs are incurred.

The medium requirements for maximum yield per cell (expression level) and maximum yield of biomass are often in conflict. As an example the culture conditions which are required for induction of a high level of gene expression may differ from those required for the production of high biomass. The degree of this problem depends on which expression system is chosen. Selection or design of the appropriate system can help to avoid some of these issues. In certain systems (e.g. constitutive expression) the accumulation of biomass and the expression of the product occur simultaneously and requirements for induction of product and cell growth can conflict. In other systems the growth and production phases can be separated and this conflict may be avoided. The latter type of process can be especially useful if the recombinant protein compromises the growth rate of the host. The wide range of available fungal promoters allows some flexibility in the design of an expression system.

Another area of conflict in medium choice is the selection of conditions which yield a high level of product but do not compromise (and add undue expense to) the purification process. Medium which does not include a high concentration of peptides or proteins is especially important in the production of pharmaceutical proteins where high levels of purity are required and a significant proportion of production costs are incurred downstream of the bioreactor.

In any microbial recombinant fermentation certain parameters are studied in order to optimize product yield. These include temperature, agitation, media components, addition of inducers etc. Certain aspects of fungal fermentations are unique, however, and two important features to be considered are inoculum size and mycelial morphology. In addition to its effects on product yield, mycelial morphology can also affect biomass separation.

Most fungal fermentations differ from their bacterial and yeast counterparts in that the inoculum is in the form of spores. The size of the spore inoculum can have significant effects on yield of product. The appropriate amount of well-separated spores will yield the optimum number of germination sites in the fermenter. Archer *et al.* [32] examined a range of *A. niger* spore concentrations (10^3 to 10^7 spores/ml final concentration) and selected 10^6 spores per ml as giving the maximum expression level of hen egg white lysozyme. Similar results have been reported by other investigators. Mycelial morphology also has an important effect on yield. The most productive part of the mycelium is at the tip and thus the selection of conditions which favour maximum levels of mycelial branching (without compromising growth) is important.

The regulatory issues surrounding the production of a recombinant protein have a significant impact on process development and process cost. All recombinant processes are affected by containment issues and product type constraints must also be considered. Pharmaceutical products fall under the most stringent constraints but there are also guidelines to be considered in the production of products intended for food use or for other applications.

Several aspects of the regulatory concerns about systems for the production of recombinant pharmaceutical proteins from filamentous fungi pertain to the specific characteristics of these organisms. Lipopolysaccharide and other pyrogens which are endogenous to bacterial cells are avoided. Evidence suggests that fungal mycelial extracts and culture filtrates do not contain these pyrogens (Allelix Pharmaceuticals, unpublished). In addition secretion systems provide a significant preliminary 'purification' step in which many of the more important contaminants of pharmaceutical recombinant proteins (nucleic acids, proteins, carbohydrates, etc.) are removed with the filtered mycelia. Several aspects of fungal recombinant systems and products are affected by the US regulatory environment in particular the good manufacturing practices (GMPs) for pharmaceuticals. Fungal systems need to be considered with regard to several basic issues. These include the maintenance and propagation of biological tissue, special facility design considerations, equipment design, effluent disposal and occupational health and safety.

5.5.1 *Maintenance and propagation of biological tissue*

For production purposes, the strain used should be derived from a master cell seed bank, where fungal spores are suspended in a nutrient medium and stored in ampoules at −70°C. Demonstration that the storage technique maintains the master seed in a stable form is important. From this parent stock, the manufacturer's working cell bank (MWCB) is derived by culturing the frozen material onto rich nutrient agar which is formulated for maximum spore yield. The technique should be designed to avoid contamination of the production organism by other microorganisms and release of the genetically modified organism into the environment. Usually operations like this are conducted under 'class 100' conditions in a laminar flow hood fitted with the appropriate filtration membranes.

It is necessary to fully document the number and place of storage of all the master cultures and MWCBs, as with all production organisms. No other organisms may be stored in the area where these cultures are maintained.

The fermentation facilities used in the production of pharmaceutical 'recombinant proteins' are often quite small when compared to conventional full-scale industrial product plants. Recombinant production facilities are usually in the 500–3000 l working volume range. Since one of the critical parameters in culturing fungi is to maximize the number of growth centres, the use of serial seed production and transfer regimes typical for yeast and bacterial expression systems is not necessary. In most cases with the spore-forming filamentous fungi, it may be simpler to develop a validatable protocol to produce a pure spore inoculum which can be added directly into the production fermenter.

For fungal species this is not difficult to achieve, since the organisms are very often prolific spore formers. This has been the experience with various species of *Aspergillus*. The key to this type of procedure is to produce a protocol which is consistent, safe and containable.

5.5.2 *Facility design*

Guidelines for the manufacture of biological products can be found in the biologic establishment standards [33] and the regulations for good manufacturing practice [34]. While these regulations do not refer exclusively to products derived from recombinant processes, they are considered to be broad enough in scope to be applicable to all biologics.

The standards for the movement of raw materials, personnel and product flow are common to those for non-spore-forming expression systems, and stipulate that there is a smooth and (ideally) linear flow through the facility, with no points of potential cross contamination.

Given the resistance of fungal spores to sterilization by heat and the requirement of many conditions connected with process validation to demonstrate sterility and/or containment [35], all movements should be strictly controlled. Physical barriers and air flows are key factors which require attention to ensure that the facility is in compliance with regulations. In addition, guidelines detailed in chapter 21 of the Code of Federal Regulations state: 'for spore bearing organisms, manufacture should be performed in a completely separate building, or in a portion of the building constructed in such a fashion as to be completely walled-off so as to prevent contamination of other areas, and shall have an entrance to that area which is independent of the remainder of the rest of the building'.

5.5.3 *Equipment design and selection*

Appendix K-III-A of the Federal Register states that cultures of viable organisms containing recombinant DNA, in particular fungal spores which are difficult to sterilize, have to be handled in a closed system which is designed to prevent the escape of viable material. This includes all fermentation vessels, transfer lines, holding tanks and primary recovery equipment such as centrifuges, homogenizers and ultrafiltration systems. One problem arising from this directive, which affects use of filamentous fungi, is the contained recovery or separation of the biomass from the extracellular liquor. Unlike bacterial or yeast cultures, filamentous fungi may not be efficiently separated using disc stack centrifugation, and the usual method of separation for non-recombinant filamentous fungi (vacuum drum filtration or filter press) is unacceptable because of the lack of containment. Although this does cause a problem, several possible solutions exist, including a contained drum-filtration system. However, this may be costly and difficult to validate. One solution is a fully contained inverting filter centrifuge which incorporates the attributes of both fast action centrifuges and filter press systems.

5.5.4 *Effluent disposal*

Effluent disposal is monitored closely by the Environmental Protection Agency (EPA) and it is the company's responsibility to ensure compliance with basic guidelines and local laws.

For filamentous production organisms, it will be necessary to develop a validatable procedure to ensure that in-process organisms which may escape through aerosols, spillages or effluent gas streams are neutralized. Spore-forming organisms are given particular attention because of their potential for long-term survival and germination outside the process environment. Under most production conditions spore formation does not occur, however, it is important to develop procedures which avoid the formation of spores during vegetative growth of the organism.

5.5.5 *Occupational health and safety*

The safety record of traditional manufacturing industries in using microbial cultures is excellent and there have been few examples where problems have arisen. The philosophy adopted by the Organization for Economic Cooperation and Development (OECD) member countries is that cases in which traditionally safe microorganisms are modified by the insertion of DNA (to enable the production of recombinant products), safety considerations are not raised beyond those that concern the products themselves. This is helpful for companies contemplating the use of filamentous fungal systems, since the pharmaceutical industry is very familiar with their use for products such as antibiotics and their use in the production of food ingredients many of which are on the GRAS (generally regarded as safe) list. In practice each case is assessed on its own merits and where appropriate the necessary levels of containment may be added to ensure compliance with regulations. Usually primary containment to BL-1 standards is sufficient. BL-1 should be defined if product compliance can be met.

It is important that the production organism be non-pathogenic, lack toxic compounds, and have a history of safe industrial use. Ideally, the organism will have environmental limitations that permit optimum growth in the industrial setting but limited survival in the environment. This should also hold true for the recombinant derivative. These issues are monitored by the FDA and EPA and are also enforced by the Occupational Safety and Health Administration (OSHA), which deals with ensuring safety and health in the work place.

The quality assurance and safety team within a company will be responsible for the monitoring and implementation of procedures for the prevention of employee exposure to the production organism or its products. Filamentous fungi, because of the potential for lung disorders as a result of the germination of inhaled spores may be a source of concern. The use of validated containment procedures where aerosols are controlled will preclude such exposure.

Pyrogenicity of the final product is also a consideration for pharmaceutical products, and potential contamination by any fungal pyrogens will require thorough testing and removal of any which are present. Although pyrogens will be present in all preparations due to the chemicals and water used in certain stages of the manufacturing process, specific fungal pyrogens will almost certainly not add significantly to the overall 'endotoxin load' as compared to pyrogens derived from a recombinant *E. coli* process. This could prove to be a production advantage for the manufacture of certain biologics.

Regulatory concerns are also important for the production of fungal recombinant products which fall into the 'industrial enzyme' category, especially food products. The issues involving containment and occupational health and safety are similar to those described above for pharmaceutical products. Some precedent has been set for food ingredients given the status of several products derived from hosts such as *A. niger*. A process for the production of recombinant calf chymosin from *A. niger* has recently received FDA approval for Genencor International. Undoubtedly the use of an organism which has been previously used for the manufacture of a GRAS product and the demonstration that the product was equivalent to the calf enzyme had a significant effect on the approval process. Similar issues probably played a role in the recent approval of Novo's recombinant lipase derived from *A. oryzae*.

5.6 Future prospects

The validity of producing recombinant proteins from filamentous fungi has recently been established with the approval of processes for the production of calf chymosim (Genencor) and a detergent lipase (Novo). These processes have built on the technological and regulatory knowhow which has been established by the enzyme industry in the production of natural (homologous) fungal proteins (see chapter 3). Companies such as Genencor, Novo and Gist-Brocades are continuing to apply fungal expression technology to the development of processes for other industrially relevant enzymes.

The application of fungal systems to the production of recombinant pharmaceutical proteins has more potential obstacles than their application to other industrial products. No precedent exists for the use of filamentous fungi as pharmaceutical protein production hosts and the purity requirements are more stringent for proteins intended for therapeutic use. However, pharmaceutical proteins are usually high value products which have lower dosage requirements. Thus extremely high expression levels are not required as is the case for other industrial proteins. These are the issues that companies such as Allelix Biopharmaceuticals face in their efforts to develop recombinant fungal products.

Despite the issues described above, several types of technological hurdles need to be overcome before filamentous fungal systems can be universally applied to the production of recombinant proteins. The more important technological questions are quantitative and qualitative. These involve increasing the expression levels of some proteins and adapting fungal systems to the production of certain proteins which have specific post-translational modifications.

High expression yields are an important consideration since the economic production of product is intrinsic to any viable recombinant protein process. In most cases the expression of non-fungal proteins from fungal hosts has been significantly lower than that obtained for the expression of fungal products. Data in the literature concerning the expression of non-fungal genes suggest that the barriers to high yields occur post-transcriptionally. Examination of recombinant strains has demonstrated that the high efficiency promoters in current use produce high levels of messenger RNA within the cell [3, 10, 18]. Some attention must be paid to improving the way in which the transformants with the highest transcription rate might be obtained. Stable transformants in filamentous fungi are possible, for the most part, when foreign DNA is integrated into the genome. Integration usually occurs at heterologous sites [36]. A large variation in product levels between transformants is often found, suggesting that the site of integration is important for maximum transcription. Identification of fungal attachment elements, similar to those described in a mammalian system [37], that mediate increased expression levels in a position-independent manner would be very useful.

The most significant post-transcriptional barrier to high expression yield is secretion into the extracellular medium. This is confirmed by the fact that intracellular expression in most organisms is relatively easy to achieve. Eukaryotic secretion is dependent, in most cases, on an N-terminal hydrophobic secretion signal sequence. In addition, important features of the non-signal portion of the protein also have a significant effect on secretion efficiency [38]. This is reflected in the fact that fungal enzymes, which are well adapted to high efficiency secretion, are usually expressed at high level in other fungal hosts (*gla* in *A. nidulans*/lipase). As emphasized by Turner [38], little fundamental study of the secretion process in filamentous fungi has been achieved and the time and expense of such a study would not make it a viable proposition for most companies. One approach to improving the efficiency of secretion of non-fungal proteins is to use mutagenesis and screening procedures on the recombinant strain as demonstrated by the Genencor group in their development of a chymosin process [9]. Unfortunately this approach will not work for many pharmaceutical products given that the assays are complex and not readily adaptable to a rapid screening process.

Another approach to the problem of low secretion levels is to fuse, at the level of the gene, the protein of choice to a well secreted fungal protein. Fusions to *A. niger* glucoamylase in this manner have been shown to be effective in increasing the yield of chymosin [7], human epidermal growth factor (Gwynne, unpublished data) and human interleukin-6 (IL-6) [39]. Fusion proteins of this type require a processing site so that the mature protein can be released following purification. Mature chymosin is released autocatalytically from the glucoamylase fusion and does not need a separate site. Contreras and co-workers [39] used an interesting approach in their glucoamylase-IL-6 fusion work. The expression construct was designed to include a *Saccharomyces*-like KEX2 specific cleavage site between the glucoamylase sequence and the IL-6 sequence. Previous data from the Allelix group had demonstrated the presence of a KEX2 like protease activity in *A. nidulans* [40]. The results showed that the glucoamylase-IL-6 fusion protein was secreted as efficiently as glucoamylase alone and that IL-6 appears in amounts which are equimolar to the glucoamylase concentration in the extracellular medium. Complete cleavage of the precursor protein was observed and IL-6 was produced at levels of 5 mg/l (compared with microgramme amounts for unfused IL-6). Future approaches to the secretion problem must involve some basic analysis of the secretion process in filamentous fungi.

In addition to secretion barriers the secreted heterologous protein may encounter proteolytic activity either during secretion or in the extracellular environment. Fungal secreted proteins are generally resistant to the actions of these proteases. This is because most filamentous fungi, as heterotrophs, release mixtures of hydrolytic enzymes into the environment and these proteases, nucleases, lipases and carbohydratases can readily coexist. One approach is to culture the recombinant fungus under conditions of nitrogen catabolite repression which reduce the activity of most secreted proteases. Another approach is to remove specific proteolytic activity using a gene deletion approach as reported by Berka and co-workers [8].

The genetic approaches used to maximize yield must be accompanied by process development work which will optimize biomass production without sacrificing yield per cell. The extremely high biomass yields (120 grammes dry weight per litre) described by Christensen *et al.* for their recombinant aspartyl proteinase process in *Mucor miehei* undoubtedly contributed to the success of the recombinant lipase process in *Mucor* [13, 14].

In addition to the concerns about product yield the problem of 'product authenticity' must be addressed. Many non-fungal proteins, especially those that are secreted, are post-translationally modified in some way. The most important type of modification is the addition of carbohydrates (glycosylation). Carbohydrate addition may be required for full activity in certain proteins. The presence of the appropriate types of carbohydrate may also be required for other purposes. This is especially important for pharmaceutical proteins where aberrantly glycosylated proteins can be immunogenic or may be cleared very quickly from the blood [41]. Fungi typically attach high mannose carbohydrates in N-linked structures. Filamentous fungi do not tend to hyperglycosylate foreign proteins as is the case with certain strains of *Saccharomyces*. An example of this is the data reported by Upshall *et al.* which showed that *A. nidulans*-produced tPA seems to be glycosylated to the same degree as the native mammalian protein [24] unlike the recombinant product from *Saccharomyces* [42, 43]. A high degree of glycosylation is observed in filamentous fungi in some cases. *A. nidulans* and *A. niger* both add excess carbohydrate to recombinant *Mucor* aspartyl proteinase [12–14]. The endogenous *T. reesei* CBH I contains N-linked glycans which are quantitatively similar to mammalian glycans of the (Man)5(GlcNAc) 2 and (Man)9 (GlcNAc)2 type [44]. A similar situation exists for *A. oryzae* glucoamylase [45]. There is no evidence that fungi produce the complex, sialic acid-containing glycoproteins which are required for both stability and non-immunogenicity in serum. Given this fact, one of the goals of expression development work in filamentous fungi should be the analysis of secreted, recombinant proteins. It should be possible to modify the host's glycosylation mechanism to add the residues and structure for the seroprotection of recombinant proteins. This may be accomplished by expressing the appropriate cloned glycosyl transferase enzymes in the selected production host.

Filamentous fungi are beginning to be recognized as a useful group of hosts for the expression of proteins from fungal and other sources. Two products which are derived from recombinant fungal processes have recently received government regulatory approval. These organisms have been accepted alongside *Saccharomyces* and *E. coli* as viable recombinant production systems for food and detergent use at least. Some interesting fungally derived recombinant human proteins are in the pipeline and are intended for human pharmaceutical use. When the lengthy drug approval process has been completed, products derived from filamentous fungi will also begin to have some impact on the health care industry. In terms of fungal products the pharmaceutical industry has come full circle. Taka-Diastase, a product of *Aspergillus oryzae* was patented by Dr Jokichi Takamine in the late nineteenth century. This product was approved for the treatment of dyspepsia at that time and was marketed by Parke-Davis.

References

1. Tekamp-Olson, P. and Valenzuela, P. (1990) *Curr. Opin. Biotechnol.* **1**: 28–35.
2. Gwynne, D.I., Buxton, F.P., Williams, S.A., Garven, S. and Davies, R.W. (1987) *Biotechnology* **5**: 713–719.
3. Cullen, D., Gray, G.L., Wilson, L.J., Hayenga, K.J., Lamsa, M.H., Rey, M.W., Norton, S. and Berka, R.M. (1987) *Biotechnology* **5**: 369–376.
4. Ward, M., Wilson, L.J., Barnett, C.C. and Berka, C.M. (1987) *Abstracts of the 14th Fungal Biology Conference*, 253.
5. Ward, M. (1991) In: *Molecular Industrial Mycology: Systems and Applications* (Leong, S. and Berka, R. eds). Marcel Dekker, New York.
6. Finkelstein, D., Rambosek, J.A., Leach, J., Wilson, R.E., Larson, A.E., Soliday, C.L. and McAda, P.C. (1988) *Abstr. Ann. Meeting Soc. Industrial Microbiology*, p. 555.
7. Ward, M., Wilson, L.J., Kodama, K.H., Rey, M. and Berka, R. (1990) *Biotechnology* **8**: 435–440.
8. Berka, R., Ward, M., Wilson, L.J., Hayenga, K.J., Kodoma, K.H., Carlomagno, L.P. and Thompson, S.A. (1990) *Gene* **86**: 153–162.
9. Lamsa, M. and Bloebaum, P. (1990) *J. Ind. Microbiol.* **5**: 229–238.
10. Haarki, A., Uusitalo, J., Bailey, M., Penttilä, M. and Knowles, J.K.C. (1989) *Biotechnology* **7**: 596–600.
11. Pourquie, J., Warzwoda, M., Chevron, F., Théry, M., Lonchamp, D. and Vandecasteele, J.P. (1987) In: *Biochemistry and Genetics of Cellulose Degradation. FEMS Symposium No. 43.* (Aubert, J.P., Beguin, P. and Millet, J. eds). Academic Press, London.
12. Gray, G.L., Hayenga, K., Cullen, D., Wilson, L.J. and Norton, S. (1986) *Gene* **48**: 41–53.
13. Christensen, T., Boel, E., Wöldike, H.E., Hjortshorj, K., Andreasen, F., Mortensen, S.B., Thim, L. and Hansen, M.J. (1987) *Abstr. 19th Lunteren Lectures on Molecular Genetics* F23.
14. Christensen, T., Woeldike, H., Boel, E., Mortensen, S.B., Hjortshoej, K., Thim, L. and Hansen, M.T. (1988) *Biotechnology* **6**: 1419–1422.
15. Dickinson, L., Harboe, M., Van Heeswijck, R., Stroman, P. and Jepsen, L.P. (1987) *Carlsberg Res. Commun.* **52**: 243–252.

16. Barnett, C., Berka, R., Shoemaker, S., Sumner, L.M., Liard, M. and Wilson, L. (1987) *FEMS Symposium – Biochemistry and Genetics of Cellulose Degradation*, Paris.
17. Bernett, C., Berka, R., Shoemaker, S., Sumner, L., Ward, M. and Wilson, L.J. (1987) *Abstr. 14th Fungal Biology Conference.*
18. Gwynne, D.I., Buxton, F.P., Williams, S.A., Garven, S. and Davies, R.W. (1987) *Bio/Technology* **5**: 713–719.
19. Chen, C.M., Ward, M., Wilson, L., Sumner, L. and Shoemaker, S. (1987) *Abstr. Proc. Ann. Chem. Soc.* p. 194.
20. Finkelstein, D., Rambosek, J., Leach, J., Wilson, R., Larson, A., Soliday, C. and McAda, P. (1986) *Abstr. Annual Meeting for Industrial Microbiology.* p. 555.
21. Barnett, C., Berka, R. and Fowler, T. (1991) *Bio/Technology* **9**: 562–567.
22. Gwynne, D.I., Buxton, F.P., Gleason, M. and Davies, R.W. (1987) In: *Protein Purification: Micro to Macro* (Burgess, R. ed.). Alan R. Liss, New York.
23. Kozak, M. (1984) *Nuc. Acids Res.* **12**: 857–876.
24. Upshall, A., Kuonar, A., Bailey, M., Parker, M., Favreau, M., Lewison, K., Joseph, M., Maraganore, S. and McKnight, G. (1987) *Bio/Technology* **5**: 1301–1304.
25. Pohl, G., Kallstrom, G., Bergsdoff, N., Walker, P. and Jornvall, H. (1984) *Biochemistry* **23**: 3701–3707.
26. Tate, K., Higgins, D., Holmes, W., Winkler, M., Heyneker, H. and Vehar, G. (1987) *Biochemistry* **26**: 338–343.
27. Boel, E., Hjort, I., Svensson, B., Norris, F., Norris, K. and Fiil, N. (1984) *EMBO J.* **5**: 1097–1102.
28. Nunberg, J.H., Meade, J.H., Cole, G., Lawyer, F., McCabe, P., Schweikhardt, V., Tal, R., Wittman, V., Flatgacra, J. and Innes, M. (1984) *Mol. Cell. Biol.* **11**: 2306–2315.
29. Leung, W.C. (1987) *Abstr. 19th Lunteren Lectures on Molecular Genetics* F156.
30. Leung, W.C., Jing, G.Z. and Leung, M.F.K. (1987) *Abstr. 19th Lunteren Lectures on Molecular Genetics* F246.
31. Zhang, X.N., Jones, A., Kole, M., Gerson, D. and Leung, W.C. (1987) *Abstr. 19th Lunteren Lectures on Molecular Genetics* F249.
32. Archer, D.B., Jeenes, D.J., Mackenzie, D.A., Brightwell, G., Lambert, N., Lorne, G., Radford, S. and Dobson, C.M. (1990) *Bio/Technology* **8**: 741–745.
33. *Code of Federal Regulations*, Title 21, Parts 600–680. April 1, 1987. Revised.
34. *Code of Federal Regulations*, Title 21, Parts 200–011. April 1, 1987. Revised.
35. *NIH Guidelines for Research Involving Recombinant DNA Molecules.* August 24, 1987.
36. Fincham, J.R.S. (1989) *Microbiol. Rev.* **53**: 148–170.
37. Stief, A., Winter, D.M., Stratling, W.H. and Sippel, A.E. (1989) *Nature* **341**: 343.
38. Turner, G. (1990) In: *Protein Production by Biotechnology.* (Harris, T. ed.). Elsevier, New York.
39. Contreras, R., Carrez, D., Kinghorn, J.R., van den Hondel, C.A.M.J.J. and Fiers, W. (1991) *Bio/Technology* **9**: 378–381.
40. Devchand, M., Gwynne, D., Buxton, F. and Davies, R.W. (1989) *Curr. Genet.* **14**: 561–566.
41. Kornfeld, R. and Kornfeld, S. (1985) *Ann. Rev. Biochem.* **54**: 631–664.
42. MacKay, V.L. (1987) In: *Biochemistry and Molecular Biology of Industrial Yeasts* (Stewart, G.G. and Klein, R. eds.). CRC Press, Boca Raton, Florida.
43. Lemontt, J.F., Wei, C.M. and Dackowski, W.R. (1985) *DNA* **4**: 419–428.
44. Salovuori, I., Makarow, M., Rauvala, H., Knowles, J. and Kääriäiren, L. (1987) *Bio/Technology* **5**: 152–156.
45. Yamaguchi, H., Ikenaka, T. and Matsushima, Y. (1971) *J. Biochem.* **70**: 537–594.

6 Stability of recombinant strains under fermentation conditions

N.S. DUNN-COLEMAN, E.A. BODIE, G.L. CARTER
and G.L. ARMSTRONG

6.1 Introduction

Filamentous fungi have a long history of use in producing large amounts of proteins in industrial fermentations. The current world market for industrial enzymes is believed to be in the region of $650 million (J. Burr, pers. commun.). *Aspergilli*, notably *A. niger* and *A. oryzae*, have been used to produce enzymes, such as glucoamylase, which are used in generally recognized as safe (GRAS) food processing. In addition, enzymes are increasingly finding a significant role in biomass conversion, waste treatment and manufacturing of fine chemicals [1].

With the development of plasmid based transformation systems for filamentous fungi such as *A. nidulans* [2], *A. niger* [3], *A. oryzae* [4], *Trichoderma reesei* [5] and *Penicillium chrysogenum* [6] (see chapter 1), there has been considerable interest in the use of filamentous fungi as hosts for the heterologous expression of mammalian proteins (see also chapter 5) and industrially important enzymes (chapter 3). This interest is not unexpected because in addition to their ability to secrete large amounts of protein (> 25 g/l) and their long history of safe use as hosts for the production of enzymes and secondary metabolites such as antibiotics, there exist established processes for inexpensive large-scale fermentation of filamentous fungi.

The strategy employed has been to clone a highly expressed gene, for example, α-amylase [7], cellobiohydrolase I [8], or glucoamylase [9], and use its promoter for the expression of the heterologous gene. (For a detailed review of heterologous gene expression in filamentous fungi, see chapter 5.) Regrettably, in most cases the yield of protein secreted has been very low [8, 10, 11]. However, two exceptions have been reported to date: the yield of 3 g/l of the naturally secreted fungal *Rhizomucor miehei* aspartic protease by *A. oryzae* using an α-amylase promoter [7] and the expression of commercially viable levels (> 1 g/l) of bovine chymosin by *A. niger* var. *awamori* [12].

However, it should be noted that the significant levels of production required for the commercialization of a heterologously expressed industrial enzyme such as chymosin, are not as important for the commercialization of a mammalian protein such as tissue plasminogen activator (tPA). In the case of high value mammalian proteins destined for pharmaceutical use, issues of the authenticity of the recombinant protein in terms of post-translational modifications, such as altered patterns of glycosylation, are important (for a review see [13]). The ability to purify the protein to homogeneity for pharmaceutical use is significantly more important than the cost savings of large-scale inexpensive fermentations. The objectives of this chapter are to provide the reader with an introduction to industrial fermentation technology. One approach (that taken by Genencor International) to the commercialization of the production of heterologous proteins in filamentous fungi is reviewed in this chapter. In addition, issues of recombinant strain stability and procedures which should be undertaken to preserve recombinant strains are discussed.

6.2 Fermentation

6.2.1 *Introduction*

Industrial microbiologists use the term 'fermentation' to describe any process in which a product is obtained by the biochemical activity of a mass culture of microorganisms. Microorganisms have been used in fermentation processes since ancient times. Early fermentation products were mainly consumable goods such as alcohol and vinegar. After 1900, the fermentation industry began to expand and new products included bakers yeast, glycerol, and organic acids such as citric and lactic acid. The first aseptic fermentation was developed to produce acetone and butanol. In the 1940s, as a result of the wartime need to produce penicillin, the fermentation industry underwent a tremendous growth and development phase resulting in many new processes and products. It was also at this time that microbial genetics, in the form of mutation and selection, were first used for strain improvement, resulting in dramatic improvements in yields [14].

6.2.2 *Fermentation development*

An industrial fermentation is a highly controlled process in which the physical and chemical environments of an organism are manipulated to maximize product yield and minimize manufacturing costs. Environmental and physiological conditions must be optimized for a microorganism so that production will be as efficient as possible. The type of product, as well as the physiology of the organism, will determine the optimal fermentation mode, inoculum conditions and environmental conditions including the medium, pH, temperature, agitation and aeration [14].

6.2.3 *Primary and secondary metabolites*

According to Pirt [15] fermentation products are either growth-linked or non-growth-linked metabolites. Metabolites made during the log phase of growth are essential to the growth of the cells. These products, known as primary metabolites, include amino acids, proteins, vitamins, polysaccharides, and ethanol. In order to obtain primary metabolites, the regulatory mechanisms present in a cell must be overcome by manipulation of the environment or by genetic alteration, to achieve over-production of a product. When designing a fermentation for a primary metabolite, it is advantageous to extend the log phase of growth as long as possible.

Secondary metabolites include products that have antimicrobial activity, such as antibiotics and growth promoters, as well as other therapeutic agents. They are not involved in cell metabolism, are not needed for growth and are only made when a nutrient is exhausted and balanced growth becomes impossible resulting in biochemical differentiation. Consequently they are produced in non-growing or slowly growing cultures. For example, many antibiotics are made only when phosphate becomes limiting [16]. Secondary metabolites are frequently dependent on primary metabolites and therefore synthesis is frequently controlled by the same regulatory mechanisms. With a secondary metabolite, conditions are needed that will reduce the lag and exponential growth phases, and either will extend the stationary growth phase or will promote slow growth throughout the fermentation.

6.2.4 *Fermentation mode*

There are three fermentation modes that are commonly used for industrial fermentations: batch, fed-batch or continuous. The choice of mode is dictated by the type of product being made.

(i) *Batch mode.* A batch mode process is a closed system in which only the nutrients present initially are available and therefore the medium must be designed specifically for individual products. For instance, for biomass production it is important to develop conditions that will result in maximum cell population and the use of a rapidly metabolizable carbon source would be advised. Fungi are chemoorganotrophs and the carbon and energy sources most frequently supplied to them are carbohydrates. For a primary metabolite subject to catabolite repression, such as a microbial enzyme, a slowly metabolizable carbohydrate would be needed. A slowly metabolizable carbohydrate would also be advantageous in a batch mode process for secondary metabolites because a slow growth rate would enhance production. Batch fermentation times can run from about 36 hours, such as with gluconic acid production, or for several days, as with fumaric acid production. For these reasons the slowly metabolizable sugar lactose was used to produce many antibiotics [17]. There are several disadvantages of batch fermentation including exhaustion of nutrients resulting in early termination of metabolite production, product feedback inhibition, and the build-up of toxic products, resulting in shorter production times.

(ii) *Fed-batch mode.* A fed-batch mode is an open system in which nutrients are added at a steady rate during the fermentation without the removal of culture broth. Thus the volume of the fermentation increases over time. In this mode it is easier to control metabolite production and solve the problems of metabolic regulation because the concentration of individual nutrients in the fermentation medium can be controlled throughout the fermentation. The growth rate can also be controlled by adjusting the rate of carbohydrate addition. As mentioned above, antibiotics were first made in a batch mode process using lactose as the carbon and energy source but today they are produced more economically by a fed-batch mode that feeds glucose at a rate at which the concentration limits growth and reduces catabolite repression [17].

Carbon sources that cause catabolite repression can be fed into the fermenter at a slow rate, resulting in low concentrations in the fermenter broth and escaping repression. The growth rate can also be controlled in this manner. A fast growth rate is desirable for a product whose formation increases with specific growth rate and a slow growth rate is required for non-growth associated products. Other advantages of the fed-batch mode include dilution of toxic materials that may inhibit product formation and extend fermentation times, when compared to a batch process, resulting in higher productivities.

(iii) *Continuous culture mode.* A continuous culture mode of operation is an open system in which nutrients are added and removed at a steady rate that maintains product production at a constant maximum. A continuous culture mode provides more control over the production phase than any other mode. Log phase growth for primary metabolites can be extended by the addition of fresh medium components. The medium can also be designed to be limiting in nutrients, thus allowing secondary metabolites to be made for an extended period of time. Since cells are continually being drawn off, a portion of the nutrients and metabolic energy of the organism is used for growth as opposed to product formation. To address this, a popular variation of the continuous culture mode includes cell recycling where the cells are removed from the draw-off material and placed back in the fermenter resulting in more efficient fermentation. Continuous culture fermentation times can be as long as several months but there is a danger of strain degeneration over time and the extended fermentation time increases the contamination risk. Continuous fermentation is useful for studying the basic physiology and biochemistry of microorganisms.

6.2.5 *Environmental conditions*

The fermentation medium must be designed with regard to scientific as well as economic aspects. It must be designed not only for growth, but also for the production of a specific product. Therefore the medium, besides containing all the nutrients necessary for growth, will also be designed specifically with the primary and secondary biochemical characteristics of the organism in mind [14].

(i) *Medium.* In laboratory research with microorganisms it is possible to study production of a product using a defined chemical medium but in industrial fermentations complex substrates, in many cases by-products of other industries, are used for economical reasons. The medium must contain sources of carbon and energy, nitrogen, sulphur, phosphorus, trace elements, and specific growth factors that the organism cannot manufacture. Often the required growth and production factors are not known and are supplied by complex nutrient sources such as yeast extract, molasses, corn steep liquor, etc.

Media development is an empirical process. Different sources of carbohydrate, nitrogenous compounds, protein digests and other growth factors are added and the effects on growth and production are observed. Mutants can have requirements that are different from the parent strain and every medium has to be specifically adapted to every new strain. There are statistical computer-based programs that are useful tools for media optimization.

A major breakthrough for penicillin manufacturing was achieved when corn steep liquor was added to the medium. Production levels increased from 20 units/ml to over 100 units/ml. Production was also shifted from penicillin F to the more desirable penicillin G. Later it was found that the active ingredients in corn steep liquor were phenylalanine and phenethylamine, which are precursors for formation of benzylpenicillin. Another example of media optimization concerns the discovery that *A. niger* was inhibited from producing citric acid when manganese or iron were present in sufficient concentration. It used to be difficult to control the concentration of these trace elements as other medium components contained them or were contaminated by them. However, Schweiger [18] discovered that copper ion could be used as an antagonist to iron and enabled the present commercial process to be developed.

Other considerations involved in media development include the availability of raw materials at the manufacturing site, the effect the media will have on downstream processing and recovery, as well as the economical impact a nutrient will have on the final product price.

(ii) *pH.* The pH of a fermentation broth is measured by combination electrode probes that can withstand sterilization temperatures and environmental and mechanical stresses associated with the fermentation. During fermentation microorganisms synthesize metabolic products which can drastically change pH. In optimizing pH, the effects on the organism as well as on the product must be considered. For some fermentations, it is advantageous to control pH using a base that can also supply a nitrogen source to the organism. For itatartaric acid production with *Aspergillus terreus*, ammonium was found to increase productivity greatly [19].

(iii) *Pressure.* In order to minimize contamination risks, fermentations are done under positive pressure, usually between 0.2–0.07 bar. In an aerobic fermentation process, another advantage of running a fermentation under pressure is that the solubility of CO_2 and oxygen is increased.

(iv) *Temperature.* As many metabolites must work in the environment of the organism, the optimal pH and temperature for production of a metabolite are usually similar to the optimal conditions for growth. Temperature can be used to control the metabolic rate. Lower temperatures can slow down the growth rate, which can be advantageous if oxygen transfer becomes limiting due to excess growth. At higher temperatures, growth may proceed faster and larger biomass levels may be obtained. In many fermentations, it is advantageous to operate at as high a temperature as possible to achieve high rates of production. However, above the optimum temperature, production rates fall off dramatically. Optimum pH and temperature ranges can change throughout the fermentation. In some fermentations, higher temperatures are used initially to obtain a large biomass after which the temperature is lowered to the optimum level for production.

(v) *Oxygen.* Oxygen is an important nutrient for most organisms and is useful in controlling growth rate and production. Polarographic or galvanic oxygen electrodes are used to measure the dissolved oxygen in the fermentation broth. Changes in dissolved oxygen concentration depend on the oxygen transfer capacity of the fermenter and the metabolic rate of the respiring cells. The oxygen concentration, transfer rate, and uptake rate by the organism, must be optimized. Oxygen is of particular concern in filamentous fungal fermentations not only because of the large biomass that can be obtained, but also because of the way fungal morphology affects oxygen transfer rates.

(a) *Controlling growth.* The carbon source as well as the concentration can affect oxygen availability. An easily metabolizable carbon source present in high concentrations will result in an extremely vigorously growing culture and will restrict oxygen transfer rates. By using a slowly metabolizable carbon source or using a fed-batch mode and keeping the concentration low, it is possible to control the growth rate as well as the oxygen transfer rate. In the fed-batch mode, it is common to control carbohydrate feed rates based on the dissolved oxygen that is present in the medium and/or the oxygen uptake rate of the organism. Oxygen availability is often the limiting factor in obtaining higher productivities in commercial fermentations. Other fermentations, including gallic acid and lysine, can be sensitive to too much oxygen [20]. Therefore, as with all nutrients, the concentration of oxygen must be optimized for each organism.

(b) *Morphology.* Fungi are capable of growing in a pelleted form consisting of individual compact discrete masses of hyphae, or in a filamentous form. In the pelleted form, oxygen becomes unavailable toward the center of the pellet resulting in a portion of the biomass that cannot grow or make product. However, pellet growth biomass is usually restricted to 10–20 g/l while filamentous growth can achieve biomasses in excess of 50–60 g/l. The filamentous form tends to become very viscous because of the concentration of biomass and also because it grows as a network of interlocking hyphae that are structurally rigid. This results in broth viscosities that behave in a non-newtonian manner and can result in oxygen transfer problems throughout the fermentation. In many fermentations, the higher biomass resulting from the filamentous form is preferable, even though the oxygen transfer rate is much lower, because there are still more product producing cells present. Citric acid production using *Aspergillus niger* is an example of a fermentation where the organism grows in a pelleted form. This form is advantageous for citric acid production because of a high degree of sensitivity to oxygen limitation.

The determining factor for morphology in many cases is the inoculum size. A very low inoculum of either spores or vegetative growth can result in pelleting. Other factors such as medium composition, stirrer speed, shear, pH and temperature can also determine morphology [21].

6.2.6 *Inoculum*

The inoculum for a fermentation is very important as it can greatly affect the quantity of product that will be produced. The inoculum development of a culture involves the revitalization of a dormant culture from storage to a final productive stage. The productive stage needs to be in sufficient quantity to meet the needs of the particular volume of the fermentation and the culture should be physiologically adapted to produce maximum amounts of the desired product. Preparation of inocula for fungal fermentations is more involved than for bacteria or yeast because once a mycelial cell is formed it does not produce new cells through division but remains as a discrete entity that grows older. Because of this, fungal cultures are not very adaptable to changing biochemical conditions or process errors [14].

For inoculum development, spores are first transferred to shake flasks, allowing germination and vegetative growth. In order to obtain a sufficient quantity of inoculum, it is usually necessary to transfer the inoculum through several stages using increasingly large vessels (Figure 6.1). The culture is then

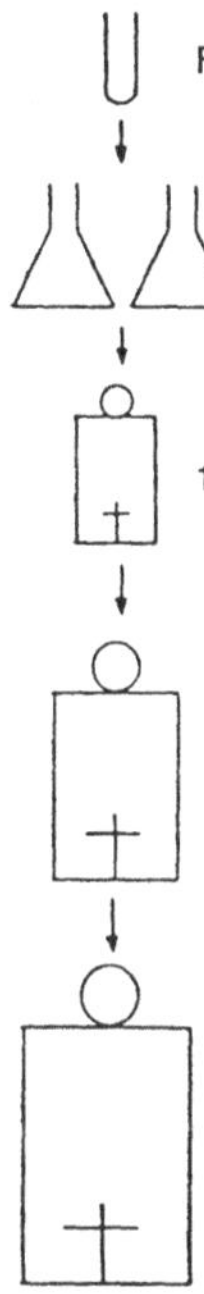

Figure 6.1 Inoculum build-up plan for industrial fermentation. Taken from Bodie [14]. Reprinted by permission of Wiley-Liss, a division of John Wiley and Sons, Inc. Copyright © 1991 Wiley-Liss.

kept in a vegetative growth stage for the remaining inoculum stages. These vegetative growth stages are advantageous as the growth rate will usually continue at a maximum rate with little interruption upon transfer from one vessel size to another. The inoculum must be efficient, predictable and be free of variants as well as contaminants in order to maintain quality through the inoculum stages.

The media used for the inoculum can be designed to obtain maximum growth rates. In many cases, a medium is used that is similar to the production medium allowing the culture to adapt to this medium through the inoculum stages resulting in a reduced lag phase in the fermentation. The inoculum size varies greatly depending on the fermentation, but it is usually important to use a large enough inoculum to reduce the lag or adaptation phase to allow production to begin as quickly as possible.

6.2.7 *Instrumentation and equipment*

A fermenter is a vessel containing some form of agitation and aeration, heating and cooling mechanisms for temperature control, ports containing probes to measure pH, temperature, dissolved oxygen, redox potential and more. It must be sterilizable, and under aseptic conditions. Aeration is supplied by pumping in air to the fermentor while agitation suspends oxygen, nutrients and cells evenly throughout the medium making them available to cells. Agitation also aids heat transfer rates. Sizes of commerical fermentors can range from 10 000 l to 200 000 l. The type and size used depends on the end-product [14].

The stirred tank fermentor was developed during the 1940 antibiotic boom for fermentations using submerged cultures. This is the basic fermentor design in use today although there have been various important modifications. The fermentor is a vertical cylindrical vessel containing a fully baffled vertical shaft driven by turbine impellers for agitation. Aeration is through air spargers located at the bottom of the fermentor. Culture vessels are usually twice as deep as they are wide. For accurate temperature control, fermentors have cooling coils and/or jackets. A variation of this impeller shaft, which does not penetrate the vessel, eliminates the need for seals.

Other fermentor designs such as the tower fermentor and the air-lift fermentor do not have mechanical agitators. Air or oxygen is pumped in to provide both aeration and agitation. The air-lift fermentor has oxygen transfer abilities similar to the stir tank fermentor. The tower fermentor is only useful for relatively non-viscous broths and is advantageous in reducing installation and operating costs when oxygen demand is low [22]. Agitation by air compression is only about 40% efficient in terms of electrical energy while agitation by turbine is about 90% efficient resulting in cost savings for many products [23]. Mixing cost may contribute a large part of the operating cost of fermentation plants [24]. There are many more types of fermentors, mainly variations of the ones described above with different mechanisms for aeration and agitation.

The fermentation conditions are frequently controlled by instrumentation overseen by a computer which is also useful for controlling fermentation parameters and collecting and summarizing data. Physical, chemical, and biological parameters are measured during a fermentation either directly or off-line in the laboratory. These measurements are used for data analysis and to control the fermentation process. Physical parameters include temperature, pressure, power consumption, viscosity, flow rates, turbidity, and the weight of the fermentor. Chemical parameters include pH, dissolved oxygen (DO), redox potential, substrate and product concentration, ionic strength, and off-gas data.

Off-gas data are collected using flow meters and sent for data analysis. There are several sensors available to measure off-gas, but a mass spectrophotometer is the most versatile. Gases such as oxygen, CO_2, and nitrogen are commonly measured and used to compute oxygen uptake rates, respiratory quotient CO_2 formation rate, specific substrate uptake rate, volume-specific energy uptake and more. These measurements provide valuable information about the metabolic status of the fermentation. In some fermentation processes valuable information about the substrate or product can be obtained through off-gas data. For example, methane has been used as a carbon and energy source in single-cell protein production [25].

6.3 Isolation and stability of chymosin-producing strains

6.3.1 *Isolation of chymosin-producing strains*

Several years ago Genencor International initiated a program to express commercially viable levels (>1 g/l) of bovine chymosin by *A. niger* var. *awamori*. Bovine chymosin is the preferred protease for cheese manufacture because it cleaves a Phe–Met bond in *k*-casein in a relatively specific manner compared to other aspartyl proteases, notably those of fungal origin, thus reducing off-flavors in the cheese as a result of non-specific cleavages. In addition, chymosin is readily heat inactivated, thereby preventing the degradation of the whey protein [26].

Chymosin is initially synthesized as a prepro enzyme, from which the 16 amino acid presequence is cleaved when the protein is secreted. Subsequently, the prosequence of 42 amino acids is autocatalytically cleaved at low pH to yield the mature enzyme [27]. Because of periodic shortages of authentic chymosin and the previously discussed disadvantages of fungal aspartyl proteases, expression of calf chymosin in a microbial host would offer several benefits.

The glucoamylase gene (*glaA*) promoter from *A. niger* var. *awamori* has been cloned for use in the expression of bovine chymosin in *Aspergillus* [9]. The *glaA* promoter and secretion signal have been used to direct chymosin production in both *A. nidulans* and in *A. niger* var. *awamori*. Several expression vectors were made incorporating prochymosin cDNA with either the glucoamylase or the chymosin secretion signal. In addition, in one construct, the first 11 codons of mature glucoamylase were fused to the prochymosin cDNA. However, the maximum yields of secreted chymosin from *A. niger* var. *awamori* were only about 15 mg/l in 50 ml shake cultures. There apparently was no correlation between the yields of secreted chymosin and the abundant chymosin mRNA being produced.

Ward *et al.* [28] greatly increased the production of secreted chymosin by fusing the chymosin cDNA to the last codon of the *A. niger* var. *awamori* *glaA* gene. Transformants obtained by using this particular construction (pGAMpR see Figure 6.2) produced 150 mg/l chymosin in 50 ml shake culture. It was also demonstrated that the low yields of chymosin produced in transformants derived from previous constructions were apparently due to inefficient secretion and probable degradation of chymosin.

As a result of mutagenesis and screening a mutant was isolated which produced substantially higher levels of secreted chymosin [29]. Subsequently, it was determined that this particular mutant produced very low amounts of

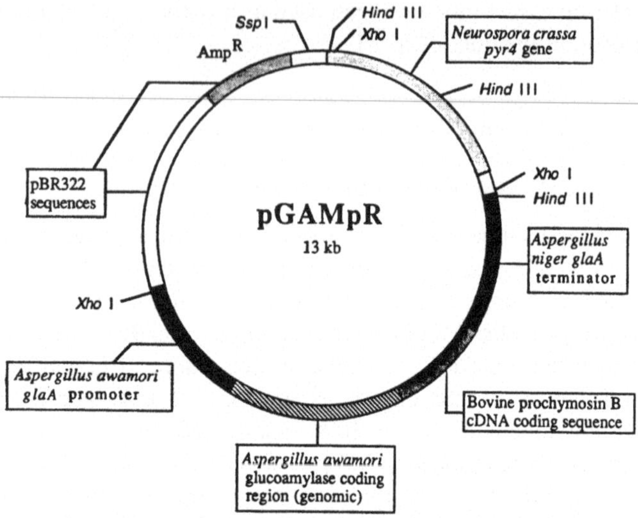

Figure 6.2 Diagram of pGAMpR. The plasmid pGAMpR contains the glucoamylase-chymosin expression cassette and has been described previously [28].

the native aspartyl protease. Furthermore, degradation of secreted chymosin by the native aspartyl protease was shown to reduce yields of chymosin in culture broths. In light of these results, it was decided to clone and delete the native aspartyl protease gene [30] and use the protease-deficient strain as a recipient for transformation using the pGAMpR plasmid (Figure 6.2). One transformant (GC4-1), was found to produce 250 mg/l extracellular chymosin. The strain GC4-1 has been used as the starting strain in all subsequent yield improvement efforts. To make the commercialization of chymosin feasible, the yields of chymosin had to be increased from 250 mg/l to greater than 1 g/l so a strain improvement program using classical mutation and selection techniques was instigated. These techniques included random mutation and screening using robotics, as well as more selective techniques such as the isolation of 2-deoxyglucose resistant (*dgr*) strains. Using a combination of these techniques, yields of greater than 1 g/l chymosin were obtained. The lineage of strains is shown in Figure 6.3. It was determined that strains which were isolated for their improved chymosin production were found to produce elevated levels of many extracellular proteins [12].

6.3.2 *Strain evaluation*

One of the key components for a successful yield improvement program to identify superior production strains is to develop a strain evaluation protocol. For example, a random mutagenesis and screening program which isolated superior chymosin producing strains found that growing the mutated spores

Strain	Method of Isolation	Chymosin (mg/l)
GC4-1	Original pGAMpR transformant	250-285
↓ GC1HD2	Isolated from GC4-1 after NTG mutagenesis and screening	377
↓ GC1HG7	NTG mutagenesis and recurrent selection	520
↓ GC1HF1-3	NTG mutagenesis and recurrent selection	650
↓ GC1HF1342	Isolated of GC1HF1 with improved growth on minimal medium	650

Deoxyglucose resistant mutants:

GC1HF1-3 →	GC1HF1-3;dgr2	1000
GC1HG7 →	GC1HG7; dgr1	780

Figure 6.3 Lineage of chymosin-producing strains.

in liquid culture using a miniaturized 96-well microtiter plate allowed the process of mutant evaluation to be automated. Robots were used to inoculate and harvest the culture supernatants for quantitative assays and comparison to the parent strains. The advantages of the semi-automated liquid culture screen were two-fold. First, the culture broths were screened with a chymosin-specific substrate and the chymosin produced by each mutant directly compared to the parent. Second, by culturing the mutants in 96-well plates, the manipulations and assays can be automated. Typically, 50 000–60 000 mutated viable spores were analysed per screen. After preliminary assays to identify the mycelial mats which produced the highest levels of chymosin, spores were purified and single spore isolates grown in shake culture. Strains which produced superior levels in shake culture would be grown in 101 fermenters. Because of the limitations of the numbers of strains which could be evaluated in a fermenter, it became essential to develop a shake culture medium which gave a reliable indication of how much chymosin the strain would produce in a fermenter. It was found that strains which displayed superior production in shake culture would also produce higher yields in a fermenter. However, each new production strain required some media and fermentor optimization to achieve maximal rates of production.

6.3.3 *Strain stability*

Although the highly expressed glucoamylase promoter was used for chymosin expression and culture conditions optimized for maximal expression of chymosin, the level of chymosin production was too low for commercialization. This finding necessitated the use of mutagenesis, screening and mutant selection to obtain superior chymosin-producing strains. As a consequence of the use of mutagenesis beneficial mutations which enhance chymosin production may be obtained but there also may be accumulation of deleterious mutations. Strains which typically have been subjected to several rounds of mutagenesis and recurrent selection display one or more of the following characteristics: growth rate is reduced, sporulation is reduced or absent and the culture is prone to produce rapidly growing sectors which typically produce lower levels of the enzyme of interest. Table 6.1 shows the effect on chymosin production of prolonged growth in liquid culture prior to transfer to production medium on chymosin production. Chymosin yields are reduced and the number of aberrant non-conidiating colonies increases the longer the strains are grown in liquid culture. These aberrant colonies were purified and grown in 10 l fermenters and the level of chymosin compared to that of 'normal' conidiating colonies isolated from the same fermenter. Aberrant colonies only produced 130–140 mg/l chymosin compared to 700–720 mg/l chymosin produced by a conidiating colony. These results indicate that the increased occurrence of the aberrant colonies was responsible for the reduced chymosin yield.

6.3.4 *Electrophoretic karyotypic of chymosin-producing strains*

Contour-clamped electric field (CHEF) gel electrophoresis has been used to obtain an electrophoretic karyotype of *A. niger* var. *awamori*. The estimated sizes of the *A. niger* var. *awamori* chromosomes were 6.2 Mb, 5.8 Mb, 4.2 Mb, 3.8 Mb, 3.3 Mb, and 2.9 Mb. The 2.9 Mb chromosome appeared to be a doublet (see Figure 6.4). This gives an estimated genome size of 29–30 Mb for the seven chromosomes identified. Previous Southern hybridization analysis indicated

Table 6.1 Effect of prolonged growth in liquid medium prior to transfer to production medium on chymosin production in the strain GC1HF1342.

Incubation time (h)	Chymosin yield (mg/l)
48	650
85	455
96	390
129	195
191	195

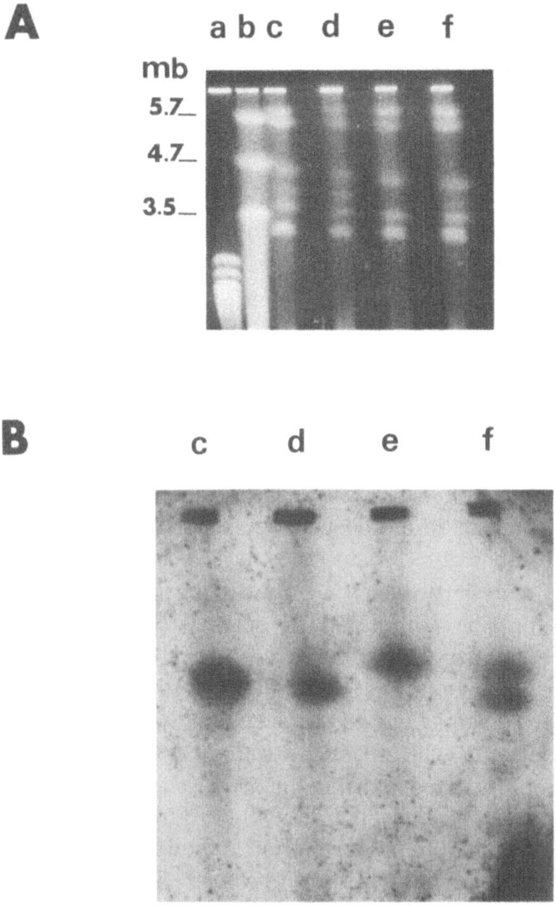

Figure 6.4 A and B CHEF gel separation of *A. niger* var. *awamori* chromosomal DNA. (A) Ethidium bromide strained chromosomal DNAs prepared from *S. pombe* and *S. cerevisiae* were used to estimate the size of the *A. niger* var. *awamori* chromosomal DNA. The estimated sizes in megabases (Mb) of the chromosomes are indicated. (B) Hybridization analysis of the CHEF gel-separated chromosomal DNAs transferred to a nylon membrane and probed with a plasmid containing the chymosin cDNA sequences. Lane a, *S. cerevisiae;* Lane b, *S. pombe*; Lane c, GC1HF1-3; *dgr2* Lane d, GC1HF1-3; *dgr2*; Lane e, GC1HF1342, Lane f, diploid strain GC1HF1342/GC1HF1-3; *dgr2*.

that several copies of the pGAMpR plasmid integrated into the genome of GC4-1, the starting strain for the yield improvement program (M. Ward, unpublished).

Four chymosin strains were subjected to CHEF gel analysis, DNA blotted to Nytran filters and probed with a plasmid containing the chymosin cDNA sequences. In all four strains the chymosin DNA sequences were found to be located on the 3.8 Mb chromosome [31]. These results indicated that the pGAMpR plasmid had integrated onto one chromosome. One unexpected outcome of using CHEF gel analysis to study the chromosomal location of the integrated chymosin DNA sequences was the identification of a chymosin-producing strain (GC1HF1342) in which there had been an alteration in the mobility of the chromosome on which the chymosin DNA sequences had integrated. Surprisingly, the chymosin DNA sequences in the strain GC1HF1342 were located in the region of the 4.2 Mb chromosome. Examination of the CHEF gel indicated that there was no longer a 3.8 Mb sized chromosome in GC1HF1342 (Figure 6.4). One possible explanation for the disappearance of the 3.8 Mb chromosome is that there had been a duplication of approximately 0.4 Mb in the 3.8 Mb chromosome and the resulting chromosome had co-migrated with the 4.2 Mb chromosome. These results indicate that *A. niger* var. *awamori* can tolerate substantial chromosomal rearrangements without any obvious loss of cell viability. It is interesting to note that GC1HF1342 is the same strain which produces large numbers of aberrant colonies upon prolonged culture in liquid medium. It is tempting to speculate that the altered mobility of the 3.8 Mb chromosome plays a role in this strain's instability.

CHEF gel analysis provides a powerful technology for the analysis of industrial strains. For example, CHEF gel analysis can be used to determine the chromosomal integration site of a transformed strain. With this knowledge it is possible to take two strains in which the transforming DNA has integrated on different chromosomes and cross them sexually (if possible) or parasexually and isolate segregants which have integrated DNA on two different chromosomes. This approach of increasing the copy number of a gene one wishes to express is a useful technique if it is difficult to re-transform the culture. CHEF gel analysis can be used to monitor strain stability by examining the presence and location on the integrated transforming DNA.

6.4 Regulatory issues

Regulatory issues should play a key role in developing the recombinant DNA technology to express foreign genes in filamentous fungi. Recently, Smart [32] has written an excellent review of regulatory issues concerning recombinant fungal fermentations.

6.4.1 *Fungal transformation and vector considerations*

Developing a selectable transformation system in which only DNA from the fungus is transformed can avoid potential regulatory problems concerning the use of antibiotic resistance genes as selectable markers for transformation (chapter 1). Nitrate reductase structural gene mutants can be readily isolated in many fungi [33] and the cloned nitrate reductase gene which is available has been successfully used to obtain large numbers of transformants [34]. Another useful selectable system which has been used relies upon the isolation of uridine auxotrophs and their complementation with the cloned orotidine-5′-monophosphate decarboxylase gene [35].

6.4.2 *Regulatory agency approval*

To date the US Food and Drug Administration (FDA), considered to be the leader of regulatory agencies, has not given approval for the use of any heterologously produced protein for pharmaceutical use or for use in the manufacture of foods produced by fungi. However, the authors believe that this delay in FDA approval for recombinant products, such as chymosin, produced by fungi, does not reflect any serious concern in this use of these host systems but reflects their rightly cautious approach to this evolving technology.

Recently, the FDA approved chymosin as the first food ingredient made via recombinant DNA technology as having GRAS (generally recognized as safe) status [36]. The approved host for the production of chymosin was *E. coli* K-12. Very recently, Flamm [37] an employee with the Office of Biotechnology at the FDA, published a report entitled *How the FDA approved chymosin: a case history*. Although approval of chymosin was given only for the *E. coli* host system and petitions using fungal hosts are awaiting approval, this article described the approval process used by the FDA. In consequence, a review of Flamm's article gives an insight to the sort of concerns companies attempting to use filamentous fungi as heterologous hosts should consider.

The GRAS petitioner was required to demonstrate that the cloned chymosin gene was identical to that of chymosin found in calf rennet. In addition, evidence was presented that the chymosin gene was properly expressed to make functional rennet. The production strain contained the beta lactamase gene which confers resistance to the antibiotic, ampicillin. To make sure that the food-grade enzyme did not contain functional copies of this antibiotic resistance gene, an additional purification step was used. The disrupted cells were treated with acid which killed intact cells and degraded their DNA. The FDA was concerned about the safety of the enzyme preparation in terms of the enzyme itself and other components in the preparation. In fact the chymosin preparation produced by *E. coli* was demonstrated to be significantly purer than rennet isolated from stomachs. The FDA concluded that the *E. coli*-produced chymosin was as safe as the calf-produced chymosin which already had GRAS status. The FDA's approval of chymosin produced by *E. coli* gives a clear indication that with suitable documentation other proteins produced by microbial hosts, e.g. fungi, will be given approval in terms of GRAS status or pronounced safe for the production of pharmaceuticals.

6.4.3 *Containment of recombinant organisms*

6.4.3.1 *Production strains.* Production vials produced from the master stock of the production strain are usually stored at −80°C at the production facility. Procedures need to be established which ensure that when a new production vial lot is produced it is checked to determine if the vial lot is free from contaminating organisms (see section 6.5.2). In addition, it is advisable to grow up a representative number of cultures from the new production vial lot in shake culture or small fermenters to determine if the new vial lot retains the master stock's productivity in terms of product yield.

6.4.3.2 *Manufacturing facility.* As Smart [32] correctly points out, a production facility should be designed not only to meet the requirements of product sterility but also to prevent the accidental release of the recombinant organism. Very recently, the US National Institute for Health (NIH) revised the guidelines concerning the physical containment for large scale uses of organisms containing recombinant DNA molecules [38]. These guidelines specifically address regulatory issues concerning fermentations of more than 10 litres of culture involving viable organisms containing recombinant DNA molecules.

6.4.3.3 *Good large scale practice (GLSP).* The most recent NIH guidelines also address GLSP. For an organism to qualify for GLSP consideration, the following factors are taken into consideration: the host organism should be non-pathogenic, the recombinant DNA-engineered organism should be non-pathogenic and the vector/insert should be well characterized and free of known harmful sequences. In addition, the insert should be as small as possible in order to perform its function. Most importantly, the vector used should not transfer any resistance markers to microorganisms not known to acquire them naturally because such an acquisition could compromise the use of the antibiotic as a disease control agent. In addition to preventing the accidental release of a recombinant organism, procedures must be developed to inactivate the organism after the fermentation has been completed.

6.5 Preservation of recombinant strains

6.5.1 *Introduction*

A wide variety of fermentation culture preservation methods are available and have been tested for significant periods of time. Simple, inexpensive methods of preservation include storage of mycelia under sterile water and storage of spores on dry soil. The most stable methods available today for long-term storage of recombinant strains are cryostorage at liquid nitrogen temperatures and lyophilization. Water crystal formation can occur down to −130°C, so

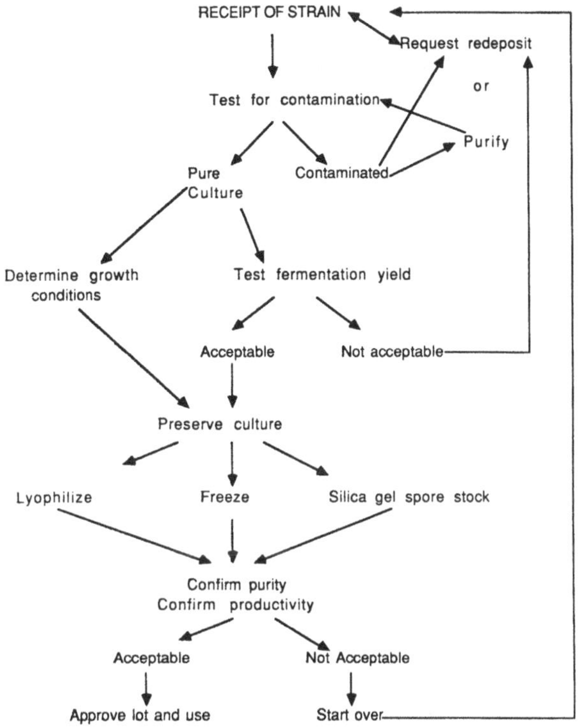

Figure 6.5 Preservation of recombinant fungal strains.

liquid nitrogen vapor is the preferred storage site for culture vials [39]. Vial storage at −80°C will preserve viability for approximately 5 years, and warmer temperatures will shorten the storage period. Lyophilized cultures are generally stored refrigerated, but will retain viability at room temperature for long periods of time [40].

6.5.2 *Initial testing*

The need for production of manufacturing vial lots may spur the deposit of a fungal strain with a culture collection. The discovery of an organism with novel properties or increased yield of a desired product may also require the generation of a vial lot for fermentation testing. Several good methods of preservation of fungal fermentation inoculum are available, and have been successfully used to produce various fungal enzymes at scales of up to 30 000 l for Genencor International. Cryopreservation and storage of the vials at liquid nitrogen vapor is the preferred method, with lyophilization and silica gel storage as 'back-up' methods.

The first task is to confirm the purity of the strain which has been deposited for preservation. The organism may be purified, or transferred onto fresh media several times, and the resulting growth observed for uniformity. Sectoring, as well as contamination is undesirable in a production strain. Another test for purity is to inoculate a liquid fungal media such as Clofine or potato dextrose broth, and observe after 24 h, 48 h and 72 h. If contamination is seen on microscopic examination or examination of samples spread onto solid media, a redeposit of the organism may be requested, or the culture repurified if it can be easily discerned from the contaminant.

After the purity of the culture is confirmed, the unique desired characteristics of the strain should be tested. For example, if the organism in question is a chymosin-producing strain, then its chymosin yield should be confirmed. This will eliminate the possibility of selection for an inferior producing mutant or revertant [41].

6.5.3 *Preservation*

In general, filamentous fungi respond very well to preservation as frozen stocks, with 10% glycerol as a cryoprotectant, and to lyophilization with 2X skim milk as the carrier. Frozen vials stored at liquid nitrogen temperatures and lyophiles in refrigerated storage can maintain viability for decades. Alternative methods are available if equipment and money are in short supply. Storage of spore suspensions on silica gel stored desiccated and refrigerated is an alternative. Non-sporulating fungi generally do not survive lyophilization and must be stored as mycelial preparations [40].

Frozen vials of sporulating fungi may be prepared from cultures grown on agar media. Agar plates are preincubated to assure sterility then inoculated with the purified, confirmed culture. The plates are incubated at the optimum temperature for the strain until heavy sporulation covers the whole plate. Plates are examined repeatedly during the growth period to eliminate any which are contaminated. A useful method of labelling cryovials is to use the date of preparation as the lot number. A spore suspension is made in an appropriate broth and sterile glycerol is added to equal 10% of the total volume. Vials are filled, with at least 10% of the vial volume left as air space to allow for expansion upon freezing. Vials are precooled on dry ice for a minimum of 30 min, then placed into a liquid nitrogen storage vessel. (If dry ice is unavailable, the vials may be cooled in a Styrofoam box in a −80°C freezer.)

Cryostorage of mycelia grown in a shaking broth culture is the preferred method of storage for non-sporulating fungi used as inoculum for fungal fermentation. For a review of the many other methods available for preservation of fungi, see [42–44].

Lyophilization of sporulating fungi is generally used for back-up and long-term storage, not as the source of inoculum for regular fermentations. Sterile Difco skim milk at double the concentration recommended on the label is an excellent carrier for fungal spores. The volume of spore suspension dispensed into an ampule for lyophilization must be less than 50% of the volume of the ampule for best results. Lyophilizers vary in size, cooling system and type of ampules or vials accepted. Each manufacturer's instructions should be inspected for details. If possible, seal the ampules while they are still under vacuum, or back fill them with nitrogen to exclude oxygen, then label and store in a refrigerator [39].

To prepare silica gel stocks of sporulating fungi, screw capped tubes are filled 1/4 full with silica gel, heated to desiccate the silica gel and cooled with the caps tight. Spores are suspended in reconstituted non-fat dry milk, then chilled. The spore suspension is added to the gel on ice, as a thermophilic reaction occurs. Tubes are kept at room temperature for 7 days, then transferred to a desiccator in a refrigerator for storage. Cultures are revived by shaking a few grains of silica gel onto an agar medium and incubating.

6.5.4 *Quality control*

Each set of inoculum preserved must be tested to assure stability of recombinant strains. One method used to determine the total number of vials to be tested from each set is to calculate the square root of the total number of vials in the set and round up to the next whole number. This will yield a statistically significant number of vials for testing. Each batch should be plated on an agar medium and the resulting growth examined, then grown in shake flask and transferred for several generations, with each generation also plated on an agar medium and examined microscopically. Purity of the culture and production of the correct level of the product should be confirmed before the lot is approved for use in fermentation.

The first batch of the organism which is confirmed for purity and production should be considered as a master stock. The first batch may be large enough to be divided into a master stock, a seed stock and a working stock. The working stock will be used for fermentation inoculum on a regular basis. One vial of the seed stock will be used to inoculate media to make each new working stock when the old one is depleted. One vial of the master stock will be used to replenish the seed stock when that is depleted. This system will assure consistency in fermentation inoculum over a long period of time. It is important to avoid continuous serial transfer of recombinant cultures to assure genetic stability.

6.6 Conclusions

The approach that Genencor International has taken to express bovine chymosin in *A. niger* var. *awamori* has been to fuse the chymosin cDNA to the last codon of the highly expressed *A. niger* var. *awamori* glucoamylase (*glaA*) gene. In the large-scale production of chymosin the mycelial inoculum for a fermentation is transferred to a production fermenter which has a carbon source such as maltose that is used to induce the glucoamylase synthesis.

The approach of Allelix, another major biotechnology company which has used filamentous fungi such as *A. nidulans* for heterologous expression of proteins, is somewhat different. For reviews of the Allelix group's results, see [32, 45] and chapter 5. The Allelix group has used the *alcA* promoter and a synthetic signal sequence to direct the expression of heterologous proteins [45]. The *alcA* gene is the structural gene for alcohol dehydrogenase. The *alcA* gene and the *aldA* gene, which encodes the aldehyde dehydrogenase gene, are under the control of the positively acting regulatory gene *alcR*. With the *alcA* system the biomass is built up in the production fermenter and then an inducer, such as a industrial ketone, is added to turn on the *alcA* promoter [32].

The Allelix group found that there was a wide range in the amount of heterologous protein made and multicopy transformants. There was, however, a correlation in the amount of heterologous protein made and copy number in transformants with 5–10 integrated copies of the plasmid. Transformants with greater than 10 copies of integrated plasmid did not show a proportional increase in the amount of protein produced. The most probable explanation for this result is that the *alcR* product required for the expression of the *alcA* promoter becomes limiting in transformants with greater than 5–10 integrated plasmids. When Allelix transformed additional copies of the *alcR* gene into a strain, if this strain was subsequently transformed with a plasmid using the *alcA* promoter, the amount of secreted protein produced was proportional up to approximately 20 copies of the integrated plasmid.

At Genencor International, the strain used for chymosin production, GC4-1, initially produced 250–270 mg/l chymosin. This strain had approximately 5–10 copies of the integrated pGAMpR plasmid. After this strain had been put through a strain improvement program, the resulting strain GC1HF1-3, *dgr246*, produced 1100 mg/l chymosin. Transformation of this strain with additional copies of pGAMpR did not result in a significant increase in chymosin production [12]. It is not known if a similar situation to that which the Allelix group found exists, but there is possibly a positively acting regulatory gene required for glucoamylase expression which is titrated out by additional copies of the *glaA* gene.

By using the *alcA* system, Allelix has reported yields of 2–3 mg/ml of fungal proteins such as glucoamylase, and 100 μg/ml human epidermal growth factor. Allelix also reported that yields of up to 500 μg/ml of intracellular proteins were possible using this system although the yields of secreted proteins are low compared with those now reported using the *glaA* system [12]. Because of the high value of the mammalian proteins destined for pharmaceutical use, the yields of proteins produced by using the *alcA* system are commercially viable. With Genencor International's success in obtaining commercially viable yields of secreted chymosin by *A. niger* var. *awamori* and

Allelix's reported results of obtaining significant levels of intracellular mammalian proteins, such as human epidermal growth factor, it is clearly evident that filamentous fungi have a significant role in biotechnology as hosts for the production of heterologous proteins.

References

1. Berka, R.M., Dunn-Coleman, N.S. and Ward, M. (1991) Industrial enzymes from *Aspergilli*. In: Aspergillus: *The Biology and Industrial Applications*. (Bennett, J.W. and Klich, M.A., eds). Butterworths, New York, (in press).
2. Ballance, D.J., Buxton, F.P. and Turner, G. (1983). *Biochem. Biophys. Res. Commun.* **112**: 284–289.
3. Goosen, T., Bloemheuvel, C., Gysler, D.A., de Bie, H.W.J., van den Broeck, H. and Swart, K. (1987) *Curr. Genet.* **11**: 499–503.
4. Mattern, I.E., Unkles, S., Kinghorn, J.R., Pouwels, P.H. and van den Hondel, C.A.M.J.J. (1987) *Mol. Gen. Genet.* **210**: 460–461.
5. Pentila, M.E., Nevalainen, M., Ratto, M., Salminen, E. and Knowles, J.K.C. (1987) *Gene.* **61**: 155–164.
6. Diez, B., Alvarez, E., Cantoral, J.M., Barredo, J.L. and Martin, J.F. (1987) *Curr. Genet.* **12**: 277–282.
7. Christensen, T., Woeldike, H., Boel, E., Mortensen, S.B., Hjortshoej, K., Thim, L. and Hansen, M.T. (1988) *Bio/Technology* **6**: 1419–1422.
8. Harkki, A., Uusitalo, J., Bailey, M., Penttila, M. and Knowles, J.K. (1989) *Bio/Technology* **7**: 596–603.
9. Cullen, D., Gary, G.L., Wilson, L.J., Hayenga, K.J., Lamsa, M.H., Rey, M.W., Norton, S. and Berka, R.M. (1987) *Bio/Technology* **5**: 369–376.
10. Upshall, A., Kumar, A.A., Bailey, M.C., Parker, M.D., Favreau, M.A., Lewison, K.P., Joeseph, M.L., Maraganore, M. and McKnight, G.L. (1987) *Bio/Technology* **5**: 1301–1304.
11. Gwynne, D.I., Buxton, F.P., Williams, S.A., Garven, S. and Davies, R.W. (1987) *Bio/Technology* **5**: 713–719.
12. Dunn-Coleman, N.S., Bloebaum, P., Berka, R.M., Bodie, E., Robinson, N., Armstrong, G., Ward, M., Przetak, M., Carter, G.L., LaCost, R., Wilson, L.J., Kodama, K.H., Baliu, E.F., Bower, B., Lamsa, M. and Heinsohn, H. (1991) *Bio/Technology* **9**: 976–981.
13. Baily, H. (1987) *Bio/Technology* **5**: 883–890.
14. Bodie, E.A. (1991) Fungal Fermentation. Appendix In: *Aspergillus nidulans* and other filamentous fungi as genetic systems (Ward, M.). In: *Modern Microbial Genetics* (Streips, U.N. and Yasbin, R.Y., eds). Wiley-Liss, New York, 491–496.
15. Pirt, S.J. (1975) *Principles of Microbe and Cell Cultivation*. Oxford Blackwell Press, Oxford.
16. Weinberg, E.D. (1974) *Dev. Industr. Microbial.* **15**: 70–81.
17. Stanbury, P.F. and Whitaker, A. (1984) *Principles of Fermentation Technology*. Pergamon Press, Oxford.
18. Schweiger, L.B. (1957) US Patent 2,970,084.
19. Nowakowska-Waszczuk, A., Zakowska, Z. and Scobocka, B. (1969) *Acta Microbiologica Polonica Ser. B.* **1**: 105–110.
20. Kristiansen, B. and Chamberlain, H.E. (1983) Fermenter design. In: *The Filamentous Fungi, Vol. IV, Fungal Technology*. (Smith, J.E., Berry, D.R. and Kristiansen, B. eds). Edward Arnold Press, London pp. 11–13.
21. Metz, B. and Kossen, N.W.F. (1977) *Biotechnol. Bioeng.* **19**: 781–799.
22. Kristiansen, B. and Bu'lock, J.D. (1980) *Biotechnol. Bioeng.* **2**: 2579–2590.
23. Solomons, G.L. (1980) Fermentor design and fungal growth. In: *Fungal Biotechnology*. (Smith, J.E., Berry, D.R. and Kristiansen, B. eds). Academic Press, London, pp. 55–80.
24. Ryu, D.Y. and Oldshue, J.Y. (1977) *Biotechnol. Bioeng.* **19**: 621–629.
25. Wang, Y. and Henry, K. (1986) Bioinstrumentation and computer control of fermentation processes. In: *Manual of Industrial Microbiology and Biotechnology*. (Demain, A.L. and Solomon, N.A. eds). American Society for Microbiology, Washington DC, pp. 308–320.

26. Foltmann, B. (1981) *Essays Biochem.* **17**: 53–84.
27. Pedersen, V.B., Christensen, K.A. and Foltmann, B. (1979) *Eur. J. Biol. Chem.* **94**: 573–580.
28. Ward, M., Wilson, J.J., Kodama, K.H., Rey, M.W. and Berka, R.M. (1990) *Bio/Technology.* **8**:435–440.
29. Lamsa, M. and Bloebaum, P. (1990) *J. Indust. Microbiol.* **5**: 229–238.
30. Berka, R.M., Ward, M., Wilson, L.J., Hayenga, K.J., Kodama, K.H., Carlomagno, L.P. and Thompson, S.A. (1990) *Gene* **86**: 2153–2162.
31. Bodie, E.A., Armstrong, G.L. and Dunn-Colerman, N.S. (in preparation).
32. Smart, N.J. (1991) Scaling-up production of recombinant DNA products using filamentous fungi as hosts. In: *Molecular Industrial Mycology.* (Leong, S.A. and Berka, R.M. eds). Dekker, pp. 251–279.
33. Cove, D.J. (1975) *Heredity* **36**: 191–203.
34. Unkles, S.E., Campbell, I.E., Carrez, D., Grieve, C., Contreras, R., Fiers, W., van den Hondel, C.A.M.J.J. and Kinghorn, J.R. (1989) *Gene* **78**: 157–166.
35. van Hartingsveldt, W., Mattern, I.E., van Zieijl, C.M.J., Pouwels, P.H. and van den Hondel, C.A.M.J.J. (1987) *Mol. Gen. Genet.* **206**: 71–75.
36. Federal Register **55** (March 23, 1990) 10932.
37. Flamm, E.L. (1991) *Bio/Technology* **9**: 349–351.
38. Federal Register **56** (No 138) 33174–33176.
39. Alexander, M.T., Drabkowski, D., Frieshtat, S., Holloway, L., Jewett, R., Pienta, P. and Simione, F.P. (1988) *Workshop on Freezing and Freeze-Drying of Microorganisms.* American Type Culture Collection, Rockville.
40. Smith, D. (1984) Maintenance of fungi. In: *Maintenance of Microorganisms: A Manual of Laboratory Methods.* (Kirsop, B.E. and Snell, J.J.S. eds). Academic Press, London, pp. 83–107.
41. Onions, A.H.S. (1983) Culture and preservation. In: *The Filamentous Fungi, Vol. IV, Fungal Technology.* (Smith, J.E., Berry, D.R. and Kristiansen, B. eds). Edward Arnold, London, pp. 373–390.
42. Ellis, J.J. (1979) *Mycologia* **71**: 1073–1075.
43. Hwang, S. (1976) *Mycologia* **68**: 378–387.
44. Smith, D. (1988) Culture and preservation. In: *Living Resources for Biotechnology: Filamentous Fungi,* (Hawksworth, D.L. and Kirsop, B.E. eds). Cambridge University Press, Cambridge, pp. 75–99.
45. Davies, R.W. (1990) Development of commercial gene expression and protein secretion systems in filamentous fungi. In: *Sixth International Symposium on Genetics of Industrial Microorganisms. Proceedings.* (Heslot, H., Davies, J., Florent, J., Bobichon, L., Durand, G. and Penasse, L. eds). **2**: 567–576.

7 Molecular biology of filamentous fungi used for biological control

J.M. CLARKSON

7.1 Introduction

Biological control offers an attractive alternative or supplement to the use of chemical pesticides. Microbial biological control agents are perceived as being less damaging to the environment and their generally complex mode of action makes it unlikely that resistance will develop to a biopesticide. To date the most successful biological control agent has been the insecticidal bacterium *Bacillus thuringiensis* [1] but there is increasing interest in the use of filamentous fungi for biological control. Enthusiasm for the use of biological control needs to be balanced by the realization that few microbial agents have been registered as commercial products, and that the likely markets for microbial agents are where effective chemical pesticides are unavailable or are limited by their toxicity.

The recent development of molecular techniques for genetic engineering of filamentous fungi provides new opportunities for the study of fungi used in biological control. The isolation of genes encoding pathogenicity or virulence determinants will allow rigorous testing of their role in pathogenesis and should provide a rational basis for strain improvement by direct genetic manipulation. The identification of molecular markers for characterizing genomes make it possible to study directly the genetic variation and structure of fungal populations. The aim of this chapter is to describe some of these molecular techniques and discuss how they might be applied to biocontrol fungi. Because of the diversity of host–pathogen interactions, it is clearly not possible to review all aspects of the molecular biology of all biocontrol fungi, and this chapter deals exclusively with fungi used for the biological control of either insect pests or plant pathogenic fungi. These two groups of biocontrol fungi show interesting parallels both in their mechanisms of pathogenesis or antagonism, and in the experimental approaches used to study them. For more general information, including commercial perspectives, the reader is referred to the several recent review articles dealing with mycopathogens [2–4] and entomopathogens [5, 6].

The broad host range of Deuteromycete fungi, and in particular *Metarhizium* and *Beauveria*, show particular promise as biological control agents and are currently used as mycoinsecticides against several pests of agricultural crops. *Trichoderma* spp., *Gliocladium virens* and *Talaromyces flavus* have been used successfully to control a number of soil-borne fungal plant pathogens, although their commercial potential has yet to be exploited. These fungi will form the focus of this review.

7.2 Mechanisms of biological control activity

A better understanding of the mechanisms of biological control activity is necessary if rational strain improvement is to become a reality. These mechanisms differ for fungi according to the host–parasite interaction, but include host location, attachment, the development of specialized infection structures, invasion, and the production of inhibitory or toxic metabolites. In addition, the growth rate, fecundity and persistence of a fungal biological control agent will significantly affect its performance in the field. Many of these characteristics are quantitative, and are likely to be determined polygenically. Genetic manipulation of polygenic characteristics is possible through sexual or parasexual hybridization. There is little scope at present, however, for the manipulation of polygenic traits by recombinant DNA techniques. Two areas where significant progress has been made in our understanding of the mechanisms of biological control activity, and where direct genetic manipulation is feasible, are the production of extracellular enzymes and toxic metabolites.

7.2.1 *Enzymes*

Filamentous fungi produce a wide range of catabolic enzymes which enable them to occupy a diversity of ecological niches. Of particular importance to both saprophytic and parasitic fungi are extracellular depolymerases, as these allow the penetration and utilization of carbohydrate and protein polymers which occur in the cell walls and integuments of plants and animals (see chapters 3 and 4). Of particular relevance here are extracellular enzymes which degrade components of fungal hyphae or insect cuticle.

The mycopathogenic fungus *Trichoderma harzianum* has been shown to produce several extracellular depolymerases including β-1,3-glucanase, protease and chitinase when grown in liquid culture on laminarin, chitin or host cell walls [7–11]. Strong circumstantial evidence for the involvement of hydrolytic enzymes in the interaction between *Trichoderma* spp. and plant pathogenic fungi has been provided by studies using scanning (SEM) and transmission (TEM) electron microscopy. Hyphae of *T. hamatum* and *T. harzianum* have been observed under SEM to attach to the hyphae of *Rhizoctonia solani* or *Sclerotium rolfsii* by coils, hooks or appressoria [12]. Lysed sites and penetration holes were found following removal of the parasitic hyphae suggesting enzymic degradation. Ultrastructural studies also indicate enzymic digestion of host cell walls [13]. The interaction between a strain of *Trichoderma* and *Fusarium oxysporum* f.sp. *radicis-lycopersici* has been investigated recently using SEM and TEM with gold cytochemistry [14]. Here, hyphal contact did not appear to be necessary for parasitism, but the observed breakdown and release of *N*-acetylglucosamine residues suggests that wall-bound chitin may be released by the action of extracellular chitinase produced by *Trichoderma*. Several glucose-repressible chitinases were detected *in vitro*.

T. harzianum (strain IMI206040) has been shown to produce at least two β-1,3-glucanases and one constitutive neutral isozyme as well as one acidic form which was subject to both carbon catabolite repression and induction by laminarin, pustulin (β-1,6-glucan) or *R. solani* cell walls [10]. Protease regulation was also studied and a constitutive neutral isozyme as well as an apparently inducible basic form were identified. The inducible isozyme was identified as a serine chymotrypsin-like protease with a p*I* of 9.2 and a molecular weight of 31 kDa. Interestingly, this is very similar to the inducible PR1 protease produced by the entomopathogen *Metarhizium anisopliae* [15] (see below). *Trichoderma harzianum* has been shown recently to be able to parasitize the elm bark beetle, *Scolytus*, [16] and it is an intriguing prospect that the species *T. harzianum*, and perhaps individual strains, may show dual pathogenicity towards both insects and fungi, mediated through the production of specific hydrolytic enzymes. *Verticillium lecanii* is able to parasitize insects and rust fungi, but a recent study has indicated that entomopathogenic and mycopathogenic isolates within the species can be separated on the basis of morphological and biochemical characteristics [17].

Pythium oligandrum is capable of penetrating fungal hyphae and produces a number of enzymes which may degrade components of the host cell wall [18]. Extracellular β-1,3-glucanase, lipase and protease, but not chitinase or cellulase activities, were detected *in vitro*. Glucanase and protease production was repressed in the presence of glucose but induced by laminarin or BSA, respectively, and by isolated host cell wall.

An extracellular enzyme produced in liquid culture by *Talaromyces flavus* that resulted in the inhibition of microsclerotial germination of *Verticillium dahliae* has been identified as glucose oxidase [19]. It was proposed that the antibiotic activity of this enzyme was caused by the action of hydrogen peroxide released on catalytic oxidation of glucose. The enzyme was purified to homogeneity and shown to consist of a dimer with a subunit molecular weight of 71 kDa and a native weight of 164 kDa [20]. Isoelectric focusing indicated that as many as five isozymes may be produced, but their similar p*I* might indicate that they result from minor variations in the glycosylation pattern, rather than being the products of different genes. Interestingly, *T. flavus* was shown to be very tolerant to hydrogen peroxide [20]. *T. flavus* has not been shown to directly penetrate the hyphae or microsclerotia of *V. dahliae* and in this interaction glucose oxidase may play a more important role than cell wall degrading enzymes. Fravel and Roberts [21] have shown recently that purified glucose oxidase from a biocontrol active strain of *T. flavus* (Tf-1) significantly reduced the growth rate of *V. dahliae* in the presence, but not absence, of eggplant roots, suggesting that glucose released from the roots was metabolized by glucose oxidase to form hydrogen peroxide. A single ascospore variant of strain Tf-1, which produced only 2% of the parent level of glucose oxidase, failed to control *Verticillium* wilt of eggplant. The wild-type parent Tf-1, but not the variant, produced an extracellular 72 kDa protein that co-migrated with purified glucose oxidase. However, the variant did produce two lower molecular weight proteins at much higher levels than the wild type.

Entomopathogenic fungi invade their hosts by direct penetration of the insect cuticle, a process that is likely to involve both mechanical pressure and enzymic hydrolysis [6, 22, 23]. Insect cuticle comprises chitin fibrils embedded in a protein matrix together with lipids/waxes and small variable amounts of phenols, inorganics and pigments [24]. The production and role of cuticle-degrading enzymes has received considerable attention and has been reviewed recently [25]. When grown in liquid cultures containing insect cuticle as the sole carbon and nitrogen source, pathogenic isolates of *M. anisopliae*, *Beauveria bassiana* and *V. lecanii* produce a range of extracellular enzymes which correspond to the major components of insect cuticle [26]. These are produced sequentially in all three fungi. Esterase activity, endoprotease, aminopeptidase and carboxypeptidase appear first (within 24 h), followed by *N*-acetylglycosaminidase. Chitinase and lipase are produced later. It has been suggested [27, 28] that the late appearance of chitinase results from induction as chitin eventually becomes available after degradation of encasing protein. Endoprotease activity from *M. anisopliae* can be resolved into three main components, a chymoelastase serine protease (PR1) (p*I* 10.3, MW 30 kDa), a trypsin-like serine protease (PR2) (p*I* 4.4, MW 28.5 kDa) and a trypsin-like cysteine protease (PR4) [15, Cole *et al.*, unpub.].

PR1-like and PR2-like enzymes are produced *in vitro* by the entomopathogenic fungi *B. bassiana*, *V. lecanii*, *Nomuraea rileyi* and *Aschersonia aleyrodis* as well as all isolates of *M. anisopliae* tested by St Leger and co-workers [29]. There is strong evidence that the penetration of insect cuticle by *M. anisopliae* (and possibly other entomopathogenic fungi) requires the action of PR1: PR1 is produced early in the sequence of cuticle-degrading enzymes when grown *in vitro* on host cuticle and is a protease adapted to extensively degrade cuticular protein [15], PR1 is the only cuticle-degrading enzyme produced in high amounts by all pathogenic isolates of this and other entomopathogenic fungi tested [29], it is produced in abundance by infection structures of the fungus during host penetration [30] and specific inhibition of PR1 delays disease symptoms and mortality of larvae of the tobacco horn worm *Manduca sexta* [31].

A study of the regulation of extracellular PR1 showed previously that basal synthesis is subject to both carbon and nitrogen metabolite repression [32, 33]. It has been demonstrated recently [Paterson *et al.*, unpub.] that PR1 is also regulated by induction by cuticular protein. Growth, but not PR1 induction, occurs with elastin, BSA or gelatin. Other fungal proteases are regulated by either metabolite repression alone, e.g. *Aspergillus* [34], or by induction under de-repressed conditions, e.g. *Neurospora* [35] and PR2 of *M. anisopliae* [Paterson *et al.*, unpub.]. However, these proteases are induced non-specifically by any protein. Presumably the unique regulation and efficacy towards insect cuticle reflects the adaptation of *M. anisopliae* to insect parasitism. The *pr1* gene has been cloned from strain ME1 using RACE–PCR to amplify a DNA fragment from cDNA made from mRNA isolated from PR1 de-repressed fungal cultures [Paterson *et al.*, unpub.]. This employed a first round of amplification using outer 5′ and 3′ primers where the 5′ consisted of an oligonucleotide based on PR1 peptide sequence data (5 peptide fragments from a tryptic digest of PR1 were previously sequenced). A second round of amplification utilized an inner 5′ primer, also based on a PR1 peptide sequence. The second round of amplification produced a single product of approximately 1 kb. Direct sequencing confirmed that this consisted of part of the *pr1* gene by identification of known amino acid sequences. This *pr1* PCR fragment has been used to isolate corresponding genomic clones from an EMBL3 library of ME1.

7.2.2 *Toxic metabolites*

Gliocladium virens produces several extracellular secondary metabolites, including gliovirin [36], viridin [37] and gliotoxin [38] which may play a role in pathogenicity. Gliovirin is an epithiodiketopiperazine [36] which inhibits the growth of *Pythium ultimum* and a *Phytophthora* sp. An ultraviolet light induced mutant of *G. virens* deficient for the production (or secretion) of gliovirin did not inhibit *P. ultimum* whereas a mutant with enhanced production was more strongly inhibitory in culture, suggesting that gliovirin may be an important determinant of pathogenicity. The observation that *G. virens* does not strictly parasitize the mycelium (as evidenced by coiling of hyphae around the host) of *P. ultimum* as it did *R. solani* [39] suggests that diffusible metabolites may play a key role in the interaction between *G. virens* and *P. ultimum.*

It has been shown previously that the antibiotics of *G. virens* can kill *R. solani* sclerotia [40]. Howell [41] isolated mutants of *G. virens* which were unable to parasitize (show hyphal coiling) *R. solani in vitro.* These mutants retained the same gliotoxin or viridin complement as the parent strain and had similar efficacy as biocontrol agents for cotton seedling disease induced by *R. solani,* indicating that mycoparasitism is not essential in the biocontrol process. Roberts and Lumsden [42] have shown recently that gliotoxin, but not extracellular enzymes, is responsible for inhibition of sporangial germination and mycelial growth of *P. ultimum*. It is possible, however, that the action of antibiotics produced by *G. virens in vivo* may be enhanced by hyphal damage caused by cell wall degrading enzymes (CWDE), and/or close proximity during hyphal coiling. *P. ultimum* is more sensitive than *R. solani* to gliotoxin but the mechanism of action towards both these fungi involves the selective binding of gliotoxin to cytoplasmic membrane thiol groups [43]. Gliotoxin has a broad spectrum of activity including inhibition of various bacteria, viruses and fungi. While showing no mutagenic effects in standard Ames/*Salmonella* assays [44], the broad spectrum of activity of this antibiotic may limit the potential for strain improvement by manipulation of genes involved in the biosynthesis of toxic metabolites of *G. virens*.

Trichoderma spp. produce a number of volatile and non-volatile metabolites active against fungi [45, 46] but their role in biological control is not clear. Claydon and co-workers [47] isolated alkyl pyrones from cultures of two strains of *T. harzianum* and showed these to be potent inhibitors of a wide range of fungi. Pyrone antibiotics have also been isolated from a strain of *T. harzianum* effective in suppressing take-all disease of wheat caused by *Gaeumannomyces graminis* var. *tritici* [48].

Entomopathogenic fungi produce a number of low molecular weight insecticidal toxins *in vitro*, including the cyclic peptides beauvericin [49] and the destruxins [50]. The destruxins are a family of closely related cyclic peptides produced by *M. anisopliae*, consisting of five amino acids, *β*-alanine, *N*-methylalanine, *N*- methylvaline, isoleucine and proline, and one hydroxy acid [51, 52]. There is good evidence for the involvement of destruxins in pathogenesis. Injected destruxins cause paralysis and death in Lepidopteran larvae and adult Diptera [53] and lethal quantities of destruxins have been isolated from *Metarhizium* infected insects [54].

Beauvericin, produced by *B. bassiana*, is a hexadepsipeptide containing three repeat units of D-2-hydroxyisovaleric acid linked to *N*-methyl phenylalanine. It is synthesized extraribosomally by the multifunctional enzyme, beauvericin synthetase (MW 250 kDa), using the thiotemplate mechanism now well established for the majority of other cyclic peptide antibiotics [49, 55]. Beauvericin is structurally related to another group of cyclic peptides, the enniatins, produced by several species of *Fusarium* [56]. Enniatin synthetase from *Fusarium scirpi* is a single large multifunctional enzyme of approximately 250 kDa. By immunologically screening a *F. scirpi* cDNA expression library, Zocher and co-workers (pers. commun.) isolated a clone representing the carboxyl terminus of the enzyme. This clone was used to screen a genomic library and the entire enniatin synthetase gene has now been isolated and characterized. Transcriptional mapping by Northern blot analysis and expression studies of fusion proteins encoded by subclones in *E. coli* revealed a 9.5 kb coding sequence. This and other cloned peptide antibiotic synthetases may be useful as molecular probes for the isolation of genes involved in the biosynthesis of insecticidal cyclic peptides from entomopathogenic fungi.

7.3 Strain improvement

Biological control activity is a complex process involving a variety of mechanisms which are likely to act in concert, but which may differ in individual biocontrol strains. For example, one strain may be optimal for the production of key enzymes whereas another may secrete a higher level of antibiotics. One of the main challenges in biological control is the development of genetic techniques by which desirable traits can be combined to produce superior strains.

7.3.1 *Sexual and parasexual recombination*

Hybridization through the sexual cycle may be applicable for some biocontrol fungi. For example, *Talaromyces flavus* is a homothallic Ascomycete which produces a cleistothecial perfect stage and Katan and co-workers [57] have used the sexual cycle to study the inheritance of benomyl and dicarboximide resistance in *T. flavus* by crossing fungicide resistant and sensitive strains.

Sexual stages have not been found in strains of *Trichoderma* and *Gliocladium* which have been identified as effective biocontrol agents. However, even in the absence of a sexual stage, genetic recombination may be possible via the parasexual cycle [58]. The essential features of the parasexual cycle are the formation of at least transient heterokaryon by hyphal anastomosis or protoplast fusion followed by rare nuclear fusions to produce a heterozygous diploid nucleus. These diploid nuclei are frequently unstable and break down to a stable haploid state by random loss of chromosomes during mitotic divisions. Genetic recombination can occur by mitotic crossing over between homologous chromosomes in diploid or aneuploid nuclei and also by the random segregation of chromosomes during haploidization.

Vegetative incompatibility, restricting genetic exchange between unrelated strains via hyphal anastomosis, is common in filamentous fungi and may constitute a major barrier to strain improvement in some biocontrol fungi. Incompatibility can sometimes be overcome through protoplast fusion [59]. Hybridization by protoplast fusion has been studied in some detail for *T. harzianum* by Harman and co-workers (reviewed in [60]). The general methodologies for protoplast isolation and fusion have been reviewed recently [61].

Protoplasts of *T. harzianum* were produced by digestion of fungal mycelia with the enzyme Novozyme 234 (Novo Industri A/S, Denmark) [62] and fusion between protoplasts derived from complementary auxotrophs was induced using polyethylene glycol. Fusions between auxotrophs derived from two different wild-type strains with desirable biocontrol potential produced slow-growing, unbalanced heterokaryons which gave rise to a range of progeny. All showed the isozyme phenotype of one of the parent strains but many were of novel morphology. Harman and co-workers [63] tested protoplast fusion progeny for their biocontrol ability and identified two that, when used as seed protectants, gave better control of *Pythium ultimum* than the two parental strains. One of these showed improved ability to colonize the rhizosphere of cotton and maize after seed treatment [64]. Pe'er and Chet [65] have similarly demonstrated that protoplast fusion progeny of *T. harzianum* can be effective in biocontrol.

The entomopathogens *M. anisopliae* and *B. bassiana* have no known sexual stages although genetic recombination through the parasexual cycle has been demonstrated for both [66–68]. Parasexual recombination has been considered as a possible method of genetic manipulation for strain improvement in *M. anisopliae.* Heale and co-workers [69] reported a parasexual cross between the auxotrophic mutants derived from pathogenic wild-type strains, which differed in sporulation rate and pathogenicity towards *Nilaparvata lugens.* Two of the haploid recombinants selected from this cross combined high pathogenicity and sporulation rates. This is in contrast to parasexual crosses between isolates of *Verticillium lecanii* [70] and the plant pathogenic *Verticillium* species *V. albo-atrum* and *V. dahliae* [71, 72] where recombinants were generally of low pathogenicity. A possible explanation is that parasexual recombination between isolates that are normally reproductively isolated disrupts adapted complexes of pathogenicity genes resulting in weakly pathogenic haploid recombinants.

7.3.2 *Genetic engineering*

An essential prerequisite for strain improvement by genetic engineering is the development of a method for the stable introduction of exogenous DNA (see chapter 1). DNA-mediated transformation procedures for filamentous fungi are frequently based on the addition of DNA to protoplasts in the presence of Ca^{2+} followed by the addition of polyethylene glycol [73, 74]. Transformants usually arise at low frequency (0.1–10 transformants/μg vector DNA) through the stable incorporation of DNA into the fungal chromosomes, although there are now several reports of autonomous replication of transforming DNA (75–78). Importantly, DNA sequences that promote autonomous replication also result in high frequency (100–10 000 transformants/μg vector DNA) transformation. A variety of selectable markers have been used for fungal transformation. Initially, transformation of the genetically well characterized 'laboratory' fungi *Aspergillus nidulans* [79] and *Neurospora crassa* [80] utilized genes encoding catabolic or biosynthetic enzymes to complement appropriate mutants (see chapter 1). Subsequently, dominant selectable markers such as hygromycin, bleomycin, phleomycin or benomyl resistance have been used, as these can be used to transform wild-type strains of a genetically uncharacterized fungus.

The mycopathogenic fungi *T. harzianum* and *T. viride* have been transformed [81] using the plasmid pAN7-1 [82] which carries a bacterial hygromycin resistance gene under the expression control of the *Aspergillus nidulans gpd* promoter and *trpC* terminator. Transformants were obtained at high frequency (200–800/μg DNA) from the integration of multiple copies of plasmid DNA into the genome of *Trichoderma*. However, they were unstable, the hygromycin resistance phenotype being lost after 2–3 sequential passages of conidia on non-selective media. Stability was achieved, without reducing transformation frequency, by inserting a 2.4 kb fragment from a *Trichoderma* α-amylase gene into plasmid pAN7-1. Co-transformation was investigated using a plasmid carrying a phleomycin resistance gene. Southern hybridization indicated that approximately 80% of the transformants selected for pAN7-1 mediated hygromycin resistance were also resistant to phleomycin. The same workers have also used high voltage electroporation to introduce plasmid pAN7-1 into protoplasts of *T. harzianum* [83]. Surprisingly, transformants produced by this method were mitotically more stable.

Transformation of *G. virens* has also been reported [84]. Protoplasts derived from young mycelia were transformed with plasmid pBT6 [85] which carries a β-tubulin gene from a benomyl resistant strain of *Neurospora crassa*. The frequency was low (*c.* 6/μg DNA) and many of the benomyl-resistant transformants were unstable. The pathogenicity of stable transformants was not reported although some showed a reduced growth rate on non-selective media.

Bernier and co-workers [86] demonstrated recently that the entomopathogen *Metarhizium anisopliae* could also be transformed to benomyl resistance with the cosmid pSV50 which contains the same cloned β-tubulin gene from *N. crassa* [87]. Transformants were stable under non-selective conditions and retained pathogenicity to larvae of the tobacco hornworm, *Manduca sexta*. Southern hybridization indicated that the majority of transformants arose through the random integration into the chromosomes of tandemly repeated vector sequences. Benomyl-resistant transformants of *M. anisopliae* have also been selected by Goettel and co-workers [88] using a plasmid containing the *benA3* allele from *A. nidulans*. In both cases transformants were identified only at low frequency. In *Beauveria bassiana* resistance to chlorate has been used as a positive selection for mutants lacking nitrate reductase activity. The nitrate reductase (*niaD*) *gene of A. nidulans* was then used to transform a nitrate reductase mutant [89]. Cloned heterologous nitrate reductase genes may have wide applicability for the transformation of biocontrol fungi.

The development of successful transformation protocols for an increasing number of filamentous fungi suggests that transformation is unlikely to be a major problem with other Ascomycete or Ascomycete-like Deuteromycete biocontrol fungi. This may not be the case for entomopathogenic fungi belonging to the entomophthoraceae, although another zygomycetous fungus, *Mucor circinelloides*, has been transformed and autonomous replication of plasmid DNA confirmed [75]. The Oomycete fungi, which include several mycopathogenic *Pythium* species, have also been recalcitrant, but a recent report of the transient expression of transforming DNA in *Phytophthora infestans* [90] represents significant progress. There have been no reports of the use, with entomopathogenic or mycopathogenic fungi, of transformation vectors containing sequences which promote high frequency autonomous replication. Although these vectors are normally unstable, and therefore of limited value in genetically modifying biocontrol fungi, they allow the rapid cloning of genes by direct complementation of mutants.

Although transformants arise mainly through the random integration of DNA in filamentous fungi, homologous integration can occur in a proportion of the transforming events [73]. This has been exploited to produce direct mutations by gene disruption or replacement with altered copies of cloned genes, thereby allowing a definitive analysis of the role of that gene product (chapter 1). This approach has been used recently to demonstrate that the cell wall-degrading enzyme polygalacturonase is not required for pathogenicity of the fungus *Cochliobolus carbonum* on its host plant, maize [91]. This powerful strategy could also be used to investigate the role of putative pathogenicity or virulence determinants of entomopathogenic or mycopathogenic fungi where transformation systems have been developed. For example, it should now be possible to establish a definitive role for the endoprotease PR1 of *M. anisopliae*. Protease deficient mutants could be produced by transformation mediated gene disruption with a disrupted or truncated copy of the cloned *pr1* gene, and the pathogenicity compared with that of the wild-type parental strain.

Genetic engineering offers an exciting alternative approach to strain improvement. The cloning of genes from biocontrol fungi, whose products are essential for pathogenicity or which contribute quantatively to virulence, should provide the rational basis for strain improvement. Transformation in filamentous fungi is frequently associated with the integration of multiple copies of plasmid sequences into chromosomal DNA [73] and this approach could be used to increase the copy number, and perhaps expression, of pathogenicity genes. A recent example of this approach is the demonstration that extracellular β-glucosidase activity could be increased five-fold in a transformant of *Trichoderma reesei* carrying an increased copy number of the *bgl1* gene [92]. Gene expression could also be modified by transformation with expression constructs containing the coding region of a pathogenicity gene fused to the promoter region of a highly expressed gene from the same or a related fungus. This approach may enable the regulation, as well as absolute amount, of the gene product to be altered. Genetic transformation will also provide a tool for moving genes between species of biocontrol fungi.

Although the technologies for producing recombinant biocontrol fungi exist already, and significant progress is being made in our understanding of the molecular basis of biological control activity, it is not clear how legislation governing the commercial release of genetically manipulated microorganisms will develop. However, at least 29 approved field tests of genetically engineered microorganisms have been initiated world-wide [93] and important progress has been made, particularly with bacteria, towards developing effective monitoring strategies.

7.4 Molecular markers for the characterization of biocontrol fungi

The genetic characterization of wild-type strains of biocontrol fungi can provide information about natural variation and population structure. It may also be possible to identify individual isolates by a unique set of genetic characteristics which might enable the fate of released biocontrol fungi to be monitored. Two types of naturally occurring genetic markers which have been used extensively for plant pathogenic fungi are mating type and specific virulence, but these are not generally applicable to mycopathogens or entomopathogens. Molecular markers have several advantages, including co-dominance, selective neutrality and abundance [94]. Variations in soluble protein patterns and isozyme polymorphisms detected by electrophoresis have proved useful in a range of fungi. deConti and co-workers [95] examined the electrophoretic variation in esterases and phosphatases in 11 wild-type isolates of *M. anisopliae* from Brazil and were able to distinguish five patterns for esterase. In a study of both *M. anisopliae* var. *anisopliae* and the large spored *M. anisopliae* var. *majus* (*major*), Riba and co-workers [96] demonstrated that var. *majus* isolates were relatively homogeneous for isozyme profiles but that var. *anisopliae* isolates were highly polymorphic. In a recent and extensive study [97], isozyme variation for eight enzyme systems was investigated in 114 isolates of *Metarhizium* spp. The highly variable nature of *M. anisopliae* was confirmed and some geographical clustering of genotypic classes was apparent. Interestingly, the thirteen isolates of *M. anisopliae* var. *majus* showed multiband phenotypes characteristic of heterozygous diploids, supporting previous suggestions that var. *majus* isolates are stable diploids [98]. Isozyme polymorphisms have also been detected in *B. bassiana* [99] and in the mycopathogens. *Trichoderma* and *Gliocladium* [100, 101].

Techniques which directly investigate genetic variation at the DNA level are rapidly gaining acceptance as the most appropriate tools for studying fungal populations. Several of these are discussed in the following sections.

7.4.1 *Restriction fragment length polymorphisms of nuclear and mitochondrial DNA*

Restriction fragment length polymorphisms (RFLPs) may be detected following digestion with an appropriate type II restriction endonuclease and separation of the DNA fragments by agarose electrophoresis. Differences between isolates in DNA sequence may alter the restriction sites for endonucleases resulting in DNA fragments of varying lengths. For fungal mitochondrial genomes, which are usually small [19–121 kb] and exist in relatively high copy number [102], RFLPs are usually easy to detect. Total DNA of the fungus is extracted and the mitochondrial DNA separated from nuclear DNA by centrifugation in CsCl gradients containing bis-benzimide [103]. The relatively small size of mitochondrial genomes results in a simple banding pattern when cut with restriction endonucleases. Mitochondrial RFLP markers have been used in a number of fungi although the extent of variation differs considerably.

Although variation may sometimes be detected between strains in ethidium bromide stained gels of restriction endonuclease digests of total genomic DNA, e.g. [104], labelled DNA probes are normally required. Total genomic DNA is extracted and digested with a restriction endonuclease. The complex array of DNA fragments are then size fractionated by agarose gel electrophoresis and transferred to a nylon or nitrocellulose filter (Southern blotting). The filter is then probed with a radioactive or non-radioactively labelled DNA probe and visualized by autoradiography or development of the filter respectively. The number of bands observed depends on the copy number of DNA sequences homologous to the probe and the number of restriction sites within the sequence.

Various types of DNA probe have been used to detect RFLPs, e.g. random genomic clones containing single copy [105] or repetitive sequences [106], rDNA [107] and oligonucleotide probes [108]. Jessop and colleagues, [Jessop *et al.*, unpub.] have recently used random genomic clones in the lambda vector EMBL3 to detect RFLPs in isolates of *M. anisopliae* from different geographical origins. Kosir and co-workers [109] probed DNA extracted from a wild-type isolate of *B. bassiana* and an ultraviolet light induced mutant which showed reduced virulence to grasshoppers, with a cloned β-tubulin gene from *N. crassa* and three random genomic clones. The heterologous probe, and two of the genomic clones, apparently failed to hybridize to the mutant whereas the third genomic clone revealed a RFLP between the wild-type and mutant. These results suggest that chromosomal changes in the mutant, including the deletion of some sequences, may be responsible for the reduction in virulence.

7.4.2 *Polymerase chain reaction methods*

The polymerase chain reaction (PCR) is an *in vitro* method for amplifying DNA sequences using very small amounts of target DNA [110]. The double-strand target DNA is denatured by heating briefly at high temperature (*c.* 95°C) and oligonucleotide primers are allowed to anneal to their complementary sequences at low temperature (dependent on probe length and composition). DNA extension from the annealed primers proceeds using dNTPs and a thermostable polymerase, e.g. *Taq* polymerase, at 72°C. Repeated cycles of denaturation, annealing and extension lead to the rapid amplification of target DNA sequences which can usually be resolved by conventional agarose or polyacrylamide gel electrophoresis.

It has been demonstrated recently that PCR can be used to identify DNA sequence polymorphisms by the amplification of random DNA segments with single oligonucleotide primers of arbitrary sequence [111, 112]. This randomly amplified polymorphic DNA (RAPD) can be visualized on ethidium bromide stained agarose gels as discrete bands which may provide a unique genetic 'fingerprint' for a particular genotype. This technique is being applied to a wide range of organisms including the entomopathogen *Metarhizium* (R. Millner, pers. commun.).

7.4.3 *Electrophoretic karyotyping*

Pulsed field gel electrophoresis (PFGE) is capable of separating DNA molecules ranging from about 20 kb to 12 000 kb [113, chapter 1]. Schwartz and Cantor [114] first applied this technique to the separation of the chromosomes of *Saccharomyces cerevisiae.* In conventional gel electrophoresis DNA molecules migrate in a single electric field. The migration rate of molecules less than approximately 20 kb is proportional to their size because they are sieved. Larger molecules are not sieved effectively and consequently migrate at similar rates. PFGE utilizes alternating electric fields which force the DNA molecules to change direction. The time required for a DNA molecule to change direction is size dependent with larger molecules requiring more time to reorientate. The duration of each alternating electric field (or pulse) determines the resolution. Total run times may last several days. The resolution of PFGE has been improved by modification to the original apparatus and protocol. One commonly used method is contour-clamped homogeneous electric field (CHEF) gel electrophoresis in which a hexagonal array of electrodes produces a uniform electric field across the gel resulting in straight running lanes and good resolution. DNA is usually prepared from fungi by embedding protoplasts in blocks of low melting point agarose. These are mixed with a detergent (e.g. sodium dodecyl sulphate), EDTA and proteinase K. The agarose blocks containing 0.5–20 μg DNA are then inserted into a well of the agarose gel and sealed with agarose.

PFGE has been reported for the entomopathogenic fungus *Paecilomyces fumosoroseus* [115]. Six chromosome-sized DNA molecules were resolved by CHEF electrophoresis with apparent sizes of 7.8, 6.2, 5.3, 4.4, 3.3 and 3.1 Mb. An electrophoretic karyotype has been determined for the cellulolytic *Trichoderma* sp., *T. reesei* [116]. Six chromosomes were separated with sizes ranging from 3.3 to 11.9 Mb. Pulsed field electrophoresis has also been used to study the molecular karyotypes of a wide range of plant pathogens and other fungi of commercial importance. Studies with field isolates of plant pathogenic fungi have revealed a surprising degree of variation in both chromosome number and length [e.g. 117, 118] suggesting that the electrophoretic karyotypes may prove to be valuable molecular markers for biocontrol fungi.

References

1. Hofte, H. and Whiteley, H.R. (1989) *Microbiol. Rev.* **53**: 242–255.
2. Lewis, J.A. and Papavizas, G.C. (1991) *Crop Protection* **10**: 95–105.
3. Papavizas, G.C. (1985) *Ann. Rev. Phytopathol.* **23**: 23–54.
4. Chet, I. (1987) In: *Innovative Approaches to Plant Disease Control* (Chet, I., ed.), John Wiley and Sons, Canada, Chapter 6.
5. Gillespie, A.T. and Moorhouse, E.R. (1989) In: *Biotechnology of Fungi for Improving Plant Growth* (Whipps, J.M. and Lumsden, R.D., eds), Cambridge University Press, Cambridge, Chapter 4.
6. Charnley. A.K. (1989) In: *Biotechnology of Fungi for Improving Plant Growth* (Whipps, J.M. and Lumsden, R.D. eds), Cambridge University Press, Cambridge, Chapter 5.
7. Elad, Y., Chet, I. and Henis, Y. (1982) *Can. J. Microbiol.* **28**: 719–725.
8. Ridout, C.J., Coley-Smith, J.R. and Lynch, J.M. (1986) *J. Gen. Microbiol.* **132**: 2345–2352.
9. Sivan, C.J. and Chet, I. (1989) *J. Gen. Microbiol.* **135**: 675–682.
10. Geremia, R., Jacobs, D., Goldman, G.H., Van Montagu, M. and Herrera-Estrella, A. (1991) In: *Biotic Interactions and Soil-Borne Diseases* (Beemster, A.B.R., Bollen, G.J., Gerlagh, M., Ruissen, M.A., Schippers, B. and Tempel, A., eds), Elsevier, Amsterdam, pp. 181–187.
11. Ulhoa, C.J. and Peberdy, J.F. (1991) *J. Gen. Microbiol.* **137**: 2163–2169.
12. Elad, Y., Chet, I., Boyle, P. and Henis, Y. (1983) *Phytopathol.* **73**: 85–88.
13. Elad, Y., Barak, R., Chet, I. and Henis, Y. (1983) *Phytopath. Z.* **107**: 168–175.
14. Cherif, M. and Benhamou, N. (1990) *Phytopathol.* **80**: 1406–1414.
15. St Leger, R.J., Charnley, A.K. and Cooper, R.M. (1987) *Arch. Biochem. Biophys.* **253**: 221–232.
16. Jassim, H.K., Foster, H.A. and Fairhurst, C.P. (1990) *Ann. Appl. Biol.* **117**: 187–196.
17. Jun, Y., Bridge, P.D. and Evans, H.C. (1991) *J. Gen. Microbiol.* **137**: 1437–1444.
18. Lewis, K., Whipps, J.M. and Cooke, R.C. (1989) In: *Biotechnology of Fungi for Improving Plant Growth* (Whipps, J.M. and Lumsden, R.D., eds), Cambridge University Press, Cambridge, Chapter 9.
19. Kim, K.K., Flavel, D.R. and Papavizas, G.C. (1988) *Phytopathol.* **78**: 488–492.
20. Kim, K.K., Flavel, D.R. and Papavizas, G.C. (1990) *Can. J. Microbiol.* **36**: 199–205.
21. Fravel, D.R. and Roberts, D.P. *Biocontrol. Science and Technology* (in press).
22. Charnley, A.K. (1984) In: *Invertebrate-Microbial Interactions* (Anderson, J.M., Rayner, A.D.M. and Walton, D.W.A., eds), Cambridge University Press, Cambridge, pp. 229–270.
23. Goettel, M.S., St Leger, R.J., Rizzo, N.W., Staples, R.C. and Roberts, D.W. (1989) *J. Gen. Microbiol.* **135**: 2223–2239.
24. Neville, A.C. (1984) In: *Biology of the Integument: 1 Invertebrates* (Boreiter-Hahn, J., Matoltsy, A.G. and Richards, K.S., eds), Springer-Verlag, Berlin, pp. 611–625.
25. Charnley, A.K. and St Leger, R.J. (1991) In: *The Fungal Spore and Disease Initiation in Plants and Animals* (Cole, G.T. and Hoch, H.C., eds), Plenum Press, New York and London pp. 267–286.
26. St Leger, R.J., Charnley, A.K. and Cooper, R.M. (1986) *J. Invertebr. Pathol.* **48**: 85–95.
27. St Leger, R.J., Cooper, R.M. and Charnley, A.K. (1986) *J. Gen. Microbiol.* **132**: 1509–1517.
28. St Leger, R.J., Cooper, R.M. and Charnley, A.K. (1986) *J. Invertebr. Pathol.* **46**: 166–167.
29. St Leger, R.J., Cooper, R.M. and Charnley, A.K. (1987) *Arch. Biochem. Biophys.* **258**: 121–131.
30. St Leger, R.J., Cooper, R.M. and Charnley, A.K. (1987) *J. Gen. Microbiol.* **133**: 1371–1382.
31. St Leger, R.J., Durrands, P.K., Charnley, A.K. and Cooper, R.M. (1988) *J. Invertebr. Pathol.* **52**: 285–294.
32. St Leger, R.J., Durrands, P.K., Charnler, A.K. and Cooper, R.M. (1988) *Arch. Microbiol.* **150**: 413–416.

33. St Leger, R.J., Staples, R.C. and Roberts, D.W. (1991) *J. Gen. Microbiol.* **137**: 807–815.
34. Cohen, B.L. (1973) *J. Gen. Microbiol.* **79**: 311–320.
35. Abbot, R.J. and Marzluf, G.A. (1984) *J. Bacteriol.* **159**: 505–510.
36. Howell, C.R. and Stipanovic, R.D. (1983) *Can. J. Microbiol.* **29**: 321–324.
37. Jones, R.W. and Hancock, J.G. (1987) *Can. J. Microbiol.* **33**: 963–966.
38. Weinding, R. and Emerson, O.H. (1936) *Phytopathol.* **26**: 1068–1070.
39. Howell, C.R. (1982) *Phytopathol.* **72**: 496–498.
40. Aluko, M.O. and Hering, T.F. (1970) *Trans. Br. Mycol. Soc.* **55**: 173–179.
41. Howell, C.R. (1987) *Phytopathol.* **77**: 992–994.
42. Roberts, D.P. and Lumsden, R.D. (1990) *Phytopathol.* **80**: 461–465.
43. Jones, R.W. and Hancock, J.G. (1988) *J. Gen. Microbiol.* **134**: 2067–2075.
44. Seigle-Murandi, F., Krivobok, S., Steiman, R. and Marzin, D. (1990) *J. Agric. Food Chem.* **38**: 1854–1856.
45. Dennis, C. and Webster, J. (1971) *Trans. Br. Mycol. Soc.* **57**: 25–39.
46. Dennis, C. and Webster, J. (1971) *Trans. Br. Mycol. Soc.* **57**: 41–48.
47. Claydon, N., Allan, M., Hanson, J.R. and Avent, A.G. (1987) *Trans. Br. Mycol. Soc.* **88**: 503–513.
48. Ghisalberti, E.L., Narbey, M.J., Dewan, M.M. and Sivasithamparam, K. (1990) *Plant and Soil* **121**: 287–291.
49. Peeters, H., Zocher, R., Madry, N., Oelrichs, P.B., Kleinkauf, H. and Kraepelin, G. (1983) *J. Antibiotics* **36**: 1762–1766.
50. Kodaira, Y. (1961) *Argic. Biol. Chem.* **25**: 261–262.
51. Suzuki, A., Kuyama, S., Kodaira, Y. and Tamura, S. (1966) *Agric. Biol. Chem.* **30**: 517–518.
52. Suzuki, A., Taguchi, H. and Tamura, S. (1970) *Agric. Biol. Chem.* **34**: 813–816.
53. Samuels, R.I., Charnley, A.K. and Reynolds, S.E. (1988) *Mycopathologia* **104**: 51–58.
54. Suzuki, A., Kawakami, K. and Tamura, S. (1971) *Agric. Biol. Chem.* **35**: 1641–1643.
55. Peeters, H., Zocher, R. and Kleinkauf, H. (1988) *J. Antibiotics* **41**: 552–559.
56. Zocher, R., Keller, U. and Kleinkauf, H. (1982) *Biochemistry* **21**: 43–48.
57. Katan, T., Dunn, M.T. and Papavizas, G.C. (1984) *Can. J. Microbiol.* **30**: 1079–1087.
58. Pontecorvo, G. (1956) *Ann. Rev. Microbiol.* **10**: 393–400.
59. Jackson, C.W. and Heale, J.B. (1987) *J. Gen. Microbiol.* **133**: 3537–3547.
60. Harman, G.E. (1991) *Crop, Prot.* **10**: 166–171.
61. Hocart, M. and Peberdy, J.F. (1989) In: *Biotechnology of Fungi for Improving Plant Growth*, Cambridge University Press, Cambridge, Chapter 11.
62. Stasz, T.E., Harman, G.E. and Weeden, N.F. (1988) *Mycologia* **80**: 141–150.
63. Harman, G.E., Taylor, A.G. and Stasz, T.E. (1989) *Plant Disease* **73**: 631–637.
64. Sivan, A. and Harman, G.E. (1991) *J. Gen. Microbiol.* **137**: 23–29.
65. Pe'er, S. and Chet, I. (1990) *Can. J. Microbiol.* **36**: 6–9.
66. Messias, C. and Azevedo, J. (1980) *Trans. Br. Mycol. Soc.* **75**: 473–477.
67. Al-Aidroos, K. (1980) *Can. J. Genet. Cytol.* **22**: 309–314.
68. Paccola-Meirelles, L. and Azevedo, J. (1991) *J. Invertebr. Pathol.* **57**: 172–176.
69. Heale, J.B., Isaac, J. and Chandler, D. (1989) *Pesticide Science* **26**: 79–92.
70. Drummond, J. and Heale, J.B. (1988) *J. Invertebr. Pathol.* **52**: 57–65.
71. Clarkson, J.M. and Heale, J.B. (1985) *Trans. Br. Mycol. Soc.* **83**: 345–350.
72. O'Garro, L.W. and Clarkson, J.M. *Plant Pathol.* (in press).
73. Fincham, J.R.S. (1989) *Microbiol. Rev.* **53**: 143–170.
74. Timberlake, W.E. and Marshall, M.A. (1989) *Science* **244**: 1313–1317.
75. Van Heeswijk, R. (1986) *Carlsberg Res. Commun.* **51**: 433–443.
76. Tsukuda, T., Carleton, S., Fotheringham, S. and Holloman, W.K. (1988) *Mol. Cell. Biol.* **8**: 3703–3709.
77. Powell, W.A. and Kistler, H.C. (1990) *J. Bacteriol.* **172**: 3163–3171.
78. Gems, D., Johnstone, I.L. and Clutterbuck, A.J. (1991) *Gene* **98**: 61–67.
79. Ballance, D.J., Buxton, F.P. and Turner, G. (1983) *Biochem. Biophys. Res. Commun.* **112**: 284–289.
80. Case, M.E., Schweizer, M., Cushner, S.R. and Giles, N.H. (1979) *Proc. Natl Acad. Sci. USA* **76**: 5259–5263.
81. Herrera-Estrella, A., Goldman, G.H. and Van Montagu, M. (1990) *Mol. Microbiol.* 4: 839–943.

82. Punt, P.J., Oliver, R.P., Dingermanse, M.A. Pouwels, P.H. and van den Hondel, C.A.M.J.J. (1987) *Gene* **56**: 117–124.
83. Goldman, G.H., Van Montagu, M. and Herrera-Estrella, A. (1990) *Curr. Genet.* **17**: 169–174.
84. Ossanna, N. and Mischke, S. (1990) *Appl. Environ. Microbiol.* **56**: 3052–3056.
85. McClung, C.R., Phillips, J.D., Orbach, M. and Dunlap, J.C. (1989) *Exp. Mycol.* **13**: 299–302.
86. Bernier, L., Cooper, R.M., Charnley, A.K. and Clarkson, J.M. (1989) *FEMS Microbiol. Lett.* **60**: 261–266.
87. Vollmer, S.J. and Yanofsky, C. (1986) *Proc. Natl Acad. Sci. USA* **83**: 4869–4873.
88. Goettel, M., St Leger, R.J., Bhairi, S., Jung, M., Oakley, B., Roberts, D. and Staples, R.C. (1990) *Curr. Genet.* **17**: 129–132.
89. Daboussi, M.J., Djeballi, A., Gerlinger, C., Blaiseau, P.L., Bouvier, I., Carson, M., Lebrun, M., Parisot, D. and Brygoo, Y. (1989) *Curr. Genet.* **15**: 453–456.
90. Judelson, H.S. and Michelmore, R.W. (1991) *Curr. Genet.* **19**: 453–459.
91. Scott-Craig, J.S., Panaccione, D., Cervone, F. and Walton, J.D. (1990) *The Plant Cell* **2**: 1191–1200.
92. Barnett, C.C., Berka, R.M. and Fowler, T. (1991) *Bio/Technology* **9**: 562–567.
93. Drahos, D.J. (1991) *Biotech. Adv.* **9**: 157–171.
94. Michelmore, R.W. and Hulbert, S.H. (1987) *Ann. Rev. Phyopathol.* **25**: 383–404.
95. deConti, E., Messias, C., Myrina, H., deSouza, L. and Azevedo, J.L. (1980) *Experientia* **36**: 293–294.
96. Riba, G., Bouvier-Fourcade, I. and Caudal, A. (1986) *Mycopathologia* **96**: 161–169.
97. St Leger, R.J., May, B., Allee, L.L., Frank, D.C. and Roberts, D.W. (in press).
98. Samuels, K.D.Z., Heale, J.B. and Llewellyn, M. (1989) *J. Invertebr. Pathol.* **53**: 25–31.
99. Mugnai, L., Bridge, P.D. and Evans, H.C. (1989) *Mycological Research* **92**: 199–209.
100. Stasz, T.E., Weeden, N.F. and Harman, G.E. (1988) *Mycologia* **80**: 870–874.
101. Stasz, T.E., Nixon, K., Harman, G.E., Weedon, N.F. and Kuter, G.A. (1989) *Mycologia* **81**: 391–403.
102. Grossman, L. and Hudspeth, M. (1985) In: *Gene Manipulations in Fungi* (Bennett, J.W. and Lasure, L., eds), Academic Press, Orlando, FL, pp. 65–103.
103. Garber, R. and Yoder, O.C. (1985) *Curr. Genet.* **8**: 621–628.
104. Coddington, A., Mathews, P.M., Cullis, C. and Smith, K.H. (1987) *J. Phytopathol.* **118**: 9–20.
105. Hulbert, S.H., Ilott, T.W., Legg, E.J., Lincoln, S.E., Lander, E.S. and Michelmore, R.W. (1988) *Genetics* **120**: 947–958.
106. Kistler, H.C., Momol, E.A. and Benny, U. (1991) *Phytopathol.* **81**: 331–336.
107. Garber, R., Turgeon, B.G., Selker, E.U. and Yoder, O.C. (1988) *Curr. Genet.* **14**: 573–582.
108. Meyer, W., Koch, A., Neimann, C., Beyermann, B., Epplen, J.T. and Borner, T. (1991) *Curr. Genet.* **19**: 239–241.
109. Kosir, J.M., MacPherson, J.M. and Khachatourians, G.G. (1991) *Can. J. Microbiol.* **37**: 534–541.
110. White, T.J., Arnheim, N. and Erlich, H.A. (1989) *Trends in Genetics* **5**: 185–189.
111. Williams, J., Kubelik, A., Lirak, K., Rafalski, J. and Tingey, S. (1990) *Nuc. Acids Res.* **18**: 6531–6535.
112. Welsh, J.E. and McClelland, M. (1990) *Nuc. Acids Res.* **24**: 7213–7218.
113. Mills, D. and McCluskey, K. (1990) *Mol. Plant-Microbe Interact.* **3**: 351–357.
114. Schwartz, D.C. and Cantor, C.R. (1984) *Cell* **37**: 67–75.
115. Shimizu, S., Nishida, Y., Yoshioka, H. and Matsumoto, T. (1991) *J. Invertebr. Pathol.* **58**: 461–463.
116. Gilly, J.A. and Sands, J.A. (1991) *Biotechnol. Lett.* **13**: 477–482.
117. Kinscherf, T. and Leong, S.A. (1988) *Chromosoma* **96**: 427–433.
118. McDonald, B.A. and Martinez, J.P. *Curr. Genet.* **19**: 265–271.

8 The application of biotechnology to the button mushroom, *Agaricus bisporus*

P.A. HORGEN

8.1 Introduction

Whereas filamentous fungi have been used for a number of different industrial processes, as discussed in this volume, the use of filamentous fungi as a food source has been restricted to those Basidiomycetes and a few Ascomycetes known as 'mushrooms'. Mushrooms are among the oldest of all cultivated crops. About 10 000 species are known, of which 600 are considered edible [1]. Only about 25 are produced on any type of commercial scale with *Agaricus bisporus*, the button mushroom, being by far the most economically important. This chapter deals with recent advances in mushroom science involving applications of numerous modern enabling technologies to *A. bisporus*.

8.2 Biological barriers to mushroom strain development

The development of new and better strains of mushrooms has occurred relatively infrequently over the last three centuries. The method of obtaining new strains has been mainly by phenotypic selection followed by production trials and the optimization of growth and fruiting conditions. The reason for the lack of a traditional 'crops breeding program' is the secondary homothallic nature of the life cycle of *Agaricus bisporus*. Figure 8.1 shows a simplified diagram of the life history of *A. bisporus*. The single most important impediment to the development of a systematic breeding program is the absence of any uninucleate propagules during the life history of *A. bisporus*. Whereas most gilled mushrooms produce uninucleate haploid basidiospores [2], nearly all of the basidiospores of *A. bisporus* are binucleate (Figure 8.1) and self-fertile, containing two nuclei of opposite mating types. Furthermore, there appears to be a controlled biological mechanism which ensures that the basidiospores contain nuclei of a compatible mating type [3, 4]. A very small and unpredictable number of basidiospores are uninucleate. Therefore, the isolation of homokaryons from basidiospores is a task of considerable difficulty. The standard methods that have been used to verify homokaryon

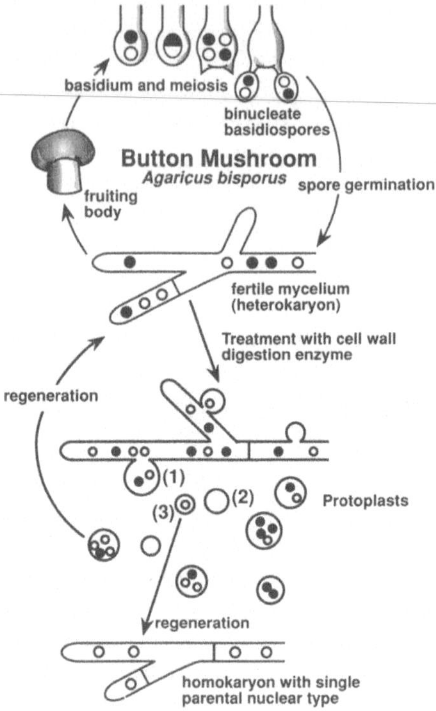

Figure 8.1 Diagrammatic representation of the secondarily homothallic life-cycle of *Agaricus bisporus* and the isolation of parental-type homokaryons by protoplast production and regeneration. The two homokaryotic nuclei are represented by open and closed circles.

status are extremely time-consuming (i.e., the ability to fruit) or somewhat problematic (i.e., slower growth rate of generating basidiospores).

8.3 Systematic isolation and verification of homokaryons

The production and regeneration of protoplasts have been utilized in a variety of different applications with higher plants and fungi [5]. In the button mushroom protoplast, production and regeneration have been developed and optimized by a number of workers [6–8]. Figure 8.1 illustrates the possible types of protoplasts that can be produced from the fertile heterokaryon. It is the type 3 protoplast that will regenerate into a homokaryotic mycelium (Figure 8.1) and this regenerate is most useful in a breeding program. In the author's laboratories, for most mushroom strains that have been examined, one out of every ten protoplasts produced is type 3 [9].

All the available evidence for the cultivated mushroom suggests that strains used today are genetically very uniform [10–13]. The author's group has been

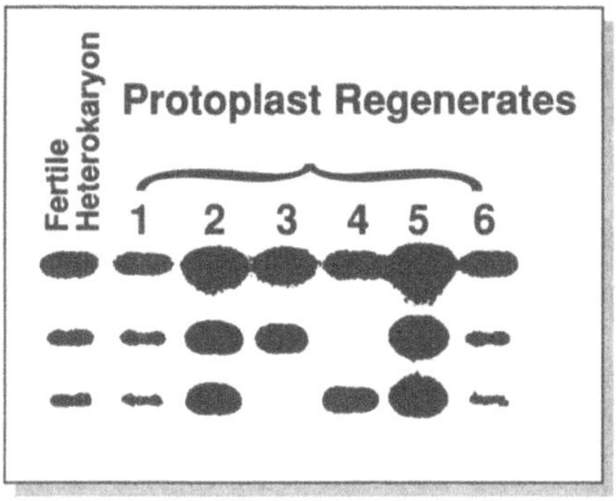

Figure 8.2 Verification of homokaryotic status by the use of RFLPs. The first lane illustrates the heteroallelic condition of DNA isolated from a fertile heterokaryon and probed with pAg33N10 [9]. Lanes 1–6 of this Southern blot illustrate the signal obtained when DNAs isolated from regenerated protoplasts are probed with pAg33N10. Lanes 3 and 4 illustrate the homoallelic condition of the two parental types; the other lanes show the typical heterokaryon pattern for this RFLP probe.

successful in using restriction fragment length polymorphisms (RFLPs) of mushroom DNA as indicators of genetic similarity (or diversity). RFLPs have become an extremely powerful technique in modern genetics [14]. When the technique is used to compare commercial isolates with wild-collected isolates of *A. bisporus*, it is nearly always found that for a given recombinant probe, there is more diversity in the wild-collected isolates [2]. This observation holds true for nuclear DNA [11, 15] and for mitochondrial DNA [12, 16]. It seems logical to attempt to utilize the genetic diversity in the wild in a breeding program for the button mushroom [9, 17].

In initial RFLP studies, it became apparent that for any specific cloned *A. bisporus* DNA fragment, multiple hybridization signals were usually observed which appeared as more than one band in hybridizations with nuclear DNAs [9, 15]. It was concluded that these multiple hybridization signals result from the heterozygous condition of the heterokaryotic (generally binucleate) mycelium which eventually produces the basidiocarp [9, 15]. Figure 8.2, lane 1 shows an example of this multiple banding phenomenon with one of the RFLP probes for a fertile heterokaryon representing the heteroallelic condition. Lanes 3 and 4 show only two bands which represent the homoallelic condition for each of the two parental nuclear types in a fertile heterokaryon. The methodology used to verify the homokaryotic condition of a regenerated protoplast is to choose one of our heteroallelic RFLP probes, and use this type of probe to screen DNA isolated from regenerated protoplasts. Figure 8.2 shows a typical experiment using RFLP probes to verify whether a regenerated protoplast has only one parental type nucleus (homokaryon) or possesses both parental type nuclei (fertile heterokaryon). The abundance of RFLPs showing a heterozygous condition has allowed easy identification of homokaryotic protoplast regenerates [2]. This procedure can be used to generate parental homokaryotic stock cultures from commercial strains, as well as wild-collected strains, and these parental stocks can be stored under liquid nitrogen for breeding purposes.

8.4 Production and verification of new mushroom strains

Mycelial plugs from homokaryotic stock cultures can be placed on petri plates and allowed to grow together. If the homokaryons are of opposite mating type, hyphal fusion will occur at the confluent zone [18]. Plugs of mycelium can be taken from this zone of anastomosis and allowed to grow on another petri plate for DNA extraction and RFLP analysis to confirm the new heterokaryotic status. Often a new 'hybrid' culture will grow very vigorously when compared to non-mated mycelium. These methodologies are described in Castle *et al.* [9] and Horgen and Anderson [2]. The approaches described above are presently being utilized by a number of spawn companies and mushroom experimental stations internationally.

8.5 A genetic linkage map and karyotype analysis for the button mushroom

Approaching mushroom genetics using classical methods has proved to be extremely difficult and for the most part relatively uninformative [4]. A limited number of genetic markers, including allozymes, auxotrophic lesions and morphological aberrations have been reported for *A. bisporus*. These markers enable mushroom scientists to carry out some initial genetic characterizations of the button mushroom. Recombination was followed with auxotrophic markers during meiosis [3]. Allozyme markers were first used to identify homokaryons with a high degree of success by Royse and May [10]. Allozyme markers were also utilized to provide limited evidence for independent assortment and genetic linkage during meiosis in *A. bisporus* [19].

DNA markers in the button mushroom were first described by Castle *et al.* [11] and these markers have been utilized in the ways described above. It has been possible with the development of the RFLP markers and more recently with randomly amplified polymorphic DNAs (RAPD markers), to begin to address more critically genetic issues in *A. bisporus* in a comprehensive manner [20]. The markers represent a large and potentially limitless set of characters which will allow mushroom scientists to address complex genetic issues, for example: What is the overall structure of the nuclear genome? How many linkage groups and chromosomes exist for the button mushroom? How do specific markers behave during meiosis? Do chromosomes assort with full independence? Do crossovers occur with a high or low frequency? Perhaps the most important issue for the mushroom industry is, can the agronomic phenotypes of the offspring be associated with the inheritance of specific genetic markers and unique chromosomal regions?

The Mushroom Research Group at the University of Toronto, with support from the mushroom industry, has produced a genetic linkage map for *A. bisporus* utilizing allozyme, RFLP and RAPD markers. This map was initially described by Kerrigan and co-workers [20] who also provided further detail later [21]. Figure 8.3 shows our linkage map with 66 genetic markers mapped to 13 different linkage groups. Two approaches to linkage analysis were followed. One involved a direct statistical test of joint segregation for all pairs of genetic markers, using the chi-square test [21]. A second approach involved an indirect statistical test, the comparison of (vanishingly small) likelihoods of exactly reproducing the actual data set given alternative hypothetical linkage relationships. A Mapmaker Macintosh (a generous gift of Scott Tingey of DuPont Corporation) was used to compute and compare the logs of these likelihood ratios (LOD scores). The results of the study suggested both joint and independent segregation [20, 21]. The frequency of crossing over was low, relative to other eukaryotes, possibly as a result of selection favoring retention of heteroallelism among heterokaryotic progeny. A number of interesting observations are discussed by Kerrigan and co-workers [21].

Associated with the linkage mapping study, protoplasting methods for *A. bisporus* have been refined to enable separation of chromosomal-sized DNA [8]. Utilizing CHEF electrophoresis, 13 chromosomes have been separated in an electrophorectically generated karyotype of the button mushroom (Figure 8.4). It has been found that the karyotype of two homokaryons may be highly polymorphic but that the fertile heterokaryon from a mating of two compatible homokaryons shows a combination of the two karyotypes [8] (Figure 8.4). A number of RFLP and RAPD genetic markers, as well as the rDNA repeat to specific chromosomes, have been located [8].

8.6 Molecular genetic studies on the mitochondria of *Agaricus*

While the size of the mitochondrial (mt) genome in eukaryotes ranges from 15 kbp (kilo base pairs) in animals to nearly 2500 kbp in higher plants, within the fungi the range is much narrower (18.9 kbp–178 kbp) with the largest organelle genome being reported for *Agaricus bitorquis* [12]. In spite of this dramatic difference in size, essentially similar sets of genes are encoded on all mitochondrial genomes [22]. Very little information exists on mitochondrial transmission and mitochondrial inheritance in basidiomycetes. Some data exist for *Armillaria* [23], *Coprinus* [24] and *Agaricus* [25]. Furthermore, a physical map for *Agaricus bisporus* has been generated [26]. While initial studies suggest very little polymorphism in the mt genome of the button

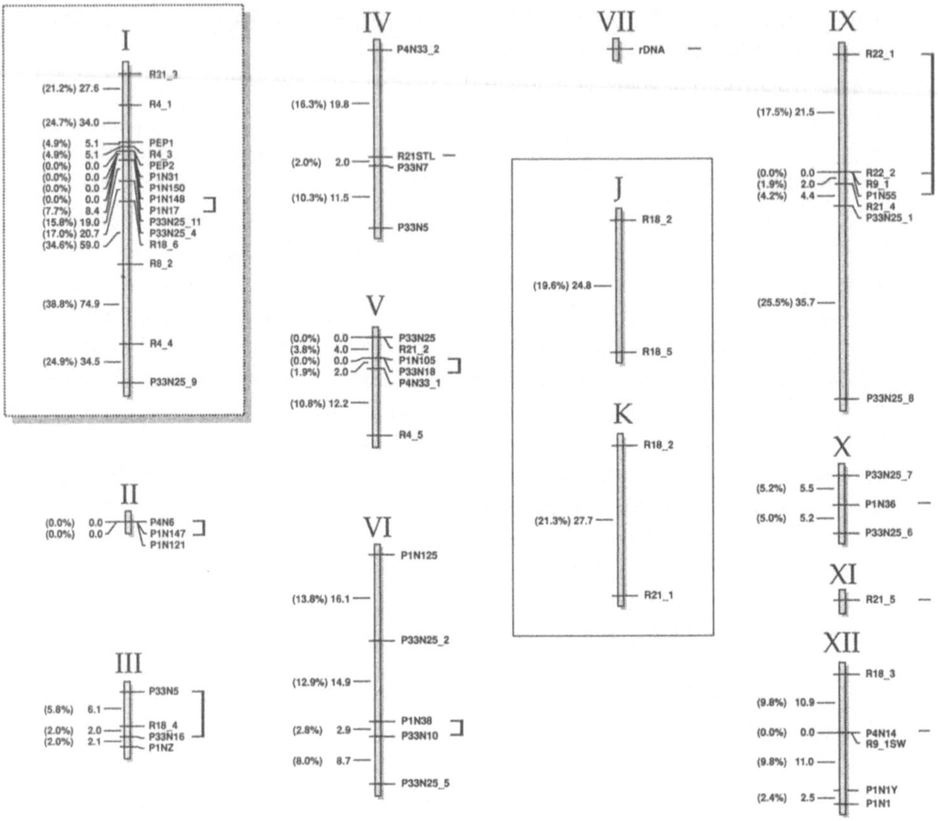

Figure 8.3 Genetic linkage map of *Agaricus bisporus*. The current linkage map shows 13 linkage groups which have been mapped to 11 chromosomes. There are two linkage groups which have not been mapped to any chromosome (J, K). Four different markers have been used: allozymes (PEP1 and PEP2) and RLFPs are designated by the letter P; RAPD markers are designated by the letter R; RAPD markers that have been cloned and used in Southern hybridizations are designated by the letters R and S. Details of this analysis are described in [20].

mushroom [12], more recently a number of polymorphic types have been reported [16]. In screening a number of wild-collected isolates of *A. bisporus*, a number of key mt polymorphisms have been identified [17].

Studies have been initiated on the mt inheritance of *A. bisporus*. It is entirely possible that mt DNA type can affect strain performance of *A. bisporus* when grown under commercial conditions. Preliminary data suggest that mt genotype can affect the phenotypic characteristics such as colony growth rates on petri plates. Three different mt genotypes, Ag50 type (Ag85) [25], the Ag2 type, which represents an approximately 20 kbp deletion of the standard Ag50 type [26], and a wild-collected mt genotype, Ag89 have been arbitrarily selected. Homokaryons were isolated from protoplasts of the three fertile

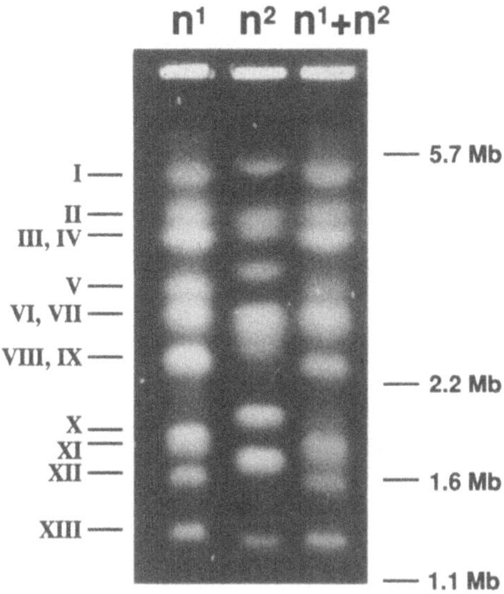

Figure 8.4 Electrophoretic karyotype of *Agaricus bisporus*. The first two lanes, n1 and n2, show separation of the chromosomal DNAs from two homokaryons. Lane 3, n1 + n2, shows a compatible mating and a new fertile heterokaryon. Separation conditions for the chromosomal DNAs are as described in [8].

heterokaryons carrying the three different mt genotypes by the methods described earlier in this chapter. Nuclear RFLP differences were used to verify homokaryon (Figure 8.5). Five of the six possible homokaryotic types from the three fertile heterokaryons used were isolated.

The results obtained to date suggest the following:

(i) Whenever an Ag85 or an Ag89 homokaryon is crossed with an Ag2 type, the new hybrid strains always carry the Ag2 mitochondrial type based on RFLP analysis (Figure 8.5) and actual fruiting trials to verify the heterokaryotic state followed by RFLP analysis.
(ii) If Ag85 homokaryons are mated with Ag89, the resultant hybrid always shows the Ag85 mt genotype.

This mitochondrial inheritance study, which is in its final stages, illustrates that the breeding approach described in this chapter is useful in both applied and basic mushroom scientific needs (Figure 8.5).

8.7 Extrachromosomal elements in *Agaricus*

Many isolates of wild *A. bitorquis* possess autonomously replicating plasmids of different sizes [27]. Two plasmids, pEM and pMPJ, have been examined in detail and have been found to be linear molecules residing within the

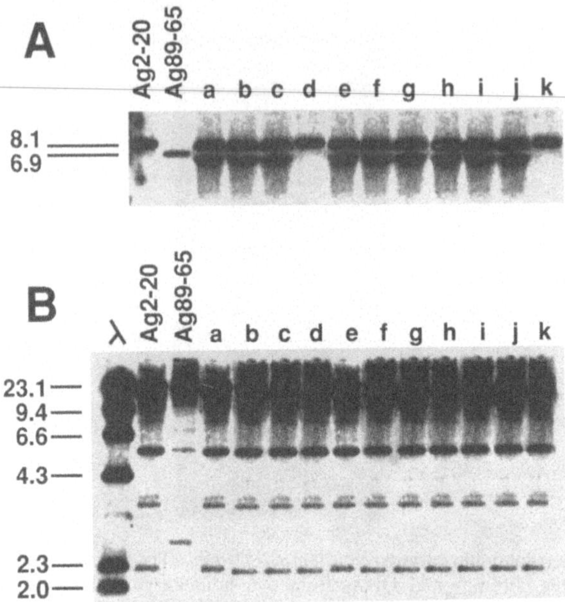

Figure 8.5 Mitochondrial inheritance in a cross of Ag2 homokaryon and an Ag89 homokaryon. (A) Verification of heterokaryon status by nuclear RFLP analysis. The two homokaryon lanes show the homoallelic signal for this nuclear probe. Lanes a–k are putative heterokaryons. All but lane d and lane k were shown to be heterokaryons. (B) Verification of mt genotype utilizing RFLP analysis. The two homokaryon lanes show the RFLP profiles of two mt genotypes followed in this cross. Lanes a–k all show the Ag2 mt genotype.

mitochondrion [27, 28]. These plasmids have structural and genetic similarities to linear plasmids found in other fungi and higher plants including the possession of two open reading frames (ORFs) that could encode virus-like DNA and RNA polymerases (Figure 8.6). To date, no plasmids have been found in *A. bisporus*. Mitochondrial genomic sequences exist within *A. bisporus* which are homologous to the ORF encoding an RNA polymerase on pEM [29]. Analysis of the pEM homologous sequence from *A. bisporus* mitochondria suggests that the mitochondrial sequence is related, but not derived from, the pEM sequence [29].

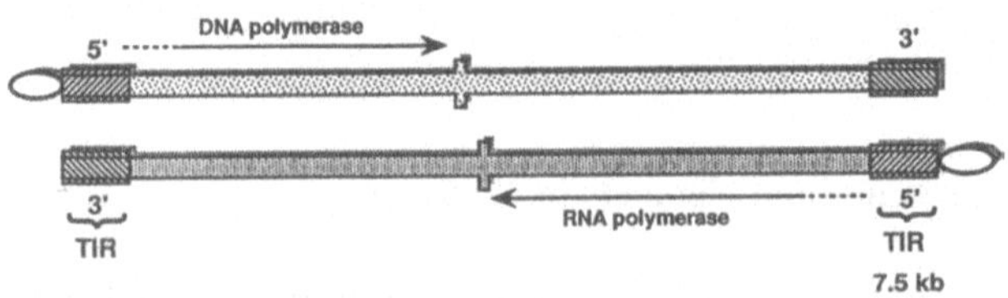

Figure 8.6 Diagrammatic representation of mitochondrial plasmid pEM. The diagram illustrates the terminal inverted repeats (TIRs) and the terminal proteins (ovals) attached at the 5′ ends of the molecule as well as the two open reading frames for the DNA and the RNA polymerases [29].

8.8 Conclusions

The cultivated mushroom, *Agaricus bisporus*, is one of the most important agronomic vegetable crops produced on a world-wide basis. Because of the life history of *A. bisporus*, the difficulty of finding and identifying it in the wild, and because of the cultural practices within the industry over the last 300 years, there is very little genetic diversity in the cultivated strains used today. Since *A. bisporus* is a difficult organism to deal with experimentally, little information has accumulated on the basic biology of this commercially important filamentous fungus. Furthermore, breeding approaches with *A. bisporus* have been based on reasonably imprecise and somewhat unreliable methodologies. Biotechnology (in the modern sense) is having and will continue to have a profound effect on mushroom science [30–32]. As described in this chapter, protoplast technology and DNA fingerprint (RFLP) technology have now made possible systematic and reproducible breeding with *A. bisporus*. As our understanding of the genetics of *A. bisporus* improves (utilizing these non-traditional approaches), the ability to handle *A. bisporus* with regard to genetic improvement will increase dramatically. Because *A. bisporus* is a filamentous fungus and is manipulated in the laboratory like other filamentous fungi, the speed at which certain experimental approaches can be applied is much greater than is possible for most green crop vegetables. It is the author's view that biotechnology significantly affects, and will continue to affect, the mushroom industry during the next decade.

References

1. Ammirati, J.P., Traquair, J.A. and Horgen, P.A. (1985) *Poisonous Mushrooms of Canada*, Fitshenry and Whiteside, Agriculture Canada, Toronto, 396 pp.
2. Horgen, P.A. and Anderson, J.B. (1991) Biotechnology and edible mushrooms. In: *Biotechnology and Filamentous Fungi* (Finkelstein, D. and Ball, C., eds). Butterworths (in press).
3. Raper, C.A., Miller, R.E. and Raper, J.R. (1972) *Mycologia* **64**: 1088–1117.
4. Elliott, T.J. (1985) The genetics and breeding in species of *Agaricus*. In: *The Biology and Technology of Cultivated Mushrooms* (Flegg, P., Spencer, D. and Wood, D.A., eds). Wiley, Chichester, pp. 110–129.
5. Peberdy, J.F. (1989) *Mycological Res.* **93**: 1–20.
6. Anderson, J.B., Petsche, D., Herr, F. and Horgen, P.A. (1984) *Can. J. Bot.* **62**: 1884–1887.
7. Sonnenberg, A.S., Wessels, J.G. and van Grensven, L.J. (1988) *Curr. Microbiol.* **17**: 285–291.
8. Royer, J., Hintz, W., Kerrigan, R. and Horgen, P. (1991) Electrophoretic karyotypic analysis of *Agaricus bisporus. Genome* (in press).
9. Castle, A.J., Horgen, P.A. and Anderson, J.B. (1988) *Appl. Environ. Microbiol.* **54**: 1643–1648.
10. Royse, D.J., and May, B. (1982) *Mycologia* **74**: 93–102.
11. Castle, A.J., Horgen, P.A. and Anderson, J.B. (1987) *Appl. Environ. Microbiol.* **53**: 816–822.
12. Hintz, W.E.A., Mohan, M., Anderson, J.B. and Horgen, P.A. (1985) *Curr. Genet.* **9**: 127–132.
13. Kerrigan, R.W. and Ross, I.K. (1989) *Mycologia* **81**: 433–443.
14. Patterson, A., Lander, E., Hewett, J., Patterson, S., Lincoln, S. and Tanksley, S. (1989) *Nature* **335**: 721–730.
15. Summerbell, R.A., Castle, A., Horgen, P.A. and Anderson, J.B (1989) *Genetics* **123**: 293–300.
16. Hintz, W.E.A., Anderson, J.B. and Horgen, P.A. (1989) *Genome* **32**: 173–178.

17. Kerrigan, R.W., Horgen, P.A. and Anderson J.B. (1991) The California *Agaricus bisporus* population comprises two interbreeding ancestral elements. *Systematic Botany* (in press).
18. Horgen, P.A., Jin, T. and Anderson, J.B. (1991) The use of protoplast production, protoplast regeneration and restriction fragment length polymorphisms in developing a systematic and highly reproducible breeding strategy for *Agaricus bisporus* In: *Proceedings for the First International Seminar on Mushroom Science.* Purdoc Wageningen, The Netherlands, pp. 62–72.
19. Spear, M.C., Royse, D.J. and May, B. (1983) *J. Heredity* **74**: 417–420.
20. Kerrigan, R.W., Horgen, P.A. and Anderson, J.B. (1991) A genetic linkage map for *Agaricus bisporus.* In: *Proceedings for the First International Seminar on Mushroom Science.* Purdoc Wageningen, The Netherlands, pp. 31–36.
21. Kerrigan, R.W., Royer, J., Baller, L., Kohli, Y., Horgen, P.A. and Anderson, J.B. (1991). A genetic map of the cultivated mushroom, *Agaricus bisporus.* Submitted to *Genetics.*
22. Dawson, A.J., Hodge, T.P., Isaac, P.G., Leaver, C.J. and Lonsdale, D.M. (1986) *Curr. Genet.* **10**: 561–564.
23. Smith, M.L., Duchesne, L.C., Bruhn, J.N. and Anderson, J.B. (1990) *Genetics* **126**: 575–592.
24. May, G. and Taylor, J. (1988) *Genetics* **118**: 213–220.
25. Hintz, W.E.A., Anderson, J.B. and Horgen, P.A. (1988) *Genetics* **119**: 35–41.
26. Hintz, W.E.A., Anderson, J.B. and Horgen, P.A. (1988) *Curr. Genet.* **14**: 43–49.
27. Meyer, R.J., Hintz, W.E.A., Mohan, M., Robison, M., Anderson, J.B. and Horgen, P.A. (1988) *Genome* **30**: 710–716.
28. Mohan, M., Meyer, R.J., Anderson, J.B. and Horgen, P.A. (1984) *Curr. Genet.* **8**: 613–619.
29. Robison, M., Royer, J. and Horgen, P.A. (1991) *Curr. Genet.* **19**: 495–500.
30. Elliott, T.J. (1987) Genetic engineering and mushrooms. *MGA Conference Proceedings 87*, pp. 526–529.
31. Horgen, P.A. and Anderson, J.B. (1989) Biotechnological advances in mushroom science. *Mushroom Science XII, Part 1*, pp. 63–73.
32. Wood, D.A. (1989) Mushroom biotechnology. *International Industrial Biotechnology* **9:1** pp. 5–8.

9 The application of molecular genetics to oriental mushrooms

K. SHISHIDO

9.1 Introduction

Mushroom is the term given to filamentous fungi (usually Basidiomycetes, but sometimes Ascomycetes) which form a fruiting body. This structure is easily visible to the naked eye and frequently the fruiting body itself is called 'mushroom'. The total number of mushrooms reported world-wide amounts to around 16 000 species. Currently, 33 mushroom species are cultivated on an economical scale and about two thirds of them are edible mushrooms such as *Agaricus bisporus*, *Lentinus edodes*, *Volvariella volvacea*, *Flammulina velutipes*, *Pleurotus* spp., *Pholiota nameko*, *Lyophyllum ulmarium*, *Auricularia* spp. *Tremella* spp., *Grifola frondosa*, *Tuber melanosporum* and *Tricholoma matsutake* (all of these are basidiomycetes except for *T. melanosporum*). Mushrooms are typical of natural fungal food and contain various chemical components such as protein, vitamins, minerals etc. A balanced diet of animal, vegetable, and fungal foods is considered to be healthy.

Agaricus bisporus (Lange) Sing. [= *A. brunnescens* Pk.] is the most popular cultivated mushroom, especially in the western hemisphere. The annual world production of this fungus is the highest among all edible mushrooms (chapter 8). *Lentinus edodes* (Berk.) Sing. ('Shiitake' in Japanese) is the second most important cultivated mushroom and is a typical oriental mushroom. This chapter provides a general description of the cultivated edible mushrooms popular in oriental countries, particularly Japan, and of current studies of their biochemistry and molecular biology.

9.2 Description of oriental edible fungi

9.2.1 Lentinus edodes (*Shiitake*) (*Figure 9.1A*)

Lentinus edodes [1] is the most common scientific name for Shiitake but *Lentinula edodes* (Berk.) Pegler is also used especially in the United States [2]. Shiitake mushrooms consist of a dark brown pileus (cap) and a cream-coloured stipe (stem). The production of the Shiitake mushroom in 1981 was around one quarter of that of *A. bisporus*; Japan alone produces around

160×10^6 kg (~US$1 billion) per year of Shiitake, corresponding to 95% of world-wide production. Japanese Shiitake cultivation began more than three hundred years ago when wild Shiitake strains were grown on fallen forest broadleaf trees of the *Fagaceae* family, e.g. the shii tree which is closely related to oak (Shiitake literally means 'shii tree mushroom'). It is distributed widely in Japan, and also in southwestern parts of China, Taiwan, Borneo, Indonesia, in the mountains of Papua and New Guinea, New Zealand, and Middle South America. Commercial large-scale cultivation of Shiitake in Japan began in the early 1940s with the development of new inoculation techniques. Small diameter hardwood logs, especially oaks, beech and chestnut are the preferred substrates for Shiitake cultivation. Trees are usually felled in the winter. In the early spring, logs are cut and inoculated with pieces of wood overgrown with the Shiitake fungus ('spawn'). After an incubation period of 1 to 2 years, mushrooms are produced for periods of 3 to 5 years, usually during the spring and autumn. Currently in Japan, however, new varieties of the Shiitake fungus, created by classical genetic breeding, permit production of mushrooms all year round. Nutritional and environmental factors critical to mycelial growth and fruiting of *L. edodes* have been extensively studied [3–6].

There are many reasons for Shiitake's popularity. For instance, when cooked it imparts a full-bodied aromatic and pleasant flavour to the dish while maintaining its own original colour and chewy texture. Shiitake mushroom is known to produce a flavour-enhancer, guanosine 5′-monophosphate [6, 7], and the aroma-bearing 'lenthionine' [8–11]. Shiitake also contains ergosterol which, when exposed to ultraviolet light (or sunlight), is converted to vitamin D_2 [12, 13]. In Japan, Shiitake is dried by heating after treatment with ultraviolet light and then marketed as a source of vitamin D. The heat treatment also enhances its flavour. If treated with sufficient ultraviolet light, 1 g of dried Shiitake can supply 400 IU (International Units), the United States Department of Agriculture (USDA) adult minimum daily requirement of vitamin D [4]. Dried Shiitake rehydrates well and the colour, shape and texture of fresh mushrooms are recovered. Dried Shiitake is exported to Hong Kong, Singapore and the US. However, China recently began commercial cultivation of Shiitake mushroom ('shiang-gu', which means 'fragrant mushroom') and exports dried Shiitake to foreign countries including, surprisingly, Japan. Shiitake cultivation is now beginning to expand rapidly throughout the western hemisphere. There is scientific evidence that Shiitake strains may synthesize chemical compounds with potential medical value, such as polysaccharides which have antitumour activity [14, 15] and an adenine derivative which has hypocholesterolemic activity [16–18].

9.2.2 Flammulina velutipes ('*Enokitake*') (*Figure 9.1 B1 and B2*)

The annual world-wide production of the Enokitake mushroom in 1981 was 65×10^6 kg. Japan alone produced 51×10^6 kg of this species. In Japan, unlike the almost constant level of Shiitake production, the annual production of Enokitake has steadily increased, reaching 78×10^6 kg ($\sim$US$370 million) in 1988. Enokitake mushrooms, cultivated on hardwood logs of dead broadleaf trees such as persimmon, hackberry, mulberry and poplar, consist of a yellowish dark brown cap and dark brown stem with woolly hairs (Figure 9.1 B1). Large-scale Enokitake cultivation is done in darkness in a culture bottle containing sawdust-rice bran (3:1) mixture with 60–65% moisture. Interestingly, the fungus forms cream-coloured slender fruit bodies in the bottle (Figure 9.1 B2). Enokitake has a distinctive aromatic flavour and texture, and is often used in soup. Since it is rich in vitamin B_1, niacin, vitamin C and ergosterol, and suggested to have potential antitumour activity [19], the Enokitake mushroom is used widely in folk medicine.

9.2.3 Volvariella volvacea *Sing.* (*'Fukurotake'*) (*Figure 9.1C*)

The world-wide production of the Fukurotake mushroom was 65×10^6 kg in 1981, similar to Enokitake production levels. Fukurotake grows well on rice straw-cotton waste, cotton waste-rice bran and rice straw-dried banana leaves at relatively high temperatures (> 30°C). Fukurotake literally means small pouched mushroom; the initial stage of fruit body has the appearance of a small white egg. The egg splits open, resulting in an appearance of a greyish dark brown mature fruit body. The Fukurotake mushroom is produced abundantly in the south of China as well as Taiwan, Hong Kong, the Philippines, Thailand and Indonesia. Large-scale cultivation is not carried out in Japan. Fukurotake (known as 'dai-mu-gu' in China) has a pleasant flavour with a high nutritive value and it is frequently used for Chinese dishes in combination with *A. bisporus* and *Auricularia* spp.

9.2.4 Pleurotus *spp.* (*Figure 9.1D*)

The world-wide production of *Pleurotus* spp. in 1981 was 40×10^6 kg of which Japan alone produced about 14×10^6 kg. Similar to *Flammulina velutipes*, the production of *Pleurotus* in Japan has steadily increased, reaching 35×10^6 kg ($\sim$US$190 million) in 1988. *Pleurotus ostreatus* Quél. ('Hiratake') is the major species used world-wide. In European countries this species is commercially cultured on a compost of vegetable manure and heap, while in Japan it is usually cultured on sawdust-rice bran mixture and sometimes on hardwood logs of hackberry, willow or poplar. Other species are *Pleurotus* sp. Florida, *Pleurotus cornucopia* Roll., *P. eryngii* Quél. (known as 'cardarella' in Italy), *P. porrigens* Sing., *P. dryinus* Quél., *P. ulmarius* Fr., *P. sajor-caju* Sing. and *P. cystidiosus* Mill ('pau-ku' in Taiwan). *Pleurotus* mushrooms (greyish dark brown cap and cream-coloured stem) have a moderate taste, flavour and texture. Additionally, they have a high nutritive value and consequently are used in soups and stews.

9.2.5 Pholiota nameko (*I. Ito*) (*S. Ito et Imai*) ('*Nameko*') (*Figure 9.1E*)

Nameko is exclusively cultured in Japan. The production was 20×10^6 and 21×10^6 kg, in 1981 and 1988 respectively. Nameko mushrooms (glossy brown cap and cream-coloured stem) grow well on hardwood logs of broadleaf trees, but large-scale cultivation is carried out on sawdust-rice bran mixture. Nameko produces a distinctive sticky substance which imparts a pleasant flavour. Japanese people consume this mushroom frequently as a miso (soybean paste) soup.

9.2.6 Lyophyllum ulmarium ('*Shirotamogitake*') (*Figure 9.1F*)

Large-scale cultivation of Shirotamogitake is performed only in Japan. A sawdust-rice bran mixture is used for cultivation. The production was 16×10^6 kg in 1988. Shirotamogitake is an elegant fruit body of brownish cream-coloured cap with brown spotted pattern and is used widely in Japanese dishes.

9.2.7 Auricularia auricula-judae, Auricula polytricha ('*Kikurage*') *and* Tremella *spp.* ('*Shiro-Kikurage*') (*Figure 9.1 G1 and G2*)

Kikurage literally means tree (ki) and jellyfish (kurage) since its shape, taste and texture resemble a jellyfish. The world-wide production of Kikurage (violet dark brown) and Shiro-kikurage was 15×10^6 kg in 1981; the majority being produced in Taiwan. Kikurage ('mu-er' in Chinese) blends well with Chinese dishes and Shiro-kikurage ('bai-mu-er') is usually eaten as a soup. In Taiwan, tradition maintains that beauty is obtained by eating Shiro-kikurage.

9.2.8 Grifola frondosa ('*Maitake*') (*Figure 9.1H*)

Maitake (greyish dark brown) has a distinctive shape of fruit body. This species is distributed in Japan, Asia, Europe and also in North America. However, large-scale cultivation with sawdust-rice bran mixture is performed only in Japan, the annual production being 4.8×10^6 kg. Maitake has a very good flavour, taste and texture.

9.2.9 Tricholoma matsutake (*S. Ito et Imai*) *Sing.* ('*Matsutake*')

Matsutake produces the best flavour of all mushrooms, but is abnormally expensive as artificial cultivation has not yet succeeded. This species can grow on the roots of pine trees. The Japanese production was about 0.4×10^6 kg in 1988 (US$68 million). Korea also produces Matsutake mushrooms but the majority are exported to Japan.

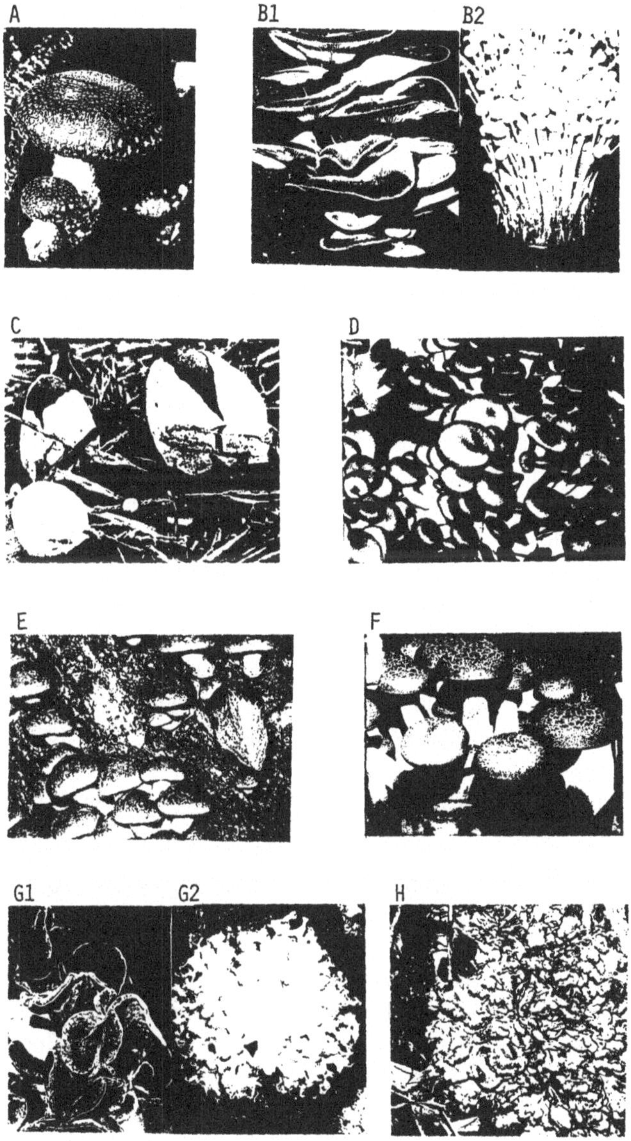

Figure 9.1 Edible oriental mushrooms. (A) *L. edodes*; (B1) and (B2) *F. velutipes*; (C) *V. volvacea*; (D) *P. ostreatus*; (E) *P. nameko*; (F) *L. ulmarium*; (G1) *A. auricula-judae*, (G2) *T. fuciformis*; (H) *G. frondosa*. Photographs courtesy of: Meiji Seika Kaisha, Japan (A), Tottori Mycol. Inst., Japan (B1, B2 to H).

9.3 Studies of the molecular biology and biochemistry of oriental mushrooms

9.3.1 *Degradative enzymes secreted by mushrooms*

L. edodes, *F. velutipes*, *P. ostreatus*, and *G. frondosa* are typical white-rot basidiomycetes. This group are highly degradative and characteristically bleach wood to lighter colours by metabolizing aromatic compounds and lignin. They also degrade the other major polymers in lignocellulose, such as cellulose and hemicellulose (chapter 4). Being capable of efficiently degrading lignocellolose, white-rot basidiomycetes are an important link in the carbon cycle. Degradative enzymes secreted by edible fungi have been extensively studied using *L. edodes* by Leatham [20] and Tokimoto and colleagues [21, 22]. These reactions include

(i) cellulose degradation to cellobiohydrolase (exocellulase), β-D-glucanase (endocellulase) and β-(1,4)-D-glucosidase (cellobiase)
(ii) hemicellulose degradation to β-D-xylanase, β-D-mannanase, α- and β-D-galactanase, β-(1,3) and (1,6)-D-glucanase (laminarinase), β-D-xylosidase, α-D-glucuronosidase, α-L-arabinosidase, α- and β-D-mannosidase, α- and β-D-galactosidase and β-D-glucosidase
(iii) lignin degradation to ligninase (Mn-dependent peroxidase) and laccase
(iv) starch and pectin degradation to (gluco)amylase and pectinase
(v) carbohydrate accessory enzymes to deacetylase (esterase) and demethylase
(vi) protein and fat degradation to acid proteinase, metallo proteinase, and lipase.

Substrates are broken down by these degradative enzymes and the nutrients liberated are absorbed and transported through hyphae. The degradative enzymes of lignin are useful for removing lignin from biomass materials and are an attractive objective of biotechnology and molecular genetics. Several ligninase genes have been cloned and sequenced from the non-edible white-rot fungus *Phanerochaete chrysosporium* [23–27 and chapter 4]. The genes have the coding capacity of 371 or 372 amino acids (interrupted by eight small introns), of which a 27 or 28 amino acid sequence corresponds to the signal peptide.

9.3.2 *Substances that regulate fruit-body formation in basidiomycetes*

The mechanism of fruiting in the basidiomycetes is an attractive scientific topic for investigation. Fruiting is known to be triggered by environmental factors such as light, temperature, humidity, nutrition and air [28]. Some chemical substances have been reported to induce fruiting in non-edible basidiomycetes, e.g. adenosine 3′,5′-cyclic monophosphate (cAMP) in *Coprinus macro-*

rhizus (=*cinereus*) [29, 30], cerebroside in *Schizophyllum commune* [31, 32], anthranilic acid and cyclooctasulfur in *Favolus arcularius* [33]. In *L. edodes*, the high level of intracellular cAMP is closely related to the onset of fruiting and/or primordium formation [34]. cAMP may be rather actively metabolized in the stage of maturing fruiting bodies [34].

Edible basidiomycetes, such as *L. edodes*, *P. ostreatus* and *F. velutipes*, produce acid proteinases and metallo proteinases as described above. The inhibitor of the former enhances growing of fruit bodies and the inhibitor of the latter stops mycelial development at the stage of primordia [35], showing the direct correlation of metallo proteinase with growing of fruit bodies.

9.3.3 *The* ras *gene of* L. edodes

The *ras* genes, which were first identified as oncogenes in Harvey (v-H-*ras*) and Kirsten (v-K-*ras*) sarcoma viruses [36], are an ubiquitous eukaryotic gene family [37]. A *ras* gene homologue has been isolated from *L. edodes* [38]. The *L. edodes ras* gene (named *Le.ras*) has a coding capacity of 217 amino acids interrupted by six small introns. The deduced *Le.*Ras protein (Figure 9.2) exhibits the highest amino acid similarity to the fission yeast *Schizosaccharomyces pombe* RAS protein (219 amino acids) [39]: 86% homology in the amino-terminal 80 amino acid sequence and 74% homology in the following 80 amino acid region [38]. The *ras* gene products of the budding yeast *S. cerevisiae*, have been shown to stimulate adenylate cyclase activity and participate in the regulation of the cell cycle via modulation of cAMP levels [40, 41]. However the *ras* gene product of *S. pombe*, appears not to be directly involved in modulation of adenylate cyclase activity [42]. In the cellular slime mould *Dictyostelium discoideum*, a transient rise in intracellular cAMP leads to a decrease in levels of *ras* gene *Dd.ras*G mRNA and simultaneously stimulates expression of another *ras* gene *Dd.ras* [43–45]. The *Le.ras* gene is transcribed at similar levels in fruit-body formation, suggesting no direct correlation of the *Le.ras* expression with intracellular cAMP levels in *L. edodes* [38] (see above section). Therefore, the *Le.ras* gene may not be directly related to the onset of fruiting and/or primordium formation.

9.3.4 *Nuclear and mitochondrial DNA*

Haploid cells of *L. edodes*, *P. nameko*, *V. volvacea*, *T. matsutake* and *P. eryngii* contain 8, 8, 9, *c.* 7 and 12 or 14 chromosomes, respectively [46]. Molecular size of nuclear genomic DNA in the *P. ostreatus* haploid cell is estimated to be 2.2×10^4 kbp, which is *c.* 6 times that of bacteria [47].

Molecular sizes of mitochondrial DNAs are 76.1 and 72.8 kbp for *L. edodes* and *P. ostreatus* [48]; for comparison those in *S. cerevisiae* and *Neurospora crassa* are 76 and 61 kbp, respectively [49]. There is no significant homology in mitochondrial DNA between *L. edodes* and *P. ostreatus*, as judged from

```
           1
Le.Ras     MAGRTTFLREYKLVVVGGGGVGKSALTIQFIQSHFVDEYDPTIEDSYRKQCVIDDEVALLDVLDTAGQEEYGAMREQYMR
H-Ras-1    -     T--------A-----------L--N-----------------V---G-TC---I---------S---D----
Sc.RAS1    -Q-NKSTI----I--------------------Y----------------V----K-SI--I---------S--------
Sc.RAS2    -PLNKSNI--------------------LT-----------------V------SI--I---------S--------
Sp.RAS     -  -S-Y----------D-----------L-----------------K-E--G-G--------------S--------
Dd.Ras     -     T-----I--------------L--N--I--------------VS----TC---I---------S---D----
Dd.RasG    -     T-----I--------------L--N--I--------------VT--E-TC---I---------S---D----

          81
Le.Ras     TGEGFLLVYSITSRNSFEEISTFHQQILRVKDQDTFPVVVVANKCDLEYERQVGMNEGRDLARHFGCKFVETSAKVRINV
H-Ras-1    ------C-FA-NNTK---D-HQYRE--K----S-DV-M-L-G-----AARTVESRQ AQ----SY-IPYI-----T-QG-
Sc.RAS1    ----------V------D-LLSYY---Q----S-YI-----G--L---N----SYED-LR--KQLNAP-L-----QA---
Sc.RAS2    N------------KS-LD-LM-YY--------T-YV-I---G--S---N-K--SYQD-LNM-KQMNAP-L-----QA---
Sp.RAS     ---------N----S--D-----Y--------K------L--------A--V-SRR--EQ--KSMH-LY------L-L--
Dd.Ras     --Q---C-------S-YD--AS-RE-------K-RV-LIL-G--A--DH----SV---QE--KDSLS -H-S---S----
Dd.RasG    --Q---C-------S--D--AS-RE-------K-RV-MIV-G------SD---TTG--Q---KS--SP-L-----I-V--

         161                                                              217
Le.Ras     DQAFQDLVREIRKYNKEQQTTGRMMTGGGGGPPGTYAGKDPNDEGAGGCCGG                    CVVL
H-Ras-1    ED--YT------QHKLRKLNPPDESGP-CMSCK                                       --LS
Sc.RAS1    -E--YS-I-LV-DDGGKYNSMN-QLDNTNEIRDSELTSSATADI-KKNNGSYV( 92 aa)-IIC
Sc.RAS2    EE--YT-A-LV-DEGGKYNKTLTENDNSKQTSQDTKGS-ANSVPRNS--HRKM(105 aa)-IIS
Sp.RAS     EE--YS---T--R---SEEKGFQNKQAVQIAQV-AST-KRASAVNNSKTEDEVSTKC              --IC
Dd.Ras     EE--YS-------EL-GD-SS-KAQKKKKQ                                          -LI-
Dd.RasG    EE--YS-------DL-GDSKPEKGKKKRPLKA                                        -TL-
```

Figure 9.2 Comparison of Ras amino acid sequences *Le* (*L. edodes*) Ras [38], human/rat H-Ras 1 [74], *Sc.* (*S. cerevisiae*) RAS1 and RAS2 (75, 76), *Sp.* (*S. pombe*) RAS [39], and *Dd.* (*D. discoideum*) Ras and Ras G [44, 45] are shown. The amino acid sequences are aligned in order to optimize matches. Identical residues to the *Le.* Ras protein are represented by a dash. Dotted box indicates the terminal Cys–A–A–X (A = aliphatic amino acid, X = any amino acid) structure present in all Ras/RAS proteins.

Southern blot hybridization analysis carried out under stringent conditions. In contrast, their nuclear genomic DNA display a degree of homology.

9.3.5 *Plasmids*

The majority of *L. edodes* and *P. ostreatus* strains have been shown to contain linear plasmid DNA elements (Table 9.1). Plasmids pLPO1 (10.0 kbp) and pLPO2 (9.4 kbp) isolated from *P. ostreatus* MS-PO.16-8 [50] have their 5′ ends blocked by association of a protein [50]. The 5′ends of pLLE1 (11.0 kbp) isolated from *L. edodes* MS-LE.1610 [51] are protected by binding of (probably) an oligopeptide [51]; pLLE1 may possess the 5′-terminal protein inherently, but it is degraded by the action of an endogenous protease during extraction of the DNA. Linear DNA plasmids observed in the bacterium *Streptomyces rochei* [52], the yeasts *Kluyveromyces lactis* [53] and *Saccharomyces kluyveri* [54], the filamentous ascomycetous fungi *Ascobolus immersus* [55] and *Claviceps purpurea* [56], the plant pathogenic fungi *Rhizoctonia solani* [57] and *Fusarium oxysporum* [58] and plants *Zea mays* [59] and *Brassica campestris* [60] all possess a terminally attached protein at each 5′ end of the duplex DNA, which is similar to mammalian adenoviruses and *Bacillus subtilis* bacteriophage ϕ29 [61]. In these viruses, DNA synthesis is shown to be primed by the 5′-terminal protein and to proceed by strand displacement [61].

Plasmids pLPO1, pLPO2, and pLLE1 are localized in the mitochondrion and are present to an extent of 1–2 copies per mitochondrial genome

Table 9.1 Natural linear DNA plasmids found in *L. edodes* and *P. ostreatus*.

Strains	Plasmids (size)
L. edodes ('Shiitake')	
FMC2	nd
MS-LE.1610	pLLE1 (11 kbp)
MS-LE.908	nd
MS-LE.1	pLLE1-like
MS-LE.2	pLLE1-like
MS-LE.3	pLLE1-like
MS-LE.4	nd
P. ostreatus ('Hiratake')	
MS-PO.16-8	pLPO1(10 kbp), pLPO2(9.4 kbp)
MS-PO.23	pLPO2-like, pLPO3(6.6 kbp)
MS-PO.3	nd
MS-PO.5	pLPO2-like
MS-PO.9	pLPO1-like, pLPO2-like
MS-PO.12	pLPO2-like
MS-PO.26	pLPO1-like
MS-PO.27	pLPO2-like
MS-PO.41	pLPO1-like, pLPO2-like
MS-PO.45	pLPO1-like, pLPO2-like
MS-PO.62	pLPO1-like
MP-1	pLPO2-like
MP-2	nd
MP-3	pLPO1-like, pLPO2-like

L. edodes FMC2, *L. edodes* MS-LE.-series, *P. ostreatus* MS-PO.-series and *P. ostreatus* MP-series are the stock strains of Forestry and Forest Products Research Institute of Japan, Meiji Seika Kaisha, Japan, and Mori Sangyo Kaisha, Japan, respectively.
nd: not detected.
'like' includes the possibility of being identical.

equivalent. They show no obvious homology with mitochondrial or nuclear genomic DNAs of plasmid-deficient strains or the plasmid-harbouring host strains [50, 51]. The biological significance of these plasmids is unclear.

The 1434 bp DNA fragment, capable of autonomous heterogous replication in the yeast *S. cerevisiae* (ARSs), was cloned from pLLE1 DNA into the yeast integration vector YIp32 (pBR322 containing yeast LEU2 DNA) [62]. The resultant vector displayed high ARS activity [51]. The cloned 1434 bp fragment lies near the end of pLLE1 DNA (nucleotides 800–2200). The fragment contains three consecutive ARS consensus sequences 5′ (A/T)-TTTAT(A/G)TTT(A/T)3′ of *S. cerevisiae* and eight dispersive ARS consensus-like sequences (Figure 9.3). The copy number of the YIp32 carrying the 1434 bp ARS fragment designated (pSKP11) is 20–30 per haploid yeast genome [51]. This is noticeably higher than the value of 6–10 for the two YIp32-ARS plasmids pSK52 and YRMp24 containing the mitochondrial genomic ARS of *L. edodes* and *S. cerevisiae*, respectively [63]. pSKP11 is

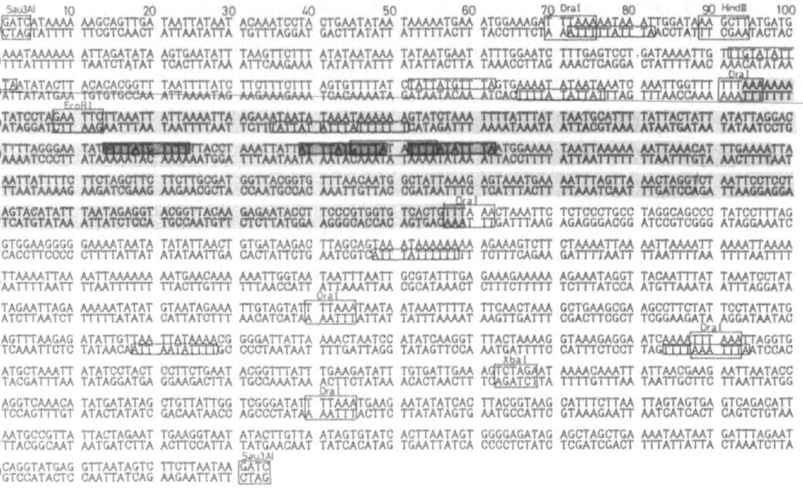

Figure 9.3 Nucleotide sequence of the cloned 1434 bp *Sau*3AI fragment with ARS activity in pLLE1 DNA [51]. ARS consensus sequence (deeply shaded) and ARS consensus-like sequence are boxed on one strand. The 366 bp *Dra*I fragment containing the three consecutive ARS consensus sequences are shaded in both strands. The cleavage sites of restriction endonucleases *Sau*3AI, *Dra*I, *Hin*dIII, *Eco*RI, and *Xba*I are boxed.

stably maintained in *S. cerevisiae* cells lacking a mitochondrial genomic DNA [51]. Therefore the plasmid is considered to be replicating in the nucleus. The 366 bp fragment (nucleotides 294–659] containing the three ARS consensus sequences (Figure 9.3) retains the replicating activity of the 1434 bp fragment, showing that the three consecutive ARSs play an important role in the expression of high ARS activity. If the ARS element does in fact function, i.e. autonomously replicate in *L. edodes*, it would be useful as a cloning vector for edible basidiomycetes.

9.3.6 *Transformation*

Protoplasts for transformation experiments can be efficiently prepared from *L. edodes* and *P. ostreatus* by digestion of the basidiospores, oidia and young mycelia by a mixture of cellulase and chitinase. The growth of *L. edodes* and *P. ostreatus* is inhibited by the antibiotic G418. The neomycin-resistance (*neo*) genes of the *E. coli* transposons Tn5 and Tn903 [64, 65], which encode aminoglycoside phosphotransferases conferring resistance to neomycin (kanamycin) and G418, was considered as a selection marker. The *neo* gene of Tn5 was inserted behind the yeast *GAL1* (galactokinase gene) promoter as well as simian virus 40 early promoter and was introduced into the protoplasts of *L. edodes* and *P. ostreatus*. However, G418-resistant transformants were not observed. For expression of the Tn5 *neo* gene in *L. edodes* and *P. ostreatus*, homologous promoters may be required. Consequently, the promoter of the *Le.ras* gene may be useful for this purpose. However, the Tn903 *neo* gene is known to be expressed in several filamentous fungal systems without the need for a eukaryotic promoter [66–69]. It may expressed in *L. edodes* and *P. ostreatus*. An alternative selection marker is a wild-type gene capable of complementing an auxotrophic mutant. So far several auxotrophic mutants have been isolated in *L. edodes* and *P. ostreatus* [70, 71]. Consequently it may be possible to isolate the corresponding genes for selection development.

Thus, the development of transformation techniques using autonomously replicating plasmid vectors and efficient chromosome-integrating vectors is under way in the edible basidiomycetes. Autonomously replicating plasmid vectors have been most recently constructed for the lignin-degrading basidiomycete *Phanerochaete chrysosporium* [72] and the phytopathogenic protobasidiomycete *Ustilago maydis* [73].

9.4 Future prospects

The molecular mechanisms of fruiting, i.e. mushroom formation in the basidiomycetes, are some of the most attractive scientific projects to be investigated. The manipulation of the genes which are considered to be related to the mushroom formation opens up new research possibilities. It is important to isolate the genes specifically (or abundantly) expressed in the initial stages of mushroom formation or during the growth of the mushroom. The genes encoding adenylate cyclase and metallo proteinase in *L. edodes* are thought to be likely candidates. Additionally there is the possibility that the *Le.ras* gene is concerned indirectly with fruiting in *L. edodes*.

The introduction of the ligninase gene of *P. chrysosporium* into the basidiomycetes, e.g. *L. edodes*, is thought to be an attractive experiment from the point of view of growing the mushroom on hardwood logs of broadleaf trees. The molecular breeding of *L. edodes* strain which can grow on hardwood logs of evergreen needle-leaved trees is of major and future interest.

Finally, the construction of expression and secretion vectors suitable for the commercial production of enzymes and other proteins by mushrooms will be important. In this regard, many proteins and enzymes are known to be glycosylated; in some cases, the sugar side chain is essential for the activity of certain enzymes. The yeast *S. cerevisiae* and animal cells differ in the type of sugar side chain attached to their proteins. Therefore animal derived protein produced in *S. cerevisiae* sometimes displays a harmful side effect. It is reasonable to assume that mushrooms may have a different type of glycosylation machinery from *S. cerevisiae*.

References

1. Singer, R. (1941) *Mycologia* **33**: 449–451.
2. Leathman, G.F. (1986) *Appl. Microbiol. Biotechnol.* **24**: 51–58.

3. Tokimoto, M. and Komatsu, M. (1978) Biological nature of *Lentinus edodes* In: *The Biology and Cultivation of Edible Mushrooms* (Chang, S.T. and Hayes, W.A., eds), Academic Press, New York, pp. 445–459.
4. Leatham, G.F. (1982) *Forest Prod. J.* **32**: 29–35.
5. Leatham, G.F. (1983) *Mycologia* **75**: 905–908.
6. Nakajima, N., Ichikawa, K., Kamada, M. and Fujita, E. (1961) *J. Agric. Chem. Soc. Jpn. (Tokyo)* **9**: 797–803.
7. Mouri, T., Hashida, W., Shiga, I. and Teramoto, S. (1969) *J. Ferment. Technol. (Jpn.)* **44**: 114–119.
8. Morita, K. and Kobayashi, S. (1966) *Tetrahedron Lett.* **6**: 573–577.
9. Morita, K. and Kobayashi, S. (1967) *Chem. Pharm. Bull.* **15**: 988–993.
10. Yasumoto, K., Iwami, K. and Mitsuda, H. (1971) *Agric. Biol. Chem.* **35**: 2059–2069.
11. Yasumoto, K., Iwami, K. and Mitsuda, H. (1971) *Agric. Biol. Chem.* **35**: 2070–2080.
12. Sumi, M. (1929) *Biochem. Z.* **204**: 397–411.
13. Fujita, A., Tokuhisa, S., Michinaka, K., Ono, T. and Sugiura, W. (1969) *Vitamins* **40**: 129–135.
14. Chihara, G., Maeda, Y., Hamuro, J., Sasaki, T. and Fukuoka, F. (1969) *Nature* **222**: 687–688.
15. Chihara, G., Hamuro, J., Maeda, Y., Arai, Y. and Fukuoka, F. (1970) *Cancer Res.* **30**: 2776–2781.
16. Chibata, I., Okumura, K., Takeyama, S. and Kotera, K. (1969) *Experientia* **15**: 1237–1238.
17. Kamiya, T., Saito, Y., Hashimoto, M. and Seki, H. (1969) *Tetrahedron Lett.* **53**: 4729–4732.
18. Kamiya, T., Saito, Y., Hashimoto, M. and Seki, H. (1972) *Tetrahedron* **28**: 899–906.
19. Mori, K. and Zennyoji, A. (1989) *Kinoko-Nenkan*, Noson Bunka-sha, Tokyo 296–302 (in Japanese).
20. Leatham, G.F. (1985) *Appl. Environ. Microbiol.* **50**: 859–867.
21. Tokimoto, K., Fukuda, M., Kishimoto, H. and Koshitani, H. (1987) *Rept. Tottori Mycol. Inst.* **25**: 24–35.
22. Tokimoto, K., Fukuda, M. and Komatsu, M. (1988) *Rept. Tottori Mycol. Inst.* **26**: 37–45.
23. Tien, M. and Tu, C.-PD. (1987) *Nature* **326**: 520–523.
24. deBoer, H.A., Zhang, Y.Z., Collins, C. and Reddy, C.A. (1987) *Gene* **60**: 93–102.
25. Smith, T.L., Schalch, H., Gaskell, J., Covert, S. and Cullen, D. (1988) *Nuc. Acids Res.* **16**: 1219.
26. Asada, Y., Kimura, Y., Kuwahara, M., Tsukamoto, A., Koide, K., Oka, A. and Takanami, M. (1988) *Appl. Microbiol. Biotechnol.* **29**: 469–473.
27. Walther, I., Kalin, M., Reiser, J., Suter, F., Fritsche, B., Saloheimo, M., Leisola, M., Teeri, T., Knowles, J.K.C. and Fiechter, A. (1988) *Gene* **70**: 127–137.
28. Manachère, G. (1980) *Trans. Br. Mycol. Soc.* **75**: 255–270.
29. Uno, I. and Ishikawa, T. (1971) *Mol. Gen. Genet.* **113**: 228–239.
30. Uno, I. and Ishikawa, T. (1974) *J. Bacteriol.* **120**: 96–100.
31. Kawai, G. and Ikeda, Y. (1982) *Biochim Biophys. Acta* **719**: 612–618.
32. Kawai, G., Ikeda, Y. and Tsubaki, K. (1985) *Agric. Biol. Chem.* **49**: 2137–2146.
33. Murao, S. (1986) *Kagaku-to-Seibutsu* **24**: 215–216 (in Japanese).
34. Takagi, Y., Katayose, Y. and Shishido, K. (1988) *FEMS Microbiol. Lett.* **55**: 275–278.
35. Terashita, T., Oda, K., Kono, M. and Murao, S. (1981) *Hakko-Kogaku* **59**: 55–57 (in Japanese).
36. Ellis, R.W., Defeo, D., Shih, T.Y., Gonda, M.A., Young, H.A., Tsuchida, N., Lowy, D.R. and Scolnick, E.M. (1981) *Nature* **292**: 506–511.
37. Barbacid, M. (1987) *Ann. Rev. Biochem.* **56**: 779–827.
38. Hori, K., Kajiwara, S., Saito, T., Miyazawa, H., Katayose, Y. and Shishido, K. (1991) *Gene* **105**: 91–96.
39. Fukui, Y. and Kaziro, Y. (1985) *The EMBO J.* **4**: 687–691.
40. Toda, T., Uno, I., Ishikawa, T., Powers, S., Kataoka, T., Broek, D., Cameron, S., Broach, J., Matsumoto, K. and Wigler, M. (1985) *Cell* **40**: 27–36.
41. Field, J., Nikawa, J., Broek, D., MacDonald, B., Rodgers, L., Wilson, I.A., Lerner, R.A. and Wigler, M. (1988) *Mol. Cell. Biol.* **8**: 2159–2165.
42. Fukui, Y., Kozasa, T., Kaziro, Y., Takeda, T. and Yamamoto, M. (1986) *Cell* **44**: 329–336.
43. Khosla, M., Robbins, S.M., Spiegelman, G.B. and Weeks, G. (1990) *Mol. Cell. Biol.* **10**: 918–922.
44. Reymond, C.D., Gomer, R.H., Mehdy, M.C. and Firtel, R.A. (1984) *Cell* **39**: 141–148.
45. Robbins, S.M., Williams, J.G., Jermyn, K.A., Spiegelman, G.B. and Weeks, G. (1989) *Proc. Natl Acad. Sci. USA* **86**: 938–942.

46. Arita, I. (1988) *Rept. Tottori Mycol. Inst.* **26**: 37–45.
47. Horgen, P.A., Arthur, R., Davy, O., Moum, A., Herr, F., Strauss, N. and Anderson, J. (1984) *Can. J. Microbiol.* **30**: 102–112.
48. Katayose, Y., Yui, Y. and Shishido, K. (1989) *Agric. Biol. Chem.* **53**: 573–575.
49. Dujon, B. (1983) In: *Mitochondria 1983* (Schweyen, R.J., Wolf, K. and Kaudewitz, K., eds), de Gruyter, Berlin, pp. 1–24.
50. Yui, Y., Katayose, Y. and Shishido, K. (1988) *Biochim Biophys. Acta* **951**: 53–60.
51. Katayose, Y., Kajiwara, S. and Shishido, K. (1990) *Nuc. Acids Res.* **18**: 1395–1400.
52. Hayakawa, T., Tanaka, T., Sakaguchi, K., Otake, N. and Yonehara, H. (1979) *J. Gen. Appl. Microbiol.* **25**: 255–260.
53. Gunge, N., Tamura, A., Ozawa, F. and Sakaguchi, K. (1981) *J. Bacteriol.* **145**: 382–390.
54. Kitada, K. and Hishinuma, F. (1987) *Mol. Gen. Genet.* **206**: 377–381.
55. Francou, F. (1981) *Mol. Gen. Genet.* **184**: 440–444.
56. Tudzynski, P., Duvell, A. and Esser, K. (1983) *Curr. Genet.* **7**: 145–150.
57. Hashiba, T., Homma, Y., Hyakumachi, M. and Matsuda, I. (1984) *J. Gen. Microbiol.* **130**: 2067–2070.
58. Kistler, H.C. and Leong, S.A. (1986) *J. Bacteriol.* **167**: 587–593.
59. Pring, D.R., Levings, C.S., III, Hu, W.W.L. and Timothy, D.H. (1983) *Proc. Natl Acad. Sci. USA* **74**: 2904–2908.
60. Erickson, L., Beversdorf, W.D. and Pauls, K.P. (1985) *Curr. Genet.* **9**: 679–682.
61. Salas, M. (1983) Curr. Top. *Microbiol. Immunol.* **109**: 89–106.
62. Botstein, D., Falco, S.C., Stewart, S.E., Brennan, M., Scherer, S., Stinchcomb, D.T., Struhl, K. and Davis, R.W. (1979) *Gene* **8**: 17–24.
63. Katayose, Y., Shishido, K. and Ohmasa, M. (1986) *Biochem. Biophys. Res. Comm.* **138**: 1110–1115.
64. Beck, E., Ludwig, G., Auerswald, E.A., Reiss, B. and Schaller, H. (1982) *Gene* **19**: 327–336.
65. Grindley, N.D.F. and Joyce, C.M. (1980) *Proc. Natl Acad. Sci. USA* **77**: 7176–7180.
66. Bull, J.H. and Wootton, J.C. (1983) *Nature* **310**: 701–704.
67. Randall, T.A., Rao, T.R. and Reddy, C.A. (1989) *Biochem. Biophys. Res. Comm.* **161**: 720–725.
68. Revuelta, J.L. and Jayaram, M. (1986) *Proc. Natl Acad. Sci. USA* **83**: 7344–7347.
69. Stahl, U., Leitner, E. and Esser, K. (1987) *Appl. Microbiol. Biotechnol.* **26**: 237–241.
70. Kawasumi, T., Baba, T. and Yanagi, O.S. (1988) *Agric. Biol. Chem.* **52**: 3197–3199.
71. Ohmasa, M. (1986) *Japan J. Breed.* **36**: 429–433.
72. Randall, T., Reddy, C.A. and Boominathan, K. (1991) *J. Bacteriol.* **173**: 776–782.
73. Kinal, H., Tao, J. and Bruenn, J.A. (1991) *Gene* **98**: 129–134.
74. Capon, D.J., Chen, E.Y., Levinson, A.D., Seebrug, P.H. and Goeddel, D.V. (1983) *Nature* **302**: 33–37.
75. DeFeo-Jones, D., Scolnick, E.M., Koller, R. and Dhar, R. (1983) *Nature* **306**: 707–709.
76. Powers, S., Kataoka, T., Fasano, O., Goldfarb, M., Strathern, J., Broach, J. and Wigler, M. (1984) *Cell* **36**: 607–612.

10 Molecular genetics of fungal secondary metabolites

J. F. MARTIN and S. GUTIERREZ

10.1 Introduction

The concept of secondary metabolism as proposed initially [1–5] is without any question an oversimplification [6] but the term is still used to distinguish between 'primary' (general) metabolites which are common to all living beings and 'secondary' (or special) types of metabolites which are restricted to certain taxonomic groups. Thus, fatty acid synthesis is common to all living cells whereas polyketide synthesis, in which the same precursors and similar biosynthetic enzymes are used, is restricted to certain taxonomic groups of microorganisms and higher plants [7, 8].

Secondary metabolites tend to have chemical structures which are infrequent in the biological world. These molecules belong to many classes of organic compounds: aminocyclitols, amino sugars, quinones, coumarins, epoxides, glutarimides, glycosides, indole derivatives, lactones, macrolides, naphthalenes, nucleosides, peptides, phenazines, polyacetylenes, polyenes, pyrroles, quinolines, terpenoids, and tetracyclines [3, 9]. In addition secondary metabolites possess unusual chemical linkages, such as β-lactam rings, cyclic peptides made of normal and modified amino acids, unsaturated bonds of polyacetylenes and polyenes, and large rings of macrolides. As a result of the broad substrate specificity of some biosynthetic enzymes, secondary metabolites are produced typically as members of a particular chemical family (e.g. there are at least 10 natural penicillins, [6]).

The precursors and the biosynthetic pathways of secondary metabolites have been reviewed in detail [10, 11]. Many enzyme proteins that carry out reactions of secondary metabolism have been now purified [12–14]. From a biochemical standpoint, there is nothing particularly special that distinguishes these reactions from ones of primary metabolism. In this regard polyketide synthases and fatty acid synthases are very similar enzymes [15]. Therefore, there is no significant biochemical base for distinguishing secondary from primary metabolism. Campbell [16] suggested that secondary metabolism exhibits a higher degree of biochemical sophistication than primary metabolism since (i) it uses a more expanded battery of precursors and biosynthetic reactions than observed in primary metabolism, and (ii) several of the basic biochemical principles that simplify the construction of primary metabolites are not conserved in generating secondary metabolites.

Another characteristic of secondary metabolites is that they are produced during the reduced growth rate of the organism after the rapid growth phase, at least in batch cultures. The evidence provided by the study of promoters of genes involved in secondary metabolism indicates that these promoters are subject to positive (or negative) regulation by intracellular effectors or DNA-interacting proteins which in turn respond to carbon, nitrogen or phosphate limitation [17, 18]. However, this behaviour is not unique to secondary metabolism promoters. Many bacterial promoters (e.g. heat shock promoters of *Escherichia coli*, stress-response gene promoters, sporulation gene promoters, extracellular enzyme gene promoters, etc.) are subject to the same type of regulation as secondary metabolism promoters.

The characteristic feature of secondary metabolites is the restricted distribution of their genetic information, and, by extension, of the enzymes required for their biosynthesis. As more information has been obtained on the distribution of genes involved in the biosynthesis of antibiotics and other secondary metabolites, it has become clear that these genes are present in restricted groups of microorganisms [19–21]. It is important, however, to note that many typical reactions of primary metabolism, e.g. carbon (from carbohydrate polymers, alcohols and hydrocarbons) utilization, nitrogen fixation, photosynthesis, are not universal among all microorganisms. The difference in terms of 'restricted distribution' of the genes is more quantitative than qualitative. It is likely that the restricted distribution of secondary metabolites is more a question of adaptation to survival in certain ecological niches than of fundamental biological differences. For example, the presence of pigments (typical secondary metabolites) in certain microorganisms is probably a protection against photodamage by light. This role is probably as important for the survival of the pigmented species as the presence of certain *Pseudomonas* hydrocarbon catabolic pathways which are considered to be primary metabolism.

Secondary metabolites are dispensable for the growth of the producer strains in the laboratory. This observation may be considered as an artifact of laboratory pure culture conditions. In nature, a secondary metabolites producer interacts with hundreds of other microorganisms and secondary metabolites are probably important signals in the communication between different members of the microbial, plant and animal communities. No molecular ecology studies are available to support or disclaim the notion of dispensability of secondary metabolites in nature. However, the amazing variety of roles found for secondary metabolites [22] suggests that removal of secondary metabolites from the producer organisms results in the long-term disappearance of those microorganisms from their usual habitats.

10.2 Fungal secondary metabolites

Fungal secondary metabolites are of major importance to mankind as drugs (e.g. penicillins, cephalosporins, griseofulvin, cyclosporin, mycophenolic acid, ergot and cyclopenin alkaloids) and toxins (e.g. aflatoxin, trichothecins, patulin, HC-toxin and other phytotoxins). Little information is available on the physiological role of these secondary metabolites in the producer organism and still less on the molecular genetics of toxin or antibiotic production. This review is limited to those groups of fungal secondary metabolites in which at least some of the biosynthetic genes have been isolated by genetic engineering technology.

10.2.1 *Polyketide-derived metabolites*

Polyketides represent a major group of secondary metabolites [23]. Only a few fungal polyketide-derived secondary metabolites have been studied in detail. These include mycotoxins [24] from species of *Aspergillus*, 6-methylsalicylic acid and patulin from *Aspergillus patulum*, ergochromes from *Claviceps* and some antimicrobial agents such as griseofulvin produced by *Penicillium griseofulvum*. Phytopathogenic fungi also synthesize polyketide-derived melanins [25] and some host-specific phytotoxins.

In spite of a remarkable variety of end products, the polyketide biosynthetic pathways follow a common basic reaction scheme that resembles fatty acid biosynthesis [7, 8].

10.2.1.1 *6-Methylsalicylic acid synthase, a model for polyketide synthases.* A significant advance in understanding fungal polyketide biosynthesis has been made through the study of 6-methylsalicylic acid biosynthesis by *Penicillium patulum*. 6-Methylsalicylic acid (6-MSA) (Figure 10.1) is one of the simplest polyketides since it is derived from one acetate starter molecule and three malonate extender units [26]. 6-MSA is later converted to patulin by this fungus.

In the literature, both synthase and synthetase are used for secondary metabolite biosynthetic enzymes. The terms synthetase and synthase imply ligase and lyase mechanisms respectively. ATP is required for peptide antibiotic biosynthesis and therefore, in those cases where a ligase mechanism is involved, such enzymes should be referred to as peptide synthetases. Fatty acid synthases and 6-MSA synthase apparently do not require ATP since they use previously activated precursors as CoA derivatives. However, in other cases not enough biochemical evidence is available and, to avoid confusion, the name proposed here is used without implying a reaction mechanism.

The 6-MSA synthase gene has been isolated and sequenced [15]. In *P. patulum* one of the genes (FAS2) encoding the fatty acid synthase is also

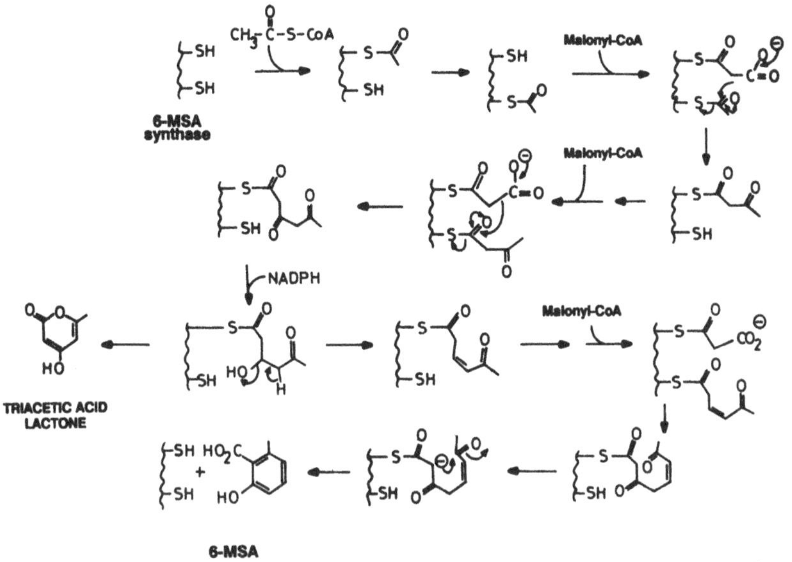

Figure 10.1 Chemical mechanisms of the reactions carried out by 6-MSA synthase. {—SH —SH represents the enzyme. Note the following sequential steps: (i) activation of acetyl and malonyl groups as enzyme-bound thioesters; (ii) two rounds of condensation of the activated monomers; (iii) reduction of the nascent polyketide; (iv) dehydration; (v) condensation with the fourth monomer (malonyl-CoA); and (vi) cyclization to form 6-MSA with the release of the free enzyme. In the presence of NADPH no free intermediates are detected whereas in the absence of NADPH triacetic acid lactone (left) is formed as the sole product. During 6-MSA biosynthesis reduction occurs only once, at a late stage, whereas during fatty acid biosynthesis reduction occurs after each condensation step. Modified from Herbert [27].

known [28]; this, of course, allows interesting comparisons at the amino acid level between the fatty acid synthase and the polyketide synthase of the same microorganism. Fatty acid synthases (FAS) and other polyketide synthases (PKS) occur either as large, multifunctional enzymes (type I PKS), or as aggregated systems of individual component enzymes (type II PKS). In *Saccharomyces cerevisiae* and *P. patulum* the fatty acid synthase (FAS) is a multifunctional protein encoded by two unlinked genes FAS1 and FAS2 that encode, respectively, a pentafunctional β-subunit and a trifunctional α-subunit which aggregate to form the $\alpha_6\beta_6$ fatty acid synthase [29]. The FAS2 proteins of *S. cerevisiae* and *P. patulum* are similar [30]. The acyl carrier protein (ACP), ketoreductase and β-ketoacyl synthase functional regions are located on the large (1887 amino acids) FAS2 polypeptide. The FAS2 gene of *P. patulum* reveals high similarity to the *S. cerevisiae* FAS2 gene (50 to 70% nucleotide similarity over the various domains) [28]. The *P. patulum* FAS2 gene is interrupted by two short introns which are absent in the *S. cerevisiae* gene.

Comparison of the 6-MSA synthase deduced protein sequence [15] with type I and type II fatty acid synthases and other polyketide synthases of actinomycetes [8] has allowed an initial assignment of functional domains and active site residues for β-ketoacyl synthase (condensing activity), acetyl/malonyl transferase, ketoreductase and ACP. Surprisingly, the overall structure of 6-MSA synthase resembles more closely that of the animal FAS multifunctional proteins than the fungal α/β subunit type I system (Figure 10.2a)

A cysteine residue (Cys-204) of 6-MSA synthase was identified as the β-ketoacyl synthase active site [8] based on the similarity with the β-ketoacyl synthase domains of the rat and chicken FAS, the *fabB* of *E. coli*, and the deduced amino acid sequences of the polyketide synthases encoded by the *gra* [31] and *tcm* [32] genes of *Streptomyces violaceusruber* and *Streptomyces glaucescens*.

Another functional domain of the 6-MSA synthase (around Ser-653) has been deduced by its similarity with the highly characteristic motif of the animal FAS acetyl/malonyl transferase domains. It seems, therefore, that chain initiation and elongation is similar in the fungal polyketide synthase and the animal FAS system. However, the central region of the 6-MSA synthase has no similarity with the central regions of the animal FAS, which appear to contain the dehydrase domain.

A putative ketoreductase domain exists near the C-terminus of 6-MSA synthase (amino acid positions 1419–1424) where a nucleotide binding site motif (Gly-Leu-Gly-Val-Leu-Gly) fits the Gly-X-Gly-X-X-Gly consensus for the ketoreductase domains in the animal FAS and also in the *act* and *gra*-encoded polyketide synthases.

Finally, a predicted ACP region occurs at the C-terminus of the 6-MSA synthase. It includes a phosphopantetheine binding Ser-1732, which is similar to the phosphopantetheine-binding sequence in several FAS, polyketide synthases and also in the oligopeptide synthetases (see section 10.2.4.1) (Figure 10.2b).

A genomic DNA fragment containing the 3′-terminus of the 6-MSA synthetase gene of a different species *Penicillium urticae*, was cloned using a

Figure 10.2 (a) Organization of the domains in the 6-MSA synthase of *P. patulum* and *P. urticae* as compared to the arrangement in the FAS of *S. cerevisiae*, rat and chicken. ACP, acylcarrier protein; AT/MT, acetyl and malonyl transferases; DH, dehydratase; ER, enoyl reductase; KR, ketoreductase; KS, β-ketoacyl synthase; TE, thioesterase. Note the similarity of the location of the KS-AT/MT and KR-ACP domains of 6-MSA synthase with those of the animal FASs. (b) Conserved amino acids in the phosphopantetheine binding site of several PKSs, FASs, peptide synthases, ACV synthetases of β-lactam-producing organisms and nodulation genes. A 13 amino acid consensus motif (D/E, I/L/V, G, I/L/V/G/A, D, S*, I/L/V, X, X, X, X, X, I/L/V) is found in the PKSs and FASs. The first amino acid of this motif is Ser (in the barley FAS I and FAS III) or Thr (in the nodulation PKSs) whereas it is Tyr or Phe in all peptide synthases. Note that the active centre Ser* at position 7, which forms a thioester bond with the phosphopantetheine molecule, is universally conserved.

(a)

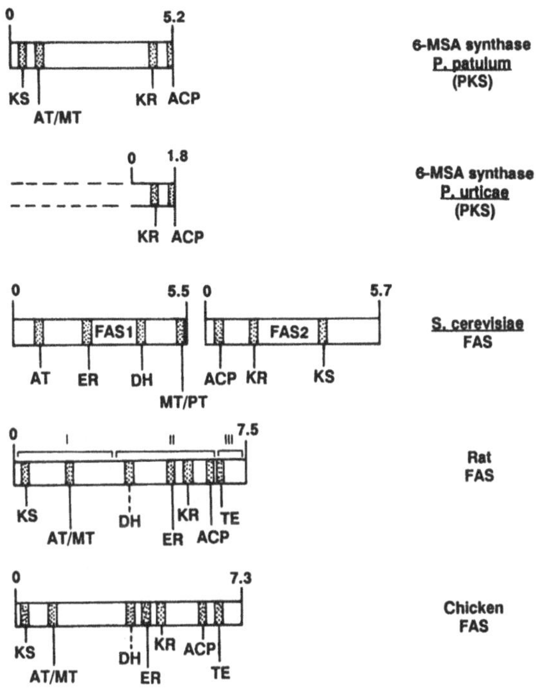

(b)

1	2	3	4	5	6	7*	8	9	10	11	12	13	14		
A	D	L	G	V	D	S	V	M	T	V	T	L	R	6-MSAS - *P. patulum*	
E	D	I	G	Y	D	S	L	A	L	M	E	T	A	*act* - *S. coelicolor*	
D	T	F	G	L	D	S	L	G	L	L	G	I	V	*whi*E - *S. coelicolor*	PKS
E	E	L	G	Y	D	S	L	A	L	M	E	S	A	*gra* - *S. violaceoruber*	
Q	D	L	G	Y	D	S	I	A	L	L	E	I	S	*tem* - *S. glaucescens*	
D	A	L	G	Y	D	S	L	A	L	L	N	T	V	*otc* - *S. rimosus*	
A	D	L	G	L	D	S	L	M	G	V	E	V	R	FAS - Rat	
A	D	L	G	L	D	S	L	M	G	V	E	V	R	FAS - Chicken	
D	L	V	G	G	K	S	T	V	Q	N	E	I	L	FAS - *S. cerevisiae*	
E	D	L	G	M	D	S	L	D	L	V	E			FAS - *Sacc. erythrea*	FAS
E	D	L	G	A	D	S	L	D	T	V	E	L	V	FAS - *E. coli*	
S	E	L	G	A	D	S	L	D	T	V	E	I	V	FASI - Barley-I	
S	D	L	G	A	D	S	L	D	T	V	E	I	V	FASIII - Barley-III	
Y	S	L	G	G	D	S	I	Q	A	I	Q	V	V	TYCA - *B. brevis*	PEPTIDE
Y	A	L	G	G	D	S	I	K	A	I	Q	V	A	GRSA - *B. brevis*	
F	R	L	G	G	H	S	I	T	C	I	Q	L	I	*pen*ACVA - *P. chrysogenum*	
F	S	L	G	G	D	S	L	K	S	T	K	L	S	*pen*ACVB - *P. chrysogenum*	
F	K	L	G	G	D	S	I	T	S	L	K	L	V	*pen*ACVC - *P. chrysogenum*	
F	R	L	G	G	H	S	I	A	C	I	Q	L	I	*cef*ACVA - *A. chrysogenum*	
F	S	L	G	G	D	S	L	K	S	T	K	L	S	*cef*ACVB - *A. chrysogenum*	β-LACTAMS
F	A	R	G	G	D	S	I	S	S	L	R	L	V	*cef*ACVC - *A. chrysogenum*	
F	R	L	G	G	Q	S	I	S	C	I	L	L	I	*cem*ACVA - *N. lactamdurans*	
F	A	L	G	G	D	S	I	R	A	M	A	L	A	*cem*ACVB - *N. lactamdurans*	
F	R	C	G	G	D	S	I	S	A	L	H	L	A	*cem*ACVC - *N. lactamdurans*	
T	A	L	G	V	D	S	L	G	L	A	D	I	I	*nod*F - *Rhizobium meliloti*	
T	S	L	G	I	D	S	L	A	L	A	D	V	C	*nod*F - *R. trifolii*	NODULATION
T	A	L	G	V	D	S	L	G	L	A	D	V	Q	*nod*F - *R. leguminosarum*	

41-mer mixed oligonucleotide probe [33]. Inspection of the nucleotide sequence revealed a large open reading frame (ORF) of 1866 bp without introns that corresponds to the carboxyl terminus of the enzyme. A putative acylcarrier protein domain was preceeded by a β-ketoreductase domain. This organization is identical to that of the 6-MSA synthetase of *P. patulum* [15].

In contrast with β-lactam tripeptide synthetases (see section 10.2.4.1), no evidence for a thioesterase domain in 6-MSA synthase is provided by amino acid sequence comparisons. An enzymatic step seems to be required for chain release from the ACP, as proposed below for the oligopeptide synthases. This enzymatic activity may be located in a small ORF separated from 6-MSA synthase.

10.2.2 *Peptide metabolites*

Of the many peptide secondary metabolites present in fungi, only a few have been studied in any detail since they have antimicrobial or other pharmacological activities (Table 10.1). These include (i) a linear peptide, alamethicin, produced by *Trichoderma viridae*, (ii) three cyclic peptides, cyclopeptin, produced by *Penicillium cyclopium*, ferrichrome, a metabolite of *Aspergillus quadricinctus*, and cyclosporin, an immunosuppressor by *Beauveria nivea*, and (iii) two depsipeptides, enniantin by *Fusarium oxysporum*, and beauvericin, synthesized by *Beauveria bassiana*. These are synthesized by a non-ribosomal thiotemplate mechanism [12]. Multienzyme synthetases involved in the formation of such compounds have been purified. Alamethicin synthetase from *Trichoderma viridae* has been partially purified [34] as has cyclopeptin synthase from *Penicillium cyclopium* [35, 36], although in the latter case there is no firm evidence for a multienzyme synthetase as has been proposed for other fungal peptides. Purification of the ferrichrome synthetase from *Aspergillus quadricinctus* [37] provided evidence indicating that this cyclic peptide is also synthesized by a multienzyme system.

One of the best known fungal peptide synthetases is one involved in the synthesis of the cyclohexadepsipeptide enniantin by *F. oxysporum*. The enniantins are cyclic hexadepsipeptides (Table 10.1) with antibiotic activity and are produced by various strains of *Fusarium*. Enniantin synthase has a molecular weight of 250 kDa [38, 39] which forms the dipeptide *N*-methyl-L-valyl-D-hydroxyisovalerate by the thiotemplate mechanism [12, 40]. The complete molecule of enniantin is formed by trimerization of activated *N*-methylvalyl-D-hydroxyisovalerate. The multienzyme accepts a variety of linear aliphatic and branched chain amino acids.

Beauvericin is an analogue of enniantin and contains phenylalanine instead of valine, with insecticidal properties. It is produced by several fungi including *Beauveria bassiana* and *Paecilomyces fumosoroseus*. The multienzyme beauvericin synthase has been purified [41, 42]. The enzyme catalyses the

Table 10.1 Several of the better characterized fungal peptide secondary metabolites. The multifunctional peptide synthases which form these metabolites have been purified but information on the genes encoding these enzymes is somewhat scant (see text).

Type	Product	Producer organism	Structure	References
Linear	Alamethicin	*Trichoderma viride*	AcAib-Pro-Aib-Ala-Aib-Ala-Gln-Aib-Val-Aib-Gly-Leu-Aib-Pro-Val-Aib-Aib-Glu-Gln-Pheol	[34]
Cyclic	Cyclopeptin	*Penicillium cyclopium*	Cyclic (Anthranilate-Phe)	[35, 36]
	Cyclosporin	*Beauveria nivea* (syn. *Tolypocladium inflatum*)	Cyclic (MeBmt-Abu-Sar-NMe-Leu-Val-NMe-Leu-Ala-D-Ala-NMe-Leu-NMe-Leu-NMe-Leu-NMe-Val)	
	Ferrichrome	*Aspergillus quadricinctus*	Cyclic (Gly_3OH-Orn_3)	[37]
Depsipeptides	Beauvericin	*Beauveria bassiana*	Cyclic $(NMe\text{-}Phe\text{-}D\text{-}Hiv)_3$	[41, 42]
	Enniantin	*Fusarium oxysporum*	Cyclic $(NMe\text{-}Val\text{-}D\text{-}Hiv)_3$	[40, 43]

AcAib, acetyl amino isobutyric acid; Aib, amino isobutyric acid; Abu, amino butyric acid; D-Hiv, D-hydroxyisovaleric acid; NMe, N-methyl; MeBmt, (4R)-4-[(E)-2-butenyl]-4-N-dimethyl-L-threonine; Pheol, phenylalaninol; Sar, sarcosine.

formation of the dipeptide *N*-methyl-L-phenylalanyl-D-hydroxyisovalerate. This multienzyme system has broad substrate specificity for aromatic amino acids (analogues of phenylalanine) but also accepts saturated hydrophobic amino acids at a reduced rate.

Replacement of analogous amino acids by enniantin and beauvericin synthetases are examples of the well-known lack of specificity of the peptide synthetases involved in secondary metabolism. Substitution of Leu/Ileu/Val/Thr or Phe/Tyr/Trp can be attributed to the broad specificity of the activating domains [12].

Cyclosporin A is a cyclic undecapeptide with anti-inflammatory, immunosuppresive, antifungal and antiparasitical activities, widely used in transplantation surgery and in the treatment of autoimmune diseases. It is produced by the fungus *Beauveria nivea* (previously designated *Tolypocladium inflatum, Tolypocladium niveum* and *Trichoderma polysporum*) in the form of a family of 25 naturally occurring compounds differing in amino acid substitution at several positions along the polypeptide.

Cyclosporins are synthesized from their precursor amino acids by a single multifunctional enzyme. The enzyme has been purified to near homegeneity and shown to have a molecular mass of approximately 800 kDa. It contains 4′-phosphopantetheine as a prosthetic group and activates all the constituent amino acids as thioesters via aminoadenylation [43]. The cyclosporin synthase is one of the largest known peptide synthases and it requires an ORF of about 20 kb in the fungal genome. Recently the gene has been isolated and characterized.

10.2.3 *Mycotoxins and phytotoxins*

10.2.3.1 *Trichothecenes and aflatoxins.* The trichothecenes are a family of sesquiterpene mycotoxins produced by members of at least eight genera of fungi [44]. Several different trichothecenes, including T2 toxin, are produced by the fungus *Fusarium sporotrichioides.* The trichothecenes play an important role in plant diseases caused by *F. sporotrichioides.* One of the enzymes involved in trichothecene biosynthesis is trichodiene synthase which catalyses the isomerization and cyclization of farnesyl pyrophosphate to form trichodiene, the first cyclic intermediate in the trichothecene biosynthetic pathway [45, 46]. Trichodiene synthase serves to branch trichodiene biosynthesis from the isoprenoid pathway. In several organisms, there are enzymes (so-called terpene cyclases) related to trichodiene synthase that function at branch points in the biosynthesis of almost all terpenes from the isoprenoid pathway [47]. The trichodiene synthase gene (*tox*5) of *F. sporotrichioides* has been cloned. The gene consists of a 1182 nucleotide ORF which contains a 60 nucleotide intron and encodes a protein of M_r 43.999. This data is in agreement with the previously established 45 kDa value determined for the purified enzyme by SDS-PAGE [48]. This is a single copy gene [49]. The trichodiene synthase is the first primary structure reported for a terpene cyclase. When genes for other terpene cyclases become available, it will be interesting to establish the relationships between these enzymes which utilize common substrates and possess similar cofactors.

Aflatoxins are mycotoxins produced by *Aspergillus flavus* and *Aspergillus parasiticus*. Various mutants of *A. flavus* and *A. parasiticus* which are unable to synthesize aflatoxins have been reported [50]. They include certain *A. flavus* mutants which are presumed to be blocked in polyketide chain formation. Such mutants might be useful for the isolation of genes involved in aflatoxin biosynthesis. Understanding of aflatoxin biosynthesis is currently being approached at the molecular level [51].

10.2.3.2 *Helminthosporium phytotoxins.* *Helminthosporium* is the commonly used name for the relatively large genus of phytopathogenic fungi that in recent taxonomic distributions has been split into the genera *Bipolaris*, *Drechslera* and *Exserohilum*. Each one of these genera has a unique perfect state: *Cochliobolus*, *Pyrenophora* and *Setosphaeria*, respectively. However, the name *Helminthosporium* is still routinely used [52]. Species of *Helminthosporium* produce many secondary metabolites including cytochalasins, ophiobolins, helminthosporal and helminthosporol [53, 54] in addition to potent host-specific phytotoxins. Among these toxins the best known are (i) victorin (or HV-toxin), a chlorinated pentapeptide produced by *H. victoriae* (syn. *Cochliobolus victoriae*) (Figure 10.3), (ii) T toxins, a family of hydrocarbons substituted with a variety of hydroxyl and carbonyl functionalities (polyketols of 37 to 43 carbon atoms) produced by *H. maydis* (syn. *Drechslera maydis*) [55], (iii) helminthosporoside (HS-toxin) a sesquiterpenoid family of compounds produced by *H. sacchari* [56] (*H. sacchari* is a plant pathogen of sugar-cane and, interestingly, addition of sugar-cane suspensions to *H. sacchari* cultures stimulates HS-toxin production) and (iv) HC-toxin, a cyclic tetrapeptide of structure cyclo(D-Pro-L-Ala-D-Ala-L-2-amino-8-oxo-9,10-epoxidecanoic acid) (Figure 10.3) produced by *H. carbonum* [57–59]. The toxicity of host-selective phytotoxins from *Helminthosporium* has been reviewed by Yoder [60] and Walton [52].

Cochliobolus, the perfect fungal state of *Bipolaris*, is a bipolar heterothallic ascomycete with eight ascospores per ascus. Since it grows quickly and is not nutritionally fastidious, a basic groundwork of classical and molecular genetics has ben developed in this fungus [61, 62]. As a result of its importance in the process of pathogenesis, interest has centred on (i) the biochemistry of host-specific toxins and (ii) the cloning and characterization of the genes involved in the biosynthesis of these secondary metabolites.

Early genetic work revealed that the production of HC-toxin by *C. carbonum* was controlled by a single genetic locus and the same held for the production of victorin by *C. victoriae* [63, 64]. Since these toxins, similar

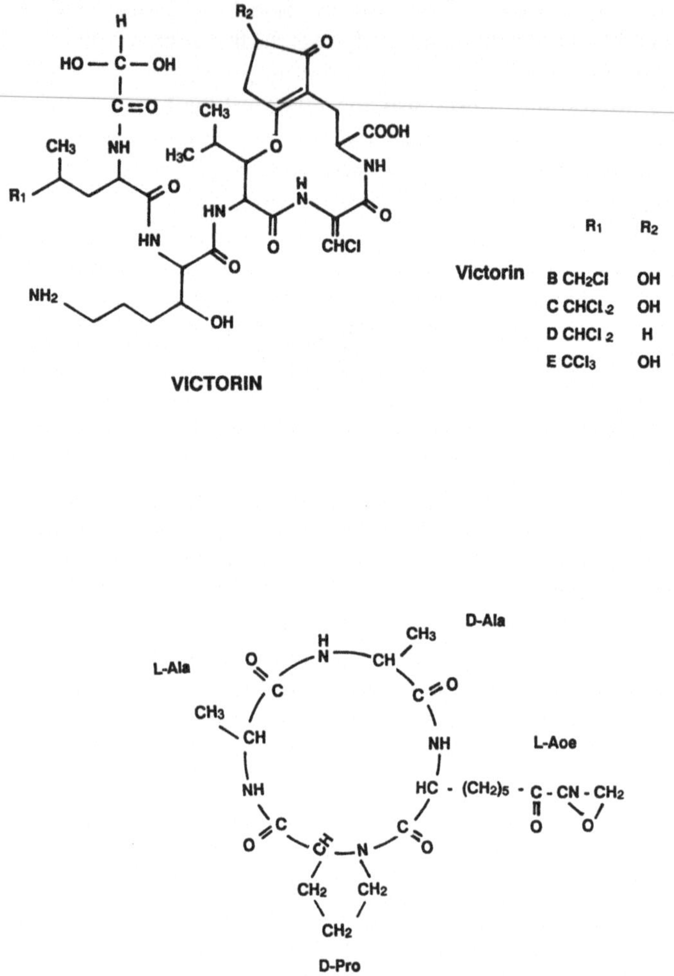

Figure 10.3 Chemical structures of the amino acid derived *Helminthosporium* phytotoxins Victorin and HC-toxin. These compounds are synthesized by multifunctional peptide synthetases (see text for details).

to other complex secondary metabolites, are synthesized by multistep pathways, it was difficult to understand at that time how the biosynthesis of these toxins might be controlled by single loci. Walton [52] has proposed several plausible explanations. The most likely explanation is that such peptide secondary metabolites are synthesized by a single multifunctional enzyme polypeptide synthetase similar to the peptide antibiotic synthetases or the α-aminoadipyl-cysteinyl-valine synthetases of *Penicillium*, *Cephalosporium* and *Aspergillus* (section 10.2.4.1). Since classical meiotic genetic analysis may not identify closely linked genes, another possibility is that several genes for toxin biosynthesis are clustered at a single locus.

The best known *Helminthosporium* phytotoxin is the HC-toxin (cyclic tetrapeptide) produced by *H. carbonum*. There are three analogues of HC-toxin produced by fungi [65, 66]. Two proteins which appear to be components of the HC-synthetase have been identified [59]. One protein, named HTS-1, catalyses an ATP/pyrophosphate exchange reaction dependent on L-alanine and D-alanine; another, HTS-2, catalyses an ATP/pyrophosphate exchange that is proline dependent. SDS-PAGE data indicate that HTS-1 has an M_r of 220 000 and HTS-2 an M_r of 169 000 [67–69]. HTS-1 also isomerizes L-proline to D-proline [67]. HTS-2 converts L-alanine to D-alanine. L-alanine appears, therefore, to be the precursor of both alanines in the HC-toxin.

Less information is available on the biosynthesis of the unusual epoxide amino acid component of HC-toxin L-2-amino-8-oxo-9,10-epoxidecanoic acid (L-AOE). Precursor feeding experiments indicate that it is derived from labelled acetate [70]. Several cyclotetrapeptides produced by different fungi contain the unusual amino acid L-AOE. Such molecules include, in addition to the HC-toxins, the compounds Cyl-1 and Cyl-2 from *Cylindrocladium scoparum* [71], chlamydocin from *Diheterospora chlamydosporia* [72, 73] and compound WF-3161 from *Petriella guttulata*. This suggests that partial homology of the template mechanisms has been conserved in the corresponding genes. An L-AOE activating domain must be present in the multienzyme synthetases of these cyclotetrapeptides.

All available biochemical evidence suggests, therefore, that the biosynthesis of HC-toxins (and probably other fungi peptide toxins) is similar to that of α-aminoadipyl-cysteinyl-valine in *Pencillium*, *Acremonium* and *Aspergillus* and enniantin in *Fusarium*. The HC-peptide synthetase is probably formed by at least two subunits as occurs with the gramicidin synthetase (GSI and GSII). It is unclear whether a third subunit activates the rare amino acid L-AOE. Confirmation of this hypothesis will be provided by the isolation of the genes involved in HC-toxin production (*tox2* in *H. carbonum*) and in T-toxin biosynthesis from *H. heterotrophus* [52].

In summary, it appears that the biosynthesis of peptide phytotoxins and other fungal secondary metabolites of peptide nature follows a similar model as the biosynthesis of bacterial peptide antibiotics [12].

10.2.4 *Amino acid-derived metabolites containing a β-lactam ring*

A group of amino acid-derived fungal secondary metabolites experience enzymatic modifications of the peptide chain that give rise to a β-lactam ring structure.

10.2.4.1 *Biosynthesis of β-lactam antibiotics: a synopsis.* Many of the biosynthetic steps of penicillins and cephalosporins have been characterized at the biochemical level [13, 74–77] but knowledge of their genes is still incomplete despite their medical and industrial importance.

Biosynthesis of penicillins. The β-lactam-thiazolidine ring (*penam*) system, which is present in all penicillins, and the β-lactam-dihydrothiazine nucleus (*cepham*) which is common to cephalosporins and cephamycins, are both derived from the condensation of L-cysteine and D-valine. A third amino acid, α-aminoadipic acid (α-AAA) is required for the biosynthesis of penicillins and cephalosporins. α-Aminoadipic acid is involved in the formation of the tripeptide δ(L-α-aminoadipyl)-L-cysteinyl-D-valine (ACV), an intermediate in the biosynthesis of all penicillins and cephalosporins [78]. ACV is then cyclized to form isopenicillin N, an intermediate which has an

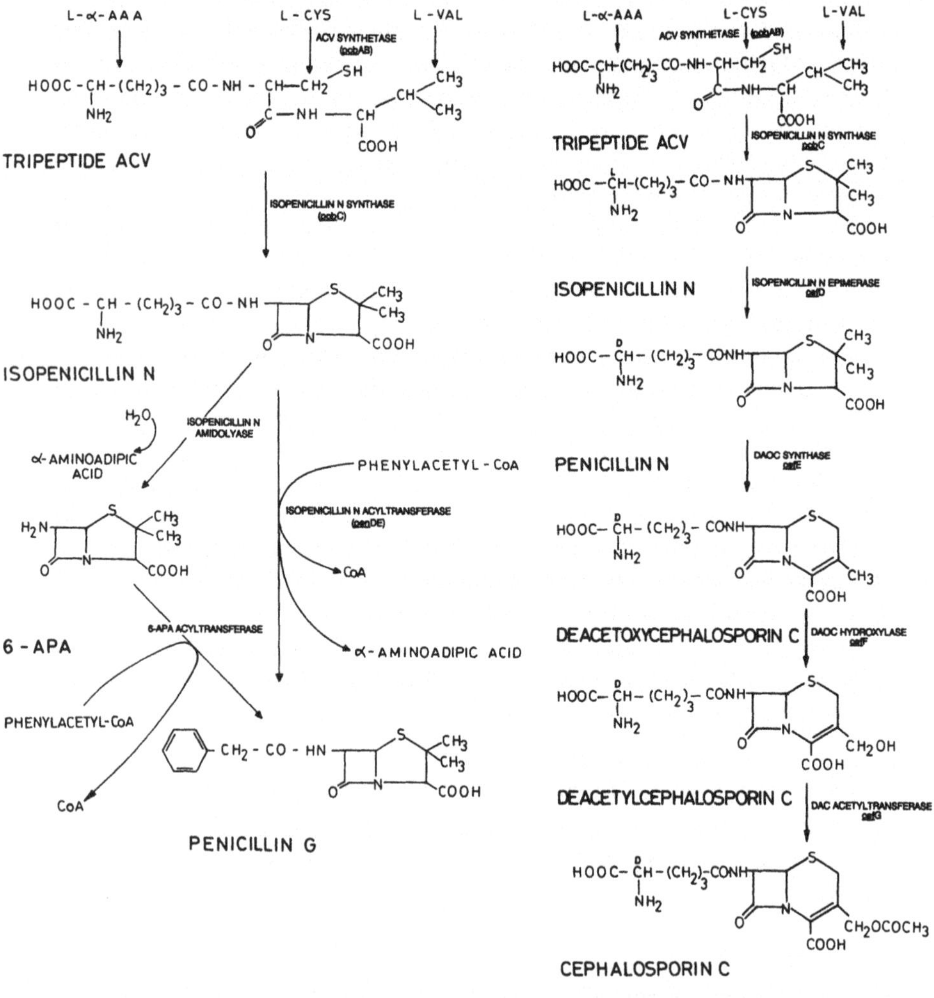

Figure 10.4 Biosynthetic pathways of penicillins (left) and cephalosporins (right). Note that the two initial steps are identical in both biosynthetic pathways. Modified from Martin and Liras [13].

L-α-aminoadipyl side chain attached to the nucleus of penicillin (Figure 10.4). This reaction was first observed with cell-free extracts of *P. chrysogenum* [78]. The isopenicillin N synthase (IPNS or cyclase) of *P. chrysogenum* has been purified to near-homogeneity and shown to have a molecular weight of 39 000 [79, 80]. The α-aminoadipyl side chain is later exchanged for phenylacetic acid during penicillin biosynthesis but not during cephalosporin biosynthesis. This reaction is carried out by an acyltransferase found in crude extracts of *P. chrysogenum* [81] which has been purified to homogeneity [82, 83]. The biochemistry of the transacylation reaction has been reviewed in detail recently [14].

Biosynthesis of cephalosporins. The biosynthetic pathways of cephalosporin and cephamycins resemble that of penicillin. Isopenicillin N is, in fact, an intermediate in the biosynthesis of both groups of compounds and is formed by cyclization of the ACV tripeptide, as in penicillin biosynthesis. The cyclases of *A. chrysogenum* (syn. *Cephalosporium acremonium*; *Acremonium strictum*) [84] and the actinomycetes, *S. clavuligerus*, *S. lipmanii* and *S. lactamdurans* have been purified. In all five microorganisms the cyclases required dithiothreitol and O_2 and were stimulated by ferrous ions and ascorbate (see review by Martin and Liras [13]).

Isopenicillin N is converted to penicillin N in cephalosporin and cephamycin-producing microorganisms but not in penicillin producers by an epimerase that isomerizes the L-α-aminoadipyl side chain to the D configuration [85]. Penicillin N is converted to deacetoxycephalosporin C by the deacetoxycephalosporin C synthetase (the so-called ring-expanding enzyme or expandase) [86]. This enzyme converts the five-membered thiazolidine ring of penicillins into the six-membered dihydrothiazine ring of cephalosporins and requires Fe^{2+} and α-ketoglutarate. Its activity is strongly stimulated by ascorbic acid and to a lesser extent by ATP, both well-known effectors of dioxygenases. The enzyme from *A. chrysogenum* has a molecular weight of 31 kDa [86]. It expands the β-lactam ring of penicillin N but does not accept the isomer isopenicillin N, penicillin G or 6-APA as substrates, i.e. the enzyme shows a high specificity for the side chain attached to the penicillin nucleus.

In the next reaction of the cephalosporin biosynthetic pathway, deacetoxycephalosporin C is hydroxylated to deacetylcephalosporin C by an α-ketoglutarate-linked dioxygenase. This enzyme catalyses the incorporation of an oxygen atom from O_2 into deacetoxycephalosporin C, and is also stimulated by α-ketoglutarate, ascorbate, dithiothreitol and Fe^{2+} [87,88]. Scheidegger and co-workers provided evidence suggesting that the two enzyme activities (deacetoxycephalosporin C synthetase and deacetoxycephalosporin C hydroxylase) from an industrial strain of *A. chrysogenum* are located on a single protein of M_r 33 000. The bifunctional role of this protein from *A. chrysogenum* has been confirmed by Dotzalf and Yeh [87] and Baldwin *et al.* [89].

Acetylation of deacetylcephalosporin C to cephalosporin C is the terminal reaction in cephalosporin-producing fungi (Figure 10.4) [90–92]. Further reactions are involved in the attachment of the C-7 methoxy group and the C-3′ carbamoyl group during cephamycin biosynthesis in the actinomycetes ([93] and see reviews by Jensen [94], Martin and Liras [13] and Martin *et al.* [14])

10.2.4.2 *Biochemical genetics of β-lactam antibiotics*

Mutants impaired in penicillin biosynthesis. Biochemical studies of impaired mutants have provided in a few cases unequivocal evidence of the steps involved in biosynthesis of β-lactam antibiotics. Nash and co-workers [95] first observed the accumulation of different intermediates of the penicillin biosynthetic pathway by blocked mutants of *P. chrysogenum.* Further studies on the genetics of penicillin biosynthesis involved the characterization of twelve mutants of *P. chrysogenum* producing 10% or less penicillin than their parental strain [96]. Analysis of heterozygous diploids formed between mutants revealed the existence of at least five complementation groups with respect to penicillin production, designated *npe V, W, X, Y* and *Z.* Mutants of the groups *V* and *W* were able to synthesize the tripeptide ACV whereas mutants of the groups *X, Y* and *Z* failed to form this intermediate. If each one of these three complementation groups corresponded to mutations in a different gene, at least three genes appeared to be involved in the synthesis of the tripeptide ACV and its control. However, it was difficult at that time to establish the exact number of genes involved in the absence of further biochemical data. Similarly, mutants impaired in the various steps of penicillin biosynthesis have been isolated by J.M. Cantoral (unpublished data) [82].

Complementation analysis of genes involved in penicillin biosynthesis. Complementation studies between mutants impaired in penicillin biosynthesis were hampered by the impermeability of fungal cell walls to intermediates of penicillin biosynthesis. Makins and co-workers [97] developed a method to overcome this limitation. Using osmotically-fragile mycelium, cosynthesis was demonstrated between pairs of the blocked mutants. This method of cosynthesis based on the passage of intermediates through enzyme-induced pores in the mycelial wall is of crucial potential for the characterization of blocked mutants in the biosynthesis of fungal secondary metabolites.

Analysis of the segregation of mutant phenotypes from heterozygous diploids, indicated that mutants type *npe W, Y* and *Z* belonged to the same haploidization group whereas types *X* and *V* mapped in the second and third linkage groups [96, 98]. This result is intriguing since recombinant DNA approaches revealed that the three structural genes required for penicillin biosynthesis in *P. chrysogenum* are linked in a cluster (see below).

Genetics of penicillin biosynthesis in Aspergillus nidulans. *A. nidulans* produces very low levels of penicillin and many wild-type strains synthesize barely detectable levels of the antibiotic. However, *A. nidulans* has the advantage of having a well developed genetic map and is a good model microorganism to perform sophisticated genetic manipulations to establish models of regulation of gene expression [99].

Despite the low level of antibiotic production, mutants of *A. nidulans* impaired in penicillin biosynthesis have been isolated [100, 101]. Using complementation tests four groups A, B, C and D, were established [101]. As in *P. chrysogenum*, over two-thirds of the isolated mutants were members of single complementation group (class A) [102] (locus *npeA*). Several of the wild-type isolates that lacked penicillin biosynthesis were altered in the same locus *npeA* [102] located on linkage group VI [100, 103]. This locus contains the cluster of penicillin biosynthetic genes in *A. nidulans* [104]. The three other mutations *npeB*, *npeC* and *npeD* were located on linkage groups III, IV and II respectively.

Mutants blocked in cephalosporin biosynthesis. The use of blocked mutants impaired in cephalosporin biosynthesis was of great importance for elucidating whether penicillin N was a precursor of cephalosporin C or whether these two types of antibiotic were distinct products of a branched pathway. Two types of mutants of *A. chrysogenum* unable to synthesize cephalosporin C or penicillin N were first studied by Lemke and Nash [95, 105]. One class produced a peptide containing α-aminoadipic acid, cysteine and valine, whereas the other did not form the peptide material.

In another laboratory, three blocked mutants of *A. chrysogenum* (N2, N31 and N79) were found which accumulate the dimer of ACV and also the *S*-methylthio derivative of the tripeptide [106, 107]. A point mutation which causes a proline to leucine change in the amino acid sequence of the IPNS is the origin of the inactivation of the enzyme in mutant N2 [108].

Similar results were obtained by Yoshida *et al.* (1978) who classified five mutants of *A. chrysogenum* blocked in β-lactam antibiotic biosynthesis in type A (which do not produce β-lactam antibiotics at all) and type B which formed penicillin N but not cephalosporin C. Cell-free extracts of type A converted penicillin N to a cephalosporin; those of type B mutants did not.

Fujisawa and co-workers characterized two distinct cephalosporin C negative mutants both involved in the final steps of cephalosporin C biosynthesis [90, 91, 109]. Certain cephalosporin-negative mutants were found to accumulate deacetylcephalosporin C (DAC). Such mutations partially inactivated the gene encoding the acetyltransferase which converts DAC into cephalosporin C [90, 92]. Queener and co-workers [110] isolated also a mutant of *A. chrysogenum* which accumulated penicillin N, deacetoxycephalosporin C (DAOC) and DAC.

10.2.4.3 *Genes involved in β-lactam antibiotic biosynthesis.* Three genes were initially assumed to be involved in penicillin biosynthesis [111]. These three genes, *pcbAB*, *pcbC* and *penDE*, encode the enzymes of the penicillin biosynthetic pathway and have been isolated in the last few years by molecular cloning techniques.

ACV is synthesized by a multifunctional peptide synthetase. ACV synthesis might be the rate-limiting step in the biosynthesis of penicillins and cephalosporins and is known to be regulated by glucose in *P. chrysogenum*, *A. chrysogenum* and *N. lactamdurans* [112–114], by phosphate in *S. clavuligerus* [115] and by ammonium in *S. clavuligerus* [115, 116]. It is also strongly affected by the oxygen transfer rate of the cultures [117].

ACV synthesis is stimulated when protein synthesis is blocked with cycloheximide or anisomycin indicating that it is synthesized by a non-ribosomal mechanism [118]. Formation of the tripeptide ACV is carried out by the enzyme ACV synthetase. This enzyme requires ATP for the reaction [119] and, therefore, should be named 'synthetase' in order to differentiate it from the isopenicillin N synthase and deacetoxycephalosporin C synthase which do not require ATP. Cell-free systems catalysing ACV formation have been described for *A. chrysogenum* [119, 120] and *S. clavuligerus* [121, 122]. A multifunctional peptide synthetase that catalyses the formation of ACV was purified from *A. nidulans* [123], *A. chrysogenum* [124, 125] and *S. clavuligerus* [122]. The molecular weight of the *A. nidulans* enzyme was originally reported to be 220 kDa [123], but upon further analysis appeared to be larger than 400 kDa [126].

An unusually massive gene (pcbAB) *encodes ACV synthetase.* The gene *pcbAB* encoding the ACV synthetase of *P. chrysogenum* was cloned using two different strategies: (i) complementation of mutants of *P. chrysogenum* blocked in penicillin biosynthesis (*npe*5 and *npe*10) and (ii) transcriptional mapping of the regions around the previously cloned (see below) *pcbC–penDE* cluster [127, 128]. *P. chrysogenum* DNA fragments, cloned in lambda EMBL3 or cosmid vectors from the upstream region of the *pcbC–penDE* cluster, carry a gene (*pcbAB*) that complements the deficiency of α-aminoadipyl-cysteinyl-valine synthetase of mutants *npe*5 and *npe*10, and restores penicillin production to mutant *npe*5. A protein of at least 250 kDa was observed in sodium dodecylsulfate- polyacrylamide gel electrophoresis (SDS-PAGE) gels of cell-free extracts of complementing strains. This protein is absent in the *npe*5 and *npe*10 mutants but exists in the parental strain from which the mutants were obtained. Transcriptional mapping studies showed the presence of one long transcript of about 11.5 kb, which hybridized with several probes internal to the *pcbAB* gene, and two small transcripts of 1.15 kb, which hybridized with the *pcbC* or the *penDE* gene, respectively. The transcription initiation and termination regions of the *pcbAB* gene were mapped by hybridization with several small probes prepared from the cloned DNA. The *pcbAB* is linked to the *pcbC* and *penDE* genes and is transcribed in the opposite orientation to them (Figure 10.5). The region has been completely sequenced. It includes an ORF of 11 376 nucleotides which encodes a protein with a deduced M_r of 425 971. Similar findings were reported independently by Smith *et al.* [129]. No introns were detected in the *pcbAB* gene. Three repeated domains were found in the α-aminoadipyl-cysteinyl-valine synthetase which have high homology with the gramicidin S synthetase I and tyrocidine synthetase I (Figure 10.6).

Initially, two *loci*, *pcbA* and *pcbB*, were allocated for the enzymatic steps that form the α-aminoadipyl-cysteine (AC) dipeptide and the ACV tripeptide. However, genetic evidence [128, 129] indicated that a single gene encodes an unusually large polypeptide which activates the three amino acids in the L form, racemizes L to D valine and carries out the polymerization steps to form the ACV tripeptide.

Corresponding *pcbAB* genes encoding ACV synthetases have been cloned from *A. chrysogenum* [130], *A. nidulans* (designated *acvA*) [126] and *Nocardia lactamdurans* [19]. A 24 kb region of *A. chrysogenum* C10 DNA was cloned by hybridization with the *pcbAB* and *pcbC* genes of *P. chrysogenum*. The

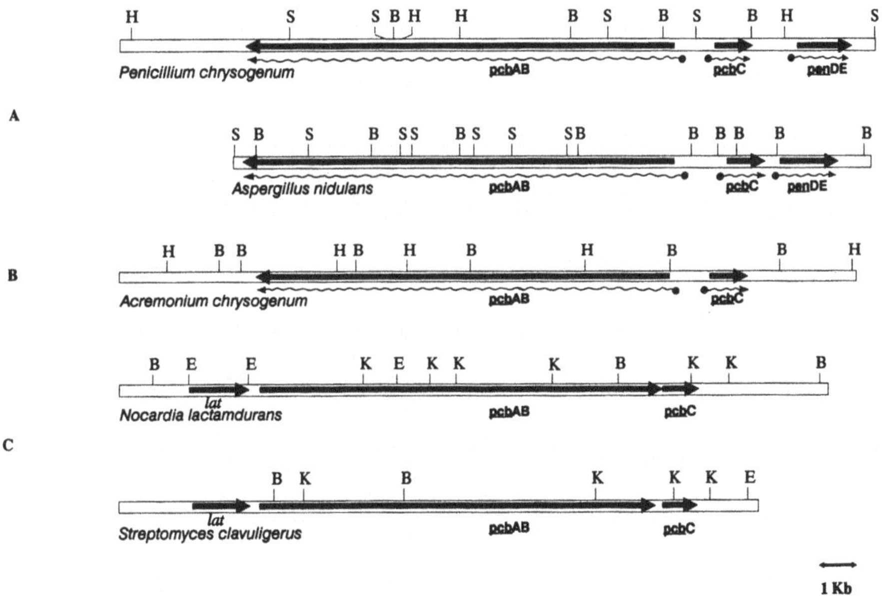

Figure 10.5 Organization of the clusters of genes involved in (A) penicillin biosynthesis in *P. chrysogenum* and *A. nidulans*; (B) early steps of cephalosporin biosynthesis in *A. chrysogenum*; and (C) early steps of cephamycin biosynthesis in *N. lactamdurans* and *S. clavuligerus*. Note the different orientation of the *pcbAB* gene with respect to the *pcbC* gene in procaryotic and eucaryotic β-lactam producers. The known transcripts formed from the penicillin and cephalosporin cluster of genes are indicated by wavy lanes. In *A. chrysogenum* a second cluster includes the *cefEF* and *cefG* genes [92].

```
N.lactamdurans ACV Synthetase  (1,3649)  MTSARHLKSAADWCARIDAIAGQRCDLEMLLKDEWRHRVAVRD-SDTAVRATQEKELTI    58
A.chrysogenum  ACV Synthetase  (1,3712)          VALEQWKTTVQSVS-ERCDLSGLSQHPTDYQLASTGVKGAGGSSIEERSAIV    51
P.chrysogenum  ACV Synthetase (62,3792)     RVRFRGGIERWKECVNQVP-ERCDLSGLTTDSTRYQLASTGF-GDASAAYQERLMTV   116
A.nidulans     ACV Synthetase (59,3770)     QVCFAGGLEGWKAGISKIT-ERCDLSSIATNSTKYQLAVTGFS-DGPDDYNEYSVPF   113

SGQDYTALKQALGAMPLE--AFALATLHSVLHAYGHGHQTVVA----------FLRDGKVLPVVVDHLE------QAGLTCAEAAEQLEDAVAREDMYLP   140
SDELFSSLRDVCSQRQLDPRSLMLFSVHQMLKRFGNGSHTVVASLVTSSEGCPSTSAWRAIPSVIHHIEGGDNNNTVASAVEQAANLLNSEGSGQDLLIP   151
PVDVHAALQELCLERRVSVGSVINFSVHQMLKGFGNGTHTITASLHREQNLQNSSPSWVVSPTIVTH--ENRDGWSVAQAVE---SIEAGRGSEKESVTA   211
PSEVLVAMEEMCLARDISMRSVIQFAVHYVLKGFGGGSHTVAASIDVGDDPNNIATSYTITPSIVCH--ESRQGQTVMQEIQ---SMEKLNQLRKQEMHP   208

PE---ELLQRGLFDALLVLADGHLGFTELP-PAPLVTIVRDDPAAGCLHWRIAYAGEFFEDKIIAGVLDVAREVLG-QFIGRPEQLVADIDLVSAEQELQ   235
IGL-TELVKSELID-LLVIFDDETNNIRLP-QD-FPLILRIHQRQDHWQLSVRYPSPLFDTMVIDSFLSALHNLLS-A-VTKPSQLVRDIELLPEYQVAQ   245
IDSGSSLVKMGLFD-LLVSFVD-ADDARIPCFD-FPLAVIVRECDANLSLTLRFSDCLFNEETICNFTDALNILLAEA-VIGRVTPVADIELLSAEQKQQ   307
GEAGLSLIRMGLFD-ILVIFAD-ANKCEGL-IAGLPLAVMVCEGGGRLQVRIHFSGSLFRQKTLVDIAEALNVLFAKA-ASGGATPVRDLELLSAEQKQQ   304

LHQWNGTDGEFDEDKRLNELFEDVVRRAPDREAVVCGDVRLTYREVNERANQFAHWLIQGPVRVRPGALIGLYLDKSDLGVVATFGIWKSGAAYVPIDPA   335
LEKWNNTDGDYPTEKRLHHLFEEAAVRRPQHVALICGDKRITYEELNAMANRLAHHLVSSGIQT--EQLVGLFLDKTELMIATILGIWKSGAAHVPIDPG   343
LEEWNNTDGEYPSSKRLHHLIEEVVERHEDKIAVVCDERELTYGELNAQGNSLARYLRSIGILP--EQLVALFLDKSEKLIVTILGVWKSGAAYVPIDPT   405
LEEWNKTDGEYPECKRLNHLIEEATQLHEDKVAIVYKRRQLTYGELNAQANCFAHYLRSIGILP--EQLVALFLEKSENLIVTILGIWKSGAAYVPIDPT   402

YPAERIRFLVGDTGLSGIVTNRRHAERLR-EVLGDEHASVHVIEVEAVVAGPHPEQARENPG----LALSSRDRAYVVTYTSGTTGVPKGVPKYHYSVVNS   430
YPDERVKFVLNDTKAQVVIASQRHVDRLRAEAVGGQHLRIIGLESLFDNLAQQTQHSPETSGNLTHLPLNSKQLAYVVTYTSGTTGFPKGIYKEHTSVVNS   443
YPDERVRFVLDDTKARAIIASNQHVERLQREVIGDRNLCIIRLEPLLASLAQDSSKFP--AHNLDDLPLTSQQLAYVVTYTSGTTGFPKGIFKQHTNVVNS   503
YPDERVRFVLEDTQAKVIIASNHLAERLQSEVISDRELSIIRLEHCLSAIDQQPSTFP--RANLRDPSLTSKQLAYVVTYTSGTTGFPKGILKQHTNVVNS   500

ITDLSERYDMRRPGTERVALFASYVFEPHLRQTLIALINEQTLVIVPDDVRLDPDLFPEYIERHGVTYLNATGSVLQHFDLRRCASLKRLLLVGEELTAS   530
ITDLSARYGVAGEDDEVILVFSAYVFEPFVRQMLMALTTGNSLAIISDEDKFDPDTLIPFIQKHKVTYIHATSSVLQEYDFGSCPSLKRMILVGENLTEP   543
ITDLSARYGVAGQHHEAILLFSACVFEPFVRQTLMALVNGHLLAVINDVEKYDADTLLPFIRRHSITYLNGTASVLQEYDFSDCPSLNRIILVGENLTEA   603
ITDLSARYGVTGDHHEAILLFSAYVFEPFVRQMLMALVNGHLLAMVDDAEKYDAEKLIPFIREHKITYLNGTASVLQEYDFSSCPSLKRLILVGENLTES   600

GLRQLREKFAGRVVNEYAFTEAAFVTAVKEFGPGVTERRDRSIGRPLRNVKWYVLSQGLKQLPIGAIGELYIGGCGVAPGYLNRDDLTAERFTANPFQTE   630
RYEALRQRFKSRILNEYGFTESAFVTALNIFEPT-SQRKDMSLGRPVRNVKCYILDANLKRVPIGVTGELHIGGLGISRGYMNREELTRQKFLPNPYQTD   642
RYLALRQRFKNRILNEYGFTESAFVTALKIFDPE-STRKDTSLGRPVRNVKCYILNPSLKRVPIGATGELHIGGLGISKGYLNRPELTPHRFIPNPFQTD   702
RYLALRRHFKNCILNEYGFTESAFVTALNVFEPG-SARNNTSLGRPVRNVKCYILNKSLKRVPIGATGELHIGGLGISKGYLNRPDLTPQRFIPNPFQTD   699

EEKARGRNGRLYRTGDLARVLLNGEVEFMGRADFQLKLNGVRVEPGEIEAQATEFPGVKKCVVVAKE-----NATGDRHLVGYYLVEDGAEVAEADLIAF   725
KERQRGVNSTMYKTGDLARWLPSGEVEYLGRADFQIKLRGIRIEPGEIESTLAMYPGIRASIVVSKKLLSQGQETIQDHLVGYYVCDEG-HIPEGDLLSF   741
CEKQLGINSLMYKTGDLARWLPNGEVEYLGRADFQIKLRGIRIEPGEIETMLAMYPRVRTSLVVSKKLRNGPEETTNEHLVGYYVCDSA-SVSEADLLSF   801
HEKELGLNQLMYKTGDLARWLPNGEIEYLGRADFQIKLRGIRIEPGEIESTLAGYPGVRTSLVVSKRLRHGEKETTNEHLVGYYVGDNT-SVSETALLQF   798

LEQRLIRIMVPARMVRLTSIPVNVNGKVDWRALPDVSLHPAPANAMNGALLAIDGSNAPLLAITEQLRAIWSEVLGVPQNRIGERDDFFRLGGQSISCIL   825
LEKKLPRYMVPTRLVQLAQIPTNINGKADLRALPAVEVAVAPTH-------KQDGERGNQLE--SDLAAIWGNILSVPAQDIGSESNFFRLGGHSIACIQ   832
LEKKLPRYMIPTRLVQLSQIPVNVNGKADLRALPAVDIS-NSTE-------VRSDLRGDT-E--IALGEIWADVLGARQRSVSRNDNFFRLGGHSITCIQ   890
LELKLPRYMIPTRLVRVSQIPVTVNGKADLRALPSVDLIQPK---------VSSCELTDEVE--IALGKIWADVLGAHHLSISRKDNFFRLGGHSITCIQ   887

LIARV--RQRLSLSLGVEDVFALRTLDALAGHLESQ----------GHAEPEVVAEEVTTGSEPVRVLANGLQQGLLYHHLKT-AGGDDAYVVQSVHRYH   912
LIARV--RQQLGQGITLEEVFQTKTLRAMAALLSEKYTKASNGTNGVTNGTAHVNGHAANGHVSDSYVASSLQQGFVYHSLKN-ELSE-AYTMQSMIHYG   928
LIARIRQRQRLSVSISVEDVFATRTLERMADLLQNK--QQEKCDKPHEAPTELLEENAATDNI---YLANSLQQGFVYHYLKSMEQSD-AYVMQSVLRYN   984
LIARI--RQQLGVIISIEDVFSSRTLERMAELLRSK-ESNGTPDERARPQLKTVAGEVANANV---YLANSLQQGFVYQFLKNMGRSE-AYVMQSVLRYD   980
```

DOMAIN A

Figure 10.6 Deduced amino acid sequences of the ACV synthetases of *N. lactamdurans*, *A. chrysogenum*, *P. chrysogenum* and *A. nidulans*. Some gaps have been introduced to obtain maximal alignment. The location of the three repeated domains in the four ACV synthetases is indicated by brackets on the right. Identical amino acids are shaded. The position of three putative phosphopantetheine binding sequences are boxed with thin lines and the thioesterase active centre is boxed with a thick line.

```
APIRPELMKDAWQAARQTYPALRLRFDWAEEPVQIVDNDDKPFDWRFVDLSATADDAEQEARVRELQERDRTEPYDLAGGRLFRVYLIKQREDLFSLIFS   1012
VPLKRDIYQAAWQRVQGEHPALRLRFTWEAEVMQIVDPKSE-LDWRVVDWTDVSSREKQLVALEQLQTEDLAKVYHLDKGPLMRLYLILLPDSKYSCLFS   1027
TTLSPDLFQRAWKHAQQSFPALRLRFSWEKEVFQLLDQDPP-LDWRFLYFTDVAAGAVEDRKLEDLRRQDLTERFKLDVGRLFRVYLIKHSENRFTCLFS   1083
VNINPDLFKKAWKQVQHMLPTLRLRFQWGQDVLQVIDEDQP-LNWWFLHLADDSALPEE-QKLLELQRRDLAEPYDLAAGSLFRIYLIEHSSTRFSCLFS   1078

CHHIILDGWSLPVLHDEVHRNYLALRAGQPIESDVDNAYVAAQRYWEAHRNDHAAYWVEQLGRIDERGDFAGLLNEKSRYRVSLGDYDHVQRHRTRKLYL   1112
CHHAILDGWSLPLLFNNVHQAYLDLVEGTASPVEQDATYLLGQQYLQSHRDDHLDFWAEQIGRIEERCDMNALLNEASRYKVPLADYDQVREQRQQTISL   1127
CHHAILDGWSLPLLFEKVHETYLQLLHGDNLTSSMDDPYTRTQRYLHAHREDHLDFWAGVVQKINERCDMNALLNERSRYKVQLADYDQVQEQRQLTIAL   1183
CHHAILDGWSLPLLFRKTHGTYLHLLHGHSLRTL-EDPYRQSQQYLQDHREDHLRYWAGIVNQIEERCDMNALLNERSRYKIQLADYDKVEDQQQLTLTV   1177

--GADLTGALKAGCAADQVTLHSVIQFVWHKVLHAIGGGNTTVVGTIVSGRNLPVDGIENSAGLFINTLPLIVDHDQQAGQNVAEAVRDIQAAVNTMNSK   1210
PWNNSMDAGVREELSSRGITLHSILQTVWHLVLHSYGGGTHTITGTTISGRHLPVPGIERSVGLFINTLPMIFDHTVCQDMTALEAIEHVQGQVNAMNSR   1227
SGDAWL-ADLRQTCSAQGITLHSILQFVWHAVLHAYGGGTHTITGTTISGRNLPILGIERAVGPYINTLPLVLDHSTFKDKTIMEAIEDVQAKVNVMNSR   1282
PDASWL-SKLRQTCSAQGITLHSILQFVWHAVLHAYGGGTHTVTGTTISGRNLPVSGIERSVGLYINTLPLVINQLAYKNKTVLEAIRDVQAIVNGMNSR   1276

SIVELGRLQSGEMKRRLFDTLLVLENYPRLLDEEEELAHQEALRFEKAYDADKVDYPIAVVARE-EGDEL-TVTLWYAGELFDEDTIDTLLDVARTLFRQ   1308
GNVELGRMSKNDLKHGLFDTLFVLENYPNL-DTEQREKHEEKLKFTIKGGTEKLSYPLAVIAQE-DGDSGCSFTLCYAGELFTDESIQALLDTVRDTLSD   1325
GNVELGRLHKTDLKHGLFDSLFVLENYPNL-DKSRTLEHQTELGYSIEGGTEKLNYPLAVIAREVETTGGFTVSICYASELFEEVMISELLHMVQDTLMQ   1381
GNVELGRLQKNELKHGLFDSLFVLENYPIL-DKSEEMRQKSELKYTIEGNIEKLDYPLAVIAREVDLTGGFTFTICYARELFDEIVISELLQMVRDTLLQ   1375

VTEDIARPVRELDLISPSMRARFDSWNETAEEFPADKTLHAVFEEMAERWPDEIAVVYRENRLTYRELNERANRLAHYLRSVVELRPDDLVALVLDKSEL   1408
ILGNIHAPIRNMEYLSSNQTAQLDKWNATAFEYP-NTTLHAMFESEAQQKPDKVAVVYEDIRLTYRELNSRANALAFYLLSQAAIQPNKLVGLIMDKSEH   1424
VARGLNEPVGSLEYLSSIQLEQLAAWNATEAEFP-DTTLHEMFENEASQKPDKIAVVYEETSLTYRELNERANRMAHQLRSDVSPNPNEVIALVMDKSEH   1480
VAKHLDDPVRSLEYLSSAQMAQLDAWNATDAEFP-DTTLHAMFEKEAAQKPDKVAVVYEQRSLTYRQLNERANRMAHQLKSDISPKPNSIIALVVDKSEH   1474

MITAIIAAWKTGAAYVPIDSGYPDDRISFMLSDTAARVVVTNEIHSDRLRSLAETGTPVLEIEL-LHLDDQ--PAVNPVTETTSTDLAYAIYTSGTTGKP   1505
MITSILAVWKTGGAYVPIDPRYPDQRIQYILEDTAALAVITDSPHIDRLRSITNNRLPVIQSDFALQLPPS--P-VHPVSNCKPSDLAYIMYTSGTTGNP   1521
MIVNILAVWKSGGAYVPIDPGYPNDRIQYILEDTQALAVIADSCYLPRIKGMAASGTLLYPSVLPANPDSKWSV-SNPSPLSRSTDLAYIIYTSGTTGRP   1579
MIATILAVWKTGGAYVPIDPEYPDDRIRYILEDTSAIAVISDACYLSRIQELAGESVRLYRSDISTQTDGNWSV-SNPAPSSTSTDLAYIIYTSGTTGKP   1573

KAVLVEHRGVVNLQVSLAKLFGLDKAHRDEALLSFSNYIFDHFVEQMTDALLNGQKLVVLDDSMRTDPGRLCRYMNDEQVTYLSGTPSVLSLYDYSS-AT   1604
KGVMVEHHGVVNLCVSLCRLFGLRNTD-DEVILSFSNYVFDHFVEQMTDALLNGQTLVVLNDEMRGDKERLYRYIETNRVTYLSGTPSVISMYEFDRFRD   1620
KGVTVEHHGVVNLQVSLSKVFGLRDTD-DEVILSFSNYVFDHFVEQMTDAILNGQTLLVLNDGMRGDKERLYRYIEKNRVTYLSGTPSVVSMYEFSRFKD   1678
KGVMVEHHGVVNLQISLSKTFGLRDTD-DEVILSFSNYVFDHFVEQMTDAILNGQTLVMLNDAMRSDKERLYQYIETNRVTYLSGTPSVISMYEFSRFKD   1672

SLTRIDAIGEDFTEPVFAKIRGTFPGLIINGYGPTEISITSHKRPYPPDVHRVNKSIGFPVANTKCHVLNKAMKPVPVGGIGELYIGGIGVTRGYLNRED   1704
HLRRVDCVGEAFSEPVFDKIRETFPGLIINGYGPTEVSITTHKRPYPFPERRTDKSIGCQLDNSTSYVLNDDMKRVPIGAVGELYLGGDGVARGYHNRPD   1720
HLRRVDCVGEAFSEPVFDKIRETFHGLVINGYGPTEVSITTHKRLYPFPERRMDKSIGQQVHNSTSYVLNEDMKRTPIGAVGELYLGGEGVVRGYHNRAD   1778
HLRRVDCVGEAFSQPVFDQIRDTFQGLIINGYGPTEISITTHKRLYPFPERRTDKSIGQQIGNSTSYVLNADMKRVPIGAVGELYLGGEGVARGYHNRPE   1772

LTADRFVENPFQTAEERRLGENGRLYKTGDLVRWL----PNGEVEYLGRTDLQVKIRGQRVELGEVEAALSSYPGVVRSLVVAREH-AVGQKYLVGFYVG   1799
LTADRFPANPFQTEQERLEGRNARLYKTGDLVRWIHNANGDGEIEYLGRNDFQVKIRGQRIELGEIEAVLSSYPGIKQSVVLAKDRKNDGQKYLVGYFVS   1820
VTAERFIPNPFQSEEDKREGRNSRLYKTGDLVRWIPGSSG--EVEYLGRNDFQVKIRGLRIELGEIEAILSSYHGIKQSVVIAKDCREGAQKFLVGYYVA   1876
VTAERFLRNPFQTDSERQNGRNSRLYRTGDLVRWIPGSNG--EIEYLGRNDFQVKIRGLRIELGEIEAVMSSHPDIKQSVVIAKSGKEGDQKFLVGYFVA   1870

EQ-EFDEQDLKQWNRKKLPESVVPARVLRITDIPVTPSGKLDARRLPETDFGAGEGAEYVAPVSEFELKLCGIWAQVLEIAPDRIGVHDDFFALGGDSIR   1898
SAGSLSAQAIRRFMLTSLPDYMVPAQLVPIAKFPVTVSGKLDAKALPVPD-DTVED-DIVPPRTEVERILAGIWSELLEIPVDRISIYSDFFSLGGDSLK   1918
DAA-LPSAAIRRFMQSRLPGYMVPSRLILVSKFPVTPSGKLDTKALPPAE-EESEI-DVVPPRSEIERSLCDIWAELLEMHPEEIGIYSDFFSLGGDSLK   1973
SSP-LSPGAIRRFMQSRLPGYMIPSSFIPISSLPVTPSGKLDTKALPTAE-EKGAM-NVLAPRNEIESILCGISAGLLDISAQTIGSDSDFFTLGGDSLK   1967
```

DOMAIN B

Figure 10.6 *Continued*

Figure 10.6 *Continued*

```
LPEPGAAVDEADLFAALRAQLMPSMVPSLLVRLDRPLPMTITGKLDVDALPSA-DFSPKRAAYAAPRDRVEARLCHLWSAQLPGGTVGIDDDFFRCGGDS   2944
TLNAGTVCEASSIRDHLHANLPPYMVPSQIHQLEGSLPVTVNGKLDLNRLSTT-QVS-QPELYTAPRNSTEETLCQLWASLLGVDHCGIDDDLFARGGDS   2990
TTDNETVSEADIL-TFMKARLPTYMVPSHLCCLEGALPVTINGKLDVRRLPEIINDS-AQSSYSPPRNIIEAKMCRLWESALGMERCGIDDDLFKLGGDS   3058
TPDTETVSDSSIL-AHMKSKLPAYMVPKYLCRLEGGLPVTINGKLDVRKLPDIGNPQ-HQISYNPPRDVLEADLCRLWASALGTERCGIDDDLFRLGGDS   3050

ISALHLASQVQREIERKVSVKYLFDHPTVRSFVDNVL------SGLAESSGDDEPEQGRLTGECPMLPIQEWFFAKPLADRHRWNHNFAIRTPPLDPGEL   3038
ISSLRLVGDIYRALGRKVTVKDIYLHRSVRALSENVLTDQKDKGTLPASPPLQRAEQGQVEGDAPLLPIQDWFLSKPLDNPAYWNHCFTIRTGALSVEGL   3090
ITSLHLVAQIHNQVGCKITVRDIFEHRTARALHDHVFMKDSDRSNV---TQF-RTEQGPVIGEAPLLPIQDWFLSKALQHPMYWNHTFYVRTPELDVDSL   3154
ITALHLAAQIHHQIGRKVTVRDIFDHPTIRGIHDNVMVKLV-PHNYV--PQF-QAEQQTVLGDAPLLPIQIWFLSKSLQHPSHWNHTFYLRTPDLDVTTL   3145

RTALDRLVEHHDAFRLRF--PESGGEV--YAEDAAPITLHELDVRGLADA-DLRQRLVDWQRTFDLANGPTACAAYLHGFDDGTARVWFALHHLVVDTVS   3133
RGALKLLQERHDVLRLRLQRRDEGRHVQTFARDCAQPRLTVLDRRSFEDAEDVQEALCEIQSHFDLENGPLYTVAYIHGYEDGSARVWFACHHVMVDTVS   3190
SAAVRDLQQYHDVFRMRLKREEVG-FVQSFAEDFSPAQLRVLNVKDVDGSAAVNEILDGWQSGFNLENGPIGSIGYLHGYEDRSARVWFSVHHMAIDTVS   3253
STAVAELQLYHDAFRMRLR-QIDGRTVQCFADDISPVQLRVLNVKDVDGSAAIDQQLQKYQSDFDLEKGPICAAAYLHGYEDRSARVWFSVHHIIIDIVS   3244

WHILAQDLEILYNGGDLGAKTGSYRQWAQAVRDYTPAEGEREFWAETTRDM-ESAELLAQTEGTTRRREEFALTAPDTRTLLAESPWAYDTEVNDLLLTA   3232
WNIILQDLQALYHGDSLGPKSSSVQQWSLAVSDYKMPLSERAHWNVLRKTVAQSFETLPICMGGVLQCQEKFSRETTTALLSKACPALDSGMHEILLMAV   3290
WQILVRDLQTLYRNGSLGSKGSSFRQWAEAIQNYKASDSERNHWNKLVMETASSISALPTSTGSRVRLSRSLSPEKTASLIQGGIDRQDVSVYDSLLTSV   3353
WQILARDLQILYEGGTLGRKSSSVRQWAEALQSYQGSASERAYWEGLLAQTAANISALPPVTGTRTRLARTWSDDRTVILLNEA-SNQNASIQDLLLAAV   3343

TGFALRSITRQATNHLTVEGHGRE-LFEGAPDVRDTVGWFTTMHPFAVEVD-PGDLGRSVLATRANRRRVPHHGIGYGALFG-GEAPLPAVSFNYLGRLG   3329
-GSALQKAAGDVPQVVTIEGHGREDTIDATLDVSRTVGWFTSMYPFEIPKV--TDPAQGVVDVKEAMRRVPNRGVGYGPAYGYGGSCLPAVSFNYLGRLD   3387
-GLALQHIAPTGPSMVTIEGHGRE-VDQTLDVSRTMGWFTTMYPFEIPRLSTENIVQGVVAVSERFRQVPARGVGYGTLYGYTQHPLPQVTVNYLGQLA   3451
-GLALQQVTPGSPSMITLEGHGREEIVDPTLDLSRTLGWFTSMYPFEIPPLNVETLSQGIASLRECLRQVPARGIGFGSLYGYCKHQMPQVTFNYLGQLT   3442

EGDGQPTEAWQL----DPALSGSHTVDGNRLANRSSIDVTMSCTGGRLVAVVDSLLGEAATRLFASELKVWLERLVSHTATVARNE-PAREATTE-LFDP   3424
QASSGAQRDWTLVMDEDEYPVGLCTSAEDSGRSSSMVDFTFSISGGQLVMDMSSSWGHGARNEFVRTVRNTLDDLIKTTSSRDFSA-PLPPSDQESSFTP   3487
RKQSKP-KEWVLAVGDNEFEYGLMTSPEDKDRSSSAVDVTAVCIDGTMIIDVDSAWSLEESEQFISSIEEGLNKILDGRASQQTSR-FPDVPQPAETYTP   3550
SKQSIT-DQWALAVGDGEMQYGLTTSPADRDQSSFAVDITASCVNGALSVEMNSAWSLEKSMRFISRIEEVLNMILSGTLAQQATPVLTPQVFNEEMYTP   3541

YILVNEDAER--TLFVLPPGEGGAESYLSNLARQLPDLRLVLF-NNVHLHTPMGSFEELGRYYVEHIRRLQPSGPYHLLGWSFGGVLSLEISRQLARAGE   3521
YFVFEEGERHGAPLFLLPPGEGGAESYFHNIVKGLPNRNLVVFNNHYREEKTLRTIEALAEYYLSHIRSIQPEGPYHILGWSFGGILGLEAAKRLTGEGH   3587
YFEYLEPPRQGPTLFLLPPGEGGAESYFNNIVKRLRQTNMVVFNNYYLHSKRLRTFEELAEMYLDQVRGIQPHGPYHFIGWSFGGILAMEMSRRLVASDE   3650
YFEFSKTPRRGPILFLLPPGEGGAESYFNNIVKHLPTTNMVVFNNYYLHSKSLNTFEKLAEMYLGHIRQIQPDGPYHFIGWSFGGTIAMEISRQLVGLGS   3641

RIDDLLLIDPYFGMRQASANIGLPGVEDILDPINYHYRPDEADLARLAGRLGNLVLFKAGEPNDVVNGPHQPRLFEYYHGTRFNHLDLLLPAAAIEV-CD   3620
KIATLALIDPYFDIPSASKAIGQPDDACVLDPIYHVYHPSPESFRTVSSLTNHIALFKATETNDQHGNATQQALYEWFATCPLNNLDKFLAADTIKV-VP   3686
KIGFLGIIDTYFNVRGATRTIGLGDTE-ILDPIHHIYNPDPANFQRLPSATDRIVLFKAMRPNNKYESENQRRLYEYYDALDSTDWTACYQAIPTSSWSR   3750
TIGLLGIIDTYFNVPGATRAIGLGDTE-VLDPIHHISQPEPADFQCLPASTDYIILFKATRVNDKFQSENQRRLYEYYDKTLLNDLDWLLPGASNIHLVR   3740

LAGETHHSWVRNEKLVRLMCERISTSLGS                                                                          3649
LEG-THFTWVHHPEQVRSMCTMLDEWLG                                                                           3712
LR-TIHTFPGSEIHNRWSRCVRLSRNTSLAIDPSLAAQYIGRWK                                                           3791
LEEDTHFSWATNPRQIAHVCSTIEKFLARY                                                                         3770
```

Figure 10.6 *Continued*

pcbAB was found to be closely linked to the *pcbC* gene forming a cluster of early cephalosporin biosynthetic genes [130]. A 3.2 kb *Bam*HI fragment of this region complemented the mutation in the structural *pcbC* gene of the *A. chrysogenum* N2 mutant, resulting in cephalosporin production. A functional α-aminoadipyl-cysteinyl-valine (ACV) synthetase was encoded by a 15.6 kb *Eco*RI-*Bam*HI DNA fragment, as shown by complementation of the ACV synthetase-deficient mutant *npe*5 of *P. chrysogenum*. The *pcbAB* gene is linked to the *pcbC* gene, forming a cluster of early cephalosporin-biosynthetic genes (Figure 10.5). Two transcripts of 1.15 and 11.4 kb were found by Northern hybridization of *A. chrysogenum* RNA with probes internal to the *pcbC* and *pcbAB* genes, respectively. An ORF of 11 136 bp was located upstream of the *pcbC* gene and matched the 11.4 kb transcript initiation and termination regions. It encoded a protein of 3712 amino acids with a deduced M_r of 414 791. The nucleotide sequence of the gene showed 62.9% similarity to the *pcbAB* gene encoding the ACV synthetase of *P. chrysogenum*; 54.9% of the amino acids were identical in both ACV synthetases. Three highly repetitive regions occur also in the deduced amino acid sequence of *A. chrysogenum* ACV synthetase. Each is similar to the three repetitive domains in the deduced sequence of *P. chrysogenum* ACV synthetase and also to the amino acid sequence of gramicidin S synthetase I and tyrocidine synthetase I of *Bacillus brevis* [131, 132]. These regions probably correspond to amino acid activating domains in the ACV synthetase protein. In addition, a thioesterase domain was present in the ACV synthetases of both fungi [130]. Each domain includes a phosphopantetheine binding sequence that is very similar to the phosphopantetheine binding motif of the polyketide and fatty acid synthases (Figure 10.2b).

Isopenicillin N synthase gene from different penicillin and cephalosporin producers. The isolation of the isopenicillin N synthase (IPNS) gene from *A. chrysogenum* was first accomplished by using a mixed oligonucleotide probe based on the first 23 N-terminal amino acids of the purified IPNS from *C. acremonium* ATCC 11550 [133, 134]. The ORF encodes a polypeptide of M_r 38 416 (Figure 10.7).

The IPNS protein has been isolated from *E. coli* transformed with the isolated gene from *A. chrysogenum* and purified to homogeneity [135]. The protein synthesized in *E. coli* suffered a slightly different N-terminal processing from that observed in the fungal protein. In *A. chrysogenum* mature IPNS has lost the terminal methionine and glycine residues whereas in the *E. coli* IPNS the terminal methionine residue is removed but the second residue, glycine, is not. The different processing observed in *E. coli* has no apparent consequence on the biochemical properties of the enzyme. The K_m and kinetics values for the conversion of LLD-ACV to isopenicillin are identical. Recombinant IPNS converted analogue substrates into unusual β-lactam antibiotics in exactly the same way as the fungal protein [134].

The *pcb*C gene (encoding IPNS) of *P. chrysogenum* has also been cloned

```
MK***MPSAEVPTIDVSPLFGDDAQEKVRVGQEINKACRGSGFFYAANHGVDVQRLQDVVNEFHRTMSPQEKYDLAIHAYNKNNS*HVRNGYYMAIEG   94
M****NRHADVPVIDISGLSGNDMDVKKDIAARIDRACRGSGFFYAANHGVDLAALQKFTTDWHMAMSAEEKWELAIRAYNPANP*RNRNGYYMAVEG   93
MPIP*MLPAHVPTIDISPLSGGDADDKKRVAQEINKACRESGFFYASHHGIDVQLLKDVVNEFHRTMTDEEKYDLAINAYNKNNP*RTRNGYYMAVKG   96
MPIL*MPSAEVPTIDISPLSGDDAKAKQRVAQEINKAARGSGFFYASNHGVDVQLLQDVVNEFHRNMSDQEKHDLAINAYNKDNP*HVRNGYYKAIKG   96
MPVL*MPSADVPTIDISPLFGTDPDAKAHVARQINEACRGSGFFYASHHGIDVRRLQDVVSEFHRTMTDQEKHDLAIHAYNENNS*HVRNGYYMARPG   96
MPVL*MPSAHVPTIDISPLFGTDAAAKKRVAEEIHGACRGSGFFIATNHGVDVQQLQDVVNEFHGAMTDQEKHDLAIHAYNPDNP*HVRNGYYKAVPG   96
MGSVPVPVANVPRIDVSPLFGDDKEKKLEVARAIDAASRDTGFFYAVNHGVDLPWLSRETNKFHMSITDEEKWQLAIRAYNKEHESQIRAGYYLPIPG   98
MAST**PKANVPKIDVSPLFGDNMEEKMKVARAIDAASRDTGFFYAVNHGVDVKRLSNKTREFHFSITDEEKWDLAIRAYNKEHQDQIRAGYYLSIPE   96
MGSV**SKANVPKIDVSPLFGDDQAAKMRVAQQIDAASRDTGFFYAVNHGINVQRLSQKYTKFHMSITPEEKWDLAIRAYNKEHQDQVRAGYYLSIPG   96

KKAVESFCYLNPSFSEDHPEIKAGTPMHEVNSWPDEEKHPSFRPFCEEYYWTMHRLSKVL*MRGFALALGKDERFFEPELKEADTLSSVSL*IRYPYL  190
KKANESFCYLNPSFDADHATIKAGLPSHEVNIWPDEARHPGMRRFYEAYFSDVFDVAAVI*LRGFALALGREESFFERHFSMDDTLSAVSL*IRYPFL  189
KKAVESWCYLNPSFSEDHPQIRSGTPMHEGNIWPDEKRHQRFRPFCEQYYRDVFSLSKVL*MRGFALALGKPEDFFDASLSLADTLSAVTL*IHYPYL  192
KKAVESFCYLNPSFSDDHPMIKSETPMHEVNLWPDEEKHPRFRPFCEDYYRQLLRLSTVI*MRGYALALGRREDFFDEALAEADTLSSVSL*IRYPYL  192
RKTVESWCYLNPSFGEDHPMIKAGTPMHEVNVWPDEERHPDFRSFGEQYYREVFRLSKVLLLRGFALALGKPEEFFENEVTEEDTLSASVLMIRYPYL  194
RKAVESFCYLNPDFGEDHPMIAAGTPMHEVNLWPDEERHPRFRPFCEGYYRQMLKLSTVL*MRGLALALGRPEHFFDAALAEQDSLSSVSL*IRYPYL  196
KKAVESFCYLNPSFSPDHPRIKEPTPMHEVNVWPDEAKHPGFRAFAEKYYWDVFGLSSAV*LRGYALALGRDEDFFTRHSRRDTTLSSVVL*IRYPYL  194
KKAVESFCYLNPNFKPDHPLIQSKTPTHEVNVWPDEKKHPGFREFAEQYYWDVFGLSSAL*LRGYALALGKEEDFFSRHFKKEDALSSVVL*IRYPYL  198
KKAVESFCYLNPNFTPDHPRIQAKTPTHEVNVWPDETKHPGFQDFAEQYYWDVFGLSSAL*LKGYALALGKEENFFARHFKPDDTLASVVL*IRYPYL  196

EDYP*P*VKTGPDGEKLSFEDHFDVSMITVLYQTQVQNLQVETVDGWRDLPTSDTDFLVNAGTYLGHLTNDYFPSPLHRVKFYNAERLSLPFFFHAGQ  286
ENYP*P*LKLGPDGEKLSFEHHQDVSLITVLYQTAIPNLQVETAEGYLDIPVSDEHFLVNCGTYMAHITNGYYPAPVHRVKYINAERLSIPFFANLSH  285
EDYP*P*VKTGPDGTKLSFEDHLDVSMITVLFQTEVQNLQVETADGWQDLPTSGENFLVNCGTYMGYLTNDYFPAPNHRVKFINAERLSLPFFLHAGH  288
EEYP*P*VKTGADGTKLSFEDHLDVSMITVLYQTEVQNLQVETVDGWQDIPRSDEDFLVNCGTYMGHITHDYFPAPNHRVKFINAERLSLPFFLNAGH  288
DPYPEAAIKTGPDGTRLSFEDHLDVSMITVLFQTEVQNLQVETVDGWQSLPTSGENFLINCGTYLGYLTNDYFPAPNHRVKYVNAERLSLPFFLHAGQ  292
EEYP*P*VKTGPDGQLLSFEDHLDVSMITVLFQTQVQNLQVETVDGWRDIPTSENDFLVNCGTYMAHVTNDYFPAPNHRVKFVNAERLSLPFFLNGGH  288
DPYPEPAIKTADDGTKLSFEWHEDVSLITVLYQSDVQNLQVKTPQGWQDIQADDTGFLINCGSYMAHITDDYYPAPIHRVKWVNEERQSLPFFVNLGW  292
NPYPPAAIKTAEDGTKLSFEWHEDVSLITVLYQSDVANLQVEMPQGYLDIEADDNAYLVNCGSYMAHITNNYYPAPIHRVKWVNEERQSLPFFVNLGF  290
DPYPEAAIKTAADGTKLSFEWHEDVSLITVLYQSNVQNLQVETAAGYQDIEADDTGYLINCGSYMAHLTNNYYKAPIHRVKWVNAERQSLPFFVNLGY  290

HTLIEPFFPDGAEPEG**KQGN*EAVRYGDYLNHGLHSLIVKNGQT    N. lactamdurans                                 328
ASAIDPFAPPPYAPPG****GN*PTVSYGDYLQHGLLDLIRANGQT    Flavobacterium                                  326
TTVMEPFSP****EDTRGKELN*PPVRYGDYLQQASNALIAKNGQT    S. griseus                                      329
NSVIEPFVP****EGAAGTVKN*PTTSYGEYLQHGLRALIVKNGQT    S. jumonjinensis                                329
NSVMKPFHP****EDTGDRKLN*PAVTYGEYLQEGFHALIAKNVQT    S. lipmanii                                     331
EAVIEPFVP****EGASEEVRN*EALSYGDYLQHGLRALIVKNGQT    S. clavuligerus                                 329
EDTIQPWDPATAKDGAKDAAKDKPAISYGEYLQGGLRGLINKNGQT    C. acremonium                                   338
NDTVQPWDP**SKEDGKT***DQRPISYGDYLQNGLVSLINKNGQT    P. chrysogenum                                  331
DSVIDPFDP**REPNGKS***DREPLSYGDYLQNGLVSLINKNGQT    A. nidulans                                     331
```

Figure 10.7 Amino acid sequences of the isopenicillin N synthase of *N. lactamdurans*, *Flavobacterium* sp., *S. griseus*, *S. jumonjinensis*, *S. lipmanii*, *S. clavuligerus*, *A. chrysogenum*, *P. chrysogenum* and *A. nidulans*. Identical amino acids are shaded. The position of the two conserved cysteines are indicated by arrowheads. Note that the second cysteine (position 255) is not present in the *N. lactamdurans* gene. A histidine residue located 10 amino acids downstream from the conserved cysteine at Cys-106 (see text) is indicated by an open circle. The Pro-285 (in the *A. chrysogenum* IPNS) indicated by an asterisk is essential for enzyme activity. The conserved region E, R, L/Q, S, I/L, P^{285}, F, F is boxed. Two peptides Asp-40 to Arg-78 and Thr-237 to Gly-256 which are labelled by the diazirinyl analogue of ACV are underlined.

Table 10.2 Analogy of the *pcbC* genes of nine different β-lactam producing organisms.

	Nocardia lactamdurans	*Streptomyces griseus*	*Streptomyces jumonjinensis*	*Streptomyces lipmanii*	*Streptomyces clavuligerus*	*Flavobacterium* sp.	*Acremonium chrysogenum*	*Penicillium chrysogenum*	*Aspergillus nidulans*
								Amino acids homology (%)	
N. lactamdurans		72.6	77.0	71.5	75.0	59.3	57.3	57.3	59.6
S. griseus	78.8		75.7	73.9	72.0	59.8	53.6	56.0	57.2
S. jumonjinensis	80.9	82.0		69.7	81.5	60.9	60.4	58.4	59.1
S. lipmanii	78.8	82.8	82.0		70.9	55.3	55.6	54.4	57.0
S. clavuligerus	79.0	78.9	84.8	79.3		59.4	56.8	56.6	57.9
Flavobacterium sp.	69.1	69.8	70.1	66.7	69.2		53.6	54.5	55.2
A. chrysogenum	66.5	69.7	70.9	69.8	67.8	67.0		76.6	74.0
P. chrysogenum	64.3	65.5	67.1	67.2	65.2	65.0	76.1		81.3
A. nidulans	63.7	63.0	64.7	65.3	63.8	69.6	71.5	76.2	
			Nucleotides homology (%)						

in two different laboratories. Barredo *et al.* [136] cloned *Sau*3A-digested fragments of total DNA of *P. chrysogenum* AS-P-78 into the *Bam*HI site in lambda EMBL-3 arms using a probe corresponding to the N-terminal end of the enzyme. A similar strategy was followed by Carr *et al.* [137] to clone the IPNS gene of a different strain of *P. chrysogenum* (strain 23X-80-269-37-2). A gene library was screened using a heterologous hybridization probe based on the nucleotide sequence of the IPNS gene of *A. chrysogenum.* The *pcbC* gene does not contain introns and is expressed in *E. coli* minicells with the help of the P_L promoter of phage lambda [136]. *E. coli* cells transformed with the IPNS encoding gene contained IPNS activity whereas untransformed cells were devoid of it. Results of both laboratories indicate that the ORF encodes a polypeptide (Figure 10.7) of M_r 37 900 which agrees with the reported value of $39\,000 \pm 1000$ for the purified protein from *P. chrysogenum* [80]. However, there is one amino acid (Tyr-194) which is different in the IPNS proteins from strains AS-P-78 and the high producer 23X-80-269-37-2 [136].

The *pcbC* gene of a third *β*-lactam producer, *Aspergillus nidulans* (designated *ipnA*), was identified by heterologous hybridization with a DNA probe corresponding to the *A. chrysogenum* IPNS [138]. The open reading frame encodes a 331 amino acid polypeptide and an M_r of 37 480 with extensive homology with the genes of other *β*-lactam producing fungi (Figure 10.7).

In addition to the three fungal *pcbC* genes, the analogous genes have been cloned from several bacteria, i.e. actinomycetes, such as *S. clavuligerus* [139], *S. griseus* [140], *S. jumonjinensis* [141], *S. lipmanii* [142], *Nocardia lactamdurans* [19] and *Flavobacterium* sp. [143]. Attempts to draw conclusions regarding important residues in IPNS proteins were initially frustrated by the extensive homology in their amino acid sequences. Many residues are conserved in 12 separate regions and it was difficult to assign significance to conserved regions (see shaded areas in Figure 10.7).

Most IPNS proteins contain two cysteine residues (numbered Cys-255 and Cys-106 in *A. chrysogenum*) the first one is observed in highly conserved regions which extend about 5 amino acids on each side of cysteine. Also the first cysteine, Cys-106 (but not the second, contrary to what was initially assumed from the first sequenced genes) has a histidine residue within 10 amino acids downstream from the cysteine (Figure 10.7). This histidine may be involved in iron binding since histidine is known to be commonly modified by Fe^{2+} in the presence of ascorbate and DTT [144].

One of the two conserved cysteines found in all other IPNSs is absent in the *N. lactamdurans pcbC* gene (residue 249 of *N. lactamdurans* IPNS). This striking result was reconfirmed by sequencing both strands using *Taq* polymerase and sequenase. A GCC triplet corresponding to alanine substitutes the normal TGC triplet encoding cysteine in other *Streptomyces* and fungi [19].

Amino acids which are important for cyclase activity have been established by *in vitro* mutagenesis of cysteine residues to serine [144] and by characterization of a mutation in the isopenicillin N synthase gene of *A. chrysogenum* N2 [108] a mutant deficient in IPNS [80]. The C→T mutation at nucleotide 854 existing in strains N2 changes amino acid 285 from proline to leucine. This proline, which is nested in one of the highly conserved regions ER(L/Q)S(I/L)P^{285}FF, defines, therefore, one of the domains involved in the active centre of IPNS.

Substitution of a serine for Cys-104 in the IPNS polypeptide of *A. chrysogenum* reduced enzyme activity by approximately 95% whereas substitution of Cys-255 has a much less dramatic effect [144] which agrees with the lack of effect on substrate cyclization of the absence of the second cysteine in the *N. lactamdurans* enzyme, which corresponds to the Cys-255 of *A. chrysogenum.* In fact, the K_m of the *N. lactamdurans* IPNS is almost identical to that of the IPNSs of the other β-lactam producers [13, 145]. Two peptides, Asp-40 to Arg-78 and Thr-237 to Gly-256 (Figure 10.7) were shown to be labelled by a diazirinyl-containing analogue of the substrate ACV [124]. These peptides coincide with some of the conserved stretches of amino acids and are probably involved in the cyclization. In summary, cysteine at position 104 and proline at position 285 in *A. chrysogenum* cyclase (but not cysteine at position 255) have been shown to be required for enzyme activity. Recently, a new model for the mechanism of cyclization of the ACV tripeptide at the active centre of IPNS has been proposed [146] in which a histidine rather than a cysteine is used to form an initial bridge between enzyme and oxygen molecule via an Fe^{2+} atom (a known cofactor of the enzyme; [80]). The conserved His residue located 10 amino acids downstream of the conserved Cys-104 is a putative candidate for interacting with the cofactor.

The penDE gene encodes isopenicillin N: acyl-CoA acyltransferase. *P. chrysogenum* and *A. nidulans* produce penicillins with acyl side chains formed from a variety of intracellular or exogenously supplied carboxylic acids. When exogenous phenylacetic acid or phenoxyacetic acid is provided, penicillins with phenylacetyl and phenoxyacetyl side chains, e.g. penicillin G and V respectively, are produced (Figure 10.5).

The biosynthesis of penicillin G and V occurs by replacement of the L-α-aminoadipyl side chain of IPN with the phenylacetyl moiety of phenylacetyl-CoA or the phenoxyacetyl moiety of phenoxyacetyl-CoA, respectively, by the isopenicillin N:acyl-CoA acyltransferase (IAT). This enzyme occurs in penicillin-producing strains but it does not occur in *A. chrysogenum* and other cephalosporin producers [83] because of the absence of the corresponding gene in this fungus [147]. This is a rare example of restricted distribution of secondary metabolites among related fungi due to the absence of a particular gene in certain genera.

The nature of the last enzymatic step of penicillin biosynthesis and the role of 6-APA as a putative intermediate in penicillin biosynthesis has remained obscure for many years (see reviews by Queener and Neuss [148], Demain [74], and Martin and Liras [13]). A two-step model for the conversion of IPN to benzylpenicillin has been proposed [148]. This would require the involvement of two enzyme activities, IPN amidohydrolase (6-APA forming), which would cleave IPN to 6-APA and α-aminoadipic acid, and acyl-CoA: 6-APA acyltransferase (AAT) (Figure 10.4). The IAT protein of *P. chrysogenum* has been purified to homogeneity. The purified preparation that catalysed the formation of benzylpenicillin from phenylacetyl-CoA and 6-APA or (with a lower affinity) from IPN was shown to contain three proteins of 40, 29 and 11 kDa. The N-terminal sequence of the 29 kDa protein was used to isolate *P. chrysogenum* DNA which contained an ORF with three introns [149] (Figure 10.8). The deduced amino acid sequence of the ORF encodes a 39 kDa protein. A DNA sequence in the gene was found which corresponded to the N-terminal sequence of the 11 kDa protein, and downstream of this sequence the nucleotide sequence matched the N-terminal sequence of the 29 kDa protein [149]. The large protein (40 kDa) corresponded to a heterodimer formed from the 11 and 29 kDa subunits [150]. The 11 and 29 subunits are probably formed by proteolysis of the 40 kDa protein encoded in the cloned gene (E. Montenegro, F.J. Perrino and J.F. Martin, unpublished data). Whiteman and co-workers reported that separation of the 10 and 29 kDa proteins purified from *P. chrysogenum* was associated with almost complete loss of acyl-CoA: IPN acyltransferase which could be stimulated 15-fold by mixing the two proteins [150]. However, the 29 kDa subunit of *P. chrysogenum* AS-P-78 retains a high enzymatic activity [83]. The exact nature of the activation and catalytic sites existing in each of the subunits is still unknown (see review by Martin *et al.* [14]).

The analogous gene encoding the IAT protein of *A. nidulans* was isolated and identified by complementation of the *npe* mutants of *P. chrysogenum* lacking IAT activity [151]. The *A. nidulans penDE* gene is very similar to the previously cloned *penDE* gene of *P. chrysogenum* [149] (Figure 10.8). Both genes contain three introns in similar positions which were identified by comparison with the fungal consensus intron/exon splicing sequences [152] and confirmed by mRNA hybridization experiments using three oligonucleotides internal to each intron.

The presence of three introns in the *penDE* genes of *P. chrysogenum* and *A. nidulans* suggests that these genes have not originated from *Streptomyces* in a recent transfer event as proposed for the other genes of the penicillin and cephalosporin pathway (see section 10.2.4.3). These genes appear to have evolved in *P. chrysogenum*, *A. nidulans* and other benzylpenicillin-producing fungi from ancestral fungal genes.

The genes encoding the cephalosporin biosynthetic pathway are located within at least two different clusters. The genes (*pcbAB* and *pcbC*) encoding

```
      10v       20v       30v       40v       50v       60v       70v       80v       90v       100v
MLHVTCQGTPSEIGYHHGSAAKGEIAKAIDFATGLIHGKTKKTQAELEQLLRELEQVMKQRWPRYYEEICGIAKGAEREVSEIVMLNTRTEFAYGLVEARD
MLH: CQGTP EIGY:HGSAAK: IA::IDFA.:LI:GKTKKT:.EL.Q:L.:L.:V:.:RWP:YYEEI GIAKGAER:VSEIVMLNTRTEFAYGL .ARD
MLHILCQGTPFEIGYEHGSAAKAVIARSIDFAVDLIRGKTKKTDEELKQVLSQLGRVIEERWPKYYEEIRGIAKGAERDVSEIVMLNTRTEFAYGLKAARD
      10^       20^       30^       40^       50^       60^       70^       80^       90^       100^

      110v      120v      130v      140v      150v      160v      170v      180v      190v      200v
GCTTVYCKTPNGALQGQNWDFFTATKENLIQLTICQPGLPTIKMITEAGIIGKVGFNSAGVAVNYNALHLHGLRPTGLPSHLALRMALESTSPSEAYEKIV
GCTT.YC: PNGALQGQNWDFF:ATKENLI:LTI Q:GLPTIK.ITEAGIIGKVGFNSAGVAVNYNALHL:GLRPTG:PSH:ALR:ALESTSPS:AY::IV
GCTTAYCQLPNGALQGQNWDFFSATKENLIRLTIRQAGLPTIKFITEAGIIGKVGFNSAGVAVNYNALHLQGLRPTGVPSHIALRIALESTSPSQAYDRIV
      110^      120^      130^      140^      150^      160^      170^      180^      190^      200^

      210v      220v      230v      240v      250v      260v      270v      280v      290v      300v
SQGGMAASAFIMVGNAHEAYGLEFSPISLCKQVADTNGRIVHTNHCLLNHGPSAQELNPLPDSWSRHGRMEHLLSGFDGTKEAFAKLWEDEDNYPLSICRA
.QGGMAASAFIMVGN:HEA:GLEFSP.S: KQV D:NGR:VHTNHCLL:HG :.:EL:PLPDSW:RH RME LL.GFDGTK:AFA:LW.DEDNYP:SICRA
EQGGMAASAFIMVGNGHEAFGLEFSPTSIRKQVLDANGRMVHTNHCLLQHGKNEKELDPLPDSWNRHQRMEFLLDGFDGTKQAFAQLWADEDNYPFSICRA
      210^      220^      230^      240^      250^      260^      270^      280^      290^      300^

      310v      320v      330v      340v      350v
YKEGKSRGSTLFNIVFDHVGRKATVRLGRPNNPDETFVMTFSNLDTKSAIQANIX   Aspergillus nidulans ATCC 28901
Y.EGKSRG:TLFNI::DH. R.ATVRLGRP.NPDE FVM F.: D.:SA::A.:X
YEEGKSRGATLFNIIYDHARREATVRLGRPTNPDEMFVMRFDEEDERSALNARLX   Penicillium chrysogenum AS-P-78
      310^      320^      330^      340^      350^
```

Figure 10.8 Amino acid sequences of the acyl-CoA: 6-amino penicillanic acid acyltransferases of *A. nidulans* (upper line) and *P. chrysogenum* (lower line) showing the identical or functionally conserved (:.) amino acids (centre line). The conserved 11 amino acid sequences surrounding the processing site (vertical arrow) of the *P. chrysogenum* enzyme are boxed.

the first two steps of the cephalosporin pathway have been cloned from the cephalosporin-producer *A. chrysogenum* and the cephamycin producer *N. lactamdurans*. They have been considered in section 10.2.4.3. since they are analogous to the genes involved in penicillin biosynthesis. The *pcbAB* was found to be closely linked to the *pcbC* gene forming a cluster of early cephalosporin-biosynthetic genes. Two transcripts of 1.15 and 11.4 kb were found by Northern hybridization of *A. chrysogenum* mRNA with *pcbC* and *pcbAB* internal probes respectively, which are transcribed in opposite orientations as in *P. chrysogenum* [130]. However, in *N. lactamdurans* this organization is different. Both *pcbAB* and *pcbC* are clustered head to tail and appear to be expressed as a single transcript (J.J.R. Coque, P. Liras and J.F. Martin, unpublished data). The location of the ACV synthetase gene upstream of the *pcbC* gene of *A. chrysogenum* has been confirmed by gene disruption of the region carrying the *pcbAB* gene [153].

A separate cluster of genes (*cefEF-cefG*) in the genome of *A. chrysogenum* encodes the bifunctional deacetoxycephalosporin C synthetase/hydroxylase [154] and the acetyl-CoA:deacetylcephalosporin C acetyltransferase [92]. *cefD*, encoding the isopenicillin N epimerase, has not yet been cloned and it is not known whether it is linked to either of the two clusters.

The *cefEF* gene (and, therefore, also the *cefG* gene) is located on chromosome II whereas the *pcbC* gene (and also the *pcbAB*) is located on chromosome VI [155]. Chromosome rearrangements have been found in improved cephalosporin C producing strains of *A. chrysogenum* [156]. In one improved strain, chromosomal changes altered the size of the chromosome on which the *pcbC* was located.

The *cefEF* gene encoding a bifunctional penicillin N expandase/DAOC 3′-hydroxylase was isolated by Samson *et al.* [154], using DNA probes based on the amino acid sequence of internal peptide fragments of the purified bifunctional protein. A protein of M_r 36 461 was encoded by the ORF. This gene was expressed in *E. coli* and the protein formed contained both penicillin expandase and DAOC 3′-hydroxylase activities, demonstrating the bifunctional nature of the polypeptide encoded by the gene.

Surprisingly, monofunctional penicillin N expandase and DAOC 3′-hydroxylase proteins are encoded by two different genes *cefE* and *cefF* both in *S. clavuligerus* [157, 158] and *N. lactamdurans* (J. Calzada, J.J.R. Coque, P. Liras and J.F. Martin, unpublished data). Probably the fungal expandase/hydroxylase gene was derived by a fusion of separate primordial expandase and hydroxylase genes. Both these genes may have arisen first in *Streptomyces* as predicted by the horizontal transfer hypothesis (see below).

The gene (*cefG*) encoding acetyl-CoA: deacetylcephalosporin C acetyltransferase, the final step of the *A. chrysogenum* strain C10 cephalosporin pathway, has been cloned recently [92]. The gene contains two introns and encodes a protein of 444 amino acids with an M_r of 49 269 that correlates well with the M_r deduced by gel filtration. The *cefG* gene is contiguous with the *cefEF* gene but it is expressed in the opposite orientation to the *cefEF* gene. Two transcripts of 1.2 and 1.4 kb were observed in *A. chrysogenum* which correspond to the *cefEF* and *cefG* genes respectively; the degree of expression of the *cefG* gene was lower than that of the *cefEF* gene, at least in 48 h cultures. The cloned *cefG* gene complemented the deficiency of deacetylcephalosporin acetyltransferase in the non-producer mutant *A. chrysogenum* ATCC 20371 and restored cephalosporin biosynthesis in this strain. Heterologous expression of the *cefG* genes was shown to be possible in *P. chrysogenum.* The deacetylcephalosporin acetyltransferase showed a much higher degree of homology with the *O*-acetylhomoserine acetyltransferases of *Saccharomyces cerevisiae* and *Ascobolous immersus* than with other *O*-acetyltransferases.

The *cefEF-cefG* gene cluster encodes, therefore, the enzymes that carry out the three final steps of the cephalosporin biosynthetic pathway.

Origin of the genes for β-lactam antibiotics: the horizontal transfer hypothesis. The β-lactam genes of prokaryotic and eukaryotic microorganisms may serve as a model in the study of the origin of the genes for many secondary metabolites. There are several important observations in this regard.

First, there is a narrow distribution of microorganisms in nature which are able to synthesize β-lactam antibiotics. DNA hybridization studies conducted on a wide collection of organisms established that *pcbC*-related sequences, encoding the IPNS protein, are distributed in just a few groups of microorganisms specifically *Ascomycotina*, actinomycetes and several Gram-negative bacteria [21]. This is a general observation for secondary metabolites. Second, there is a remarkable sequence similarity among the amino acid sequences deduced from the gene sequences from fungal and bacterial producers of β-lactams. Third, genes are clustered, although in different arrangements for different producer organisms.

The similarity of the nine different *pcbC* genes (Table 10.2) is in the range of 64 to 85%. Analysis of these data reveals differences between the fungal cyclase genes, the Gram-positive actinomycetes and the Gram-negative *Flavobacterium*. However, in all cases the similarity in nucleotide sequence is too high to be explained simply on the base of the separate evolution of the different microbial groups. Clearly fungal and bacterial *pcbAB* or *pcbC* genes are much closer than the corresponding 5s RNA encoding genes which are considered to be a slowly evolving group of genes. Weigel *et al.* [142] proposed two explanations for the high similarity between fungal and bacterial genes. This may have resulted from a horizontal transfer event, possibly from the bacterial ancestry to the fungal lineage after the prokaryotic–eukaryotic split. Alternatively, the similarity might simply be a consequence of a slow but constant rate of evolutionary change of the *pcbAB* and *pcbC* gene, perhaps reflecting strict functional constraints on the ACVS and IPNS protein, thus precluding many amino acid changes. Weigel and co-workers favoured the first hypothesis and estimated that a gene transfer event took place about 370 million years ago, well after the proposed time of the procaryotic/eucaryotic split which occurred about 2 to 4 billion years ago.

The evolution of other genes of the β-lactam pathway, in addition to the cyclase genes, may also be compared between fungi and bacteria. The proteins encoded by the *cefEF* gene of *A. chrysogenum* and the *cefE* of *S. clavuligerus* or the ACV synthetases of *P. chrysogenum*, *A. chrysogenum*, *A. nidulans* and *N. lactamdurans* (Figure 10.6) showed almost the same percentage sequence identity as the *pcbC* genes from these organisms, i.e. they probably have also been transferred from bacteria to fungi.

However, the arrangement of the cluster of early cephalosporin biosynthetic genes is different between fungi and bacteria. The *pcbAB* and *pcbC* of *N. lactamdurans* are located head to tail, forming in fact an operon with the *lat* gene (encoding lysine-6-aminotransferase, an α-aminoadipic acid forming enzyme [20]), whereas in the three filamentous fungi *pcbAB* is oriented in the opposite direction to *pcbC* and expressed from a bidirectional promoter. If it is assumed that a gene transfer event took place from a bacterium to a filamentous fungus, then it can be postulated that a major reorganization of the cluster occurred after transfer. Moreover, two of the genes, *penDE* involved in the last step of penicillin biosynthesis and *cefG* which encodes the last enzyme of the cephalosporin pathway, contain introns [92, 149].

If a single transfer event took place from the Gram-positive actinomycetes to filamentous fungi, the gene from the Gram-negative *Flavobacterium* sp. should be less related to the actinomycetes and fungal genes since Gram-positive and Gram-negative bacteria diverged around 1 to 1.5 million years ago. Since this is not the case, Aharonowitz *et al.* [21] have proposed that the gene transfer event occurred before the Gram-positive/Gram-negative bacterial split. A simpler, but more provocative explanation, is that several transfer events occurred between different lineages. If gene transfers occurred after the different fungal genera diverged, this would easily explain why certain fungi acquired the genes and are β-lactam producers whereas others are not.

10.2.5 *Isoprene-derived metabolites: fungal carotenoids*

Carotenoids are a group of photoprotective pigments present in all photosynthetic and many non-photosynthetic organisms including fungi and bacteria [159]. They are probably the most widespread group of pigments in nature. Many of the carotenoids contain a C-40 carbon backbone derived from eight C-5 isoprene units.

Fungal carotenoids have been extensively studied in *Phycomyces* [160] and *N. crassa* [161, 162]. Carotenoids are non-essential for the growth of *N. crassa*, in contrast to plants where they are essential for photoprotection of the photosynthetic apparatus. The absorption of light is due to the presence of a chromophore of 3 to 15 conjugated double bonds which is formed by successive desaturations of the colourless precursor phytoene [161, 162]. Phytoene is desaturated to produce lycopene via neurosporene by a series of four didehydrogenations. In some organisms two dehydrogenases are required to carry out the four dehydrogenation steps whereas in others one dehydrogenase is sufficient [161]. The lack of pigment formation is likely to be due to an impairment in the biosynthesis of carotenoids. Colourless (e.g. albino) mutants are very easy to isolate. In this regard, *al-1* mutants are known to be deficient in phytoene dehydrogenase and they accumulate the colourless intermediate phytoene [164].

Nelson and co-workers [165] isolated the *al-3* gene of *N. crassa* and Schmidhauser *et al.* [166] cloned the *al-1* gene. The *al-1* gene encodes the phytoene dehydrogenase, which is similar to the phytoene dehydrogenases of *Rhodobacter capsulatus* and *glycine max* (soybean) [167]. All the phytoene dehydrogenases show conservation of several amino acid residues including a dinucleotide binding motif that could mediate binding of FAD. The cloned *al-3* gene of *N. crassa* probably encodes the geranylgeranyl pyrophosphate (GGPP) synthetase [165]. This enzyme catalyses the conversion of isopentenyl pyrophosphate to GGPP, which is very low in albino-3 mutants. Further analysis of the *al-3* gene is required to establish the nature of the protein encoded by the cloned DNA fragment.

The activity of two enzymes in the *N. crassa* carotenoid biosynthetic pathway corresponding to the products of the *al-2* and *al-3* genes were shown to increase following photoinduction [168]. Using the cloned genes it has been possible to establish that the messenger RNA levels of *al-3*, *al-1* and *al-2* increase in response to photoinduction [165, 166]. The photoinduced increase of *al-1* and *al-3* (and also *al-2*, unpublished) mRNAs was not observed in *N. crassa* mutants which are defective in all physiological photoresponses (designated *wc* mutants) This suggests that these genes may be controlled by a common blue light response regulatory mechanism, probably a component of the photosensor system.

10.3 Are most genes for secondary metabolites tightly linked to each other?

Particularly relevant to the origin of secondary metabolites is the clustering of the genes which has been observed in *P. chrysogenum* [127, 169], *A. chrysogenum* (with at least two gene clusters; [92, 130, 155]), *A. nidulans* [104, 126] and also in Gram-positive bacteria actinomycetes [17]. Late evolutionary horizontal gene transfer events may explain the restricted taxonomical distribution of many secondary metabolites. If the genes for the biosynthesis of secondary metabolites are contiguous or clustered, the transfer of large fragments of DNA containing several biosynthetic genes would be more effective than multiple single gene transfers. Unfortunately little information is yet available on clustering of genes in other secondary metabolites.

An interesting question is the linkage in a single cluster of intron-free bacterial genes for β-lactam antibiotics with intron-containing genes of fungal origin. The *penDE* gene of *P. chrysogenum* or *A. nidulans* contains three introns and is linked to the intron-free *pcbAB* and *pcbC* genes. The *cefG* gene of *A. chrysogenum* contains two introns and is closely linked to the *cefEF* gene in the cluster of 'late' biosynthetic cephalosporin genes. Such associations can be explained if the product of the biosynthetic pathway (formed at higher yield by coordinated expression and/or amplification of the cluster of genes) confers survival advantage to the organisms that acquired an adequate combination of these genes. Although molecular ecology studies to support this hypothesis have not been carried out, the conservation of the genes through many generations suggest that their products may be important for survival of the producer strains during competition in nature.

This argument emphasizes the lack of information on the physiological and ecological role of the thousands of secondary metabolites in nature. It is likely that many of the secondary metabolites may play control functions either in the producer organisms or in the communities of interacting organisms (symbiotic, parasitic, plant growth regulators, etc.) as seems to be the case for certain compounds such as antibiotics and toxins [170].

10.4 Hybrid secondary metabolites: strains producing metabolites which are not produced in nature

Taxonomically restricted production of secondary metabolites is due to the lack of genes encoding the required biosynthetic enzymes in many fungi. Confirmation of this hypothesis has been provided by Gutiérrez *et al.* [147]. No DNA sequences homologous to the *penDE* gene of *P. chrysogenum* were found in the genome of three different strains of *A. chrysogenum.* Genetic transformation systems are now available [171] to transfer genes from secondary metabolite-producing strains to other filamentous fungi which might have better fermentation characteristics. This 'directed' genetic manipulation may result in the production of hybrid secondary metabolites as a result of the new combination of genes in the recipient strain.

Since *A. chrysogenum* does not form isopenicillin N acyltransferase because of the lack of the *penDE* gene, the gene of *P. chrysogenum* was introduced in two different strains of *A. chrysogenum* [147]. The heterologous gene became integrated by non-homologous recombination. The *penDE* gene was observed to be expressed in the transformants by synthesising a transcript of 1.15 kb which was processed and translated giving rise to active acyl-CoA:isopenicillin N acyltransferase. In this way a cephalosporin-producing strain was converted to a penicillin producer. The reverse conversion (of penicillin to cephalosporin production) might be of great industrial interest because of the high levels of penicillin produced by industrial strains of *P. chrysogenum.*

The entire cluster of penicillin biosynthetic genes can also be transferred to a filamentous fungus which hitherto in nature was not able to form penicillin. In effect, transfer of a DNA fragment containing the *pcbAB–pcbC–penDE* cluster to *A. niger* and *N. crassa* has resulted in the production of penicillin by these penicillin non-producing fungi [172].

Exogenous DNA from bacteria or other fungi may be introduced into a producer of secondary metabolites to obtain modified products. Two genes encoding, respectively, a D-amino acid oxidase from *Fusarium solani* and a cephalosporin acylase from *Pseudomonas diminuta,* were introduced into the cephalosporin producer *A. chrysogenum,* and expressed using the promoter of the *A. chrysogenum* alkaline protease. The resulting transformants were able to synthesize small amounts of 7-aminocephalosporanic acid [173] demonstrating the feasibility of this strategy.

As further genes encoding enzymes required for secondary metabolite syntheses become available, the construction of hybrid products should be routine. However, nature has worked on gene engineering for millions of years and has constructed thousands of secondary metabolites that, to a large extent, remain to be explored.

References

1. Bu'Lock, J.D. (1961) *Adv. Appl. Microbiol.* **3**: 293–342.
2. Bu'Lock, J.D. (1967) *Essays in Biosynthesis and Microbial Development.* John Wiley & Sons, Inc., New York.
3. Weinberg, E.C. (1970) *Adv. Microbial. Physiol.* **4**: 1–44.
4. Weinberg, E.D. (1971) *Perspect. Biol. Med.* **14**: 565–577.
5. Bushell, M.E. (1989) The process physiology of secondary metabolite production. In: *Microbial Products: New Approaches* (Baumberg, S., Hunter, I.S. and Rhodes, P.M., eds), Cambridge University Press, Cambridge, pp. 95–120.
6. Martin, J.F. and Demain, A.L. (1980) *Microbiol. Rev.* **44**: 230–251.
7. Martin, J.F. (1976) *Dev. Ind. Microbiol.* **17**: 223–231.
8. Hopwood, D.A. and Sherman, D.H. (1990) *Ann. Rev. Genet.* **24**: 37–66.
9. Berdy, J. (1974) *Adv. Appl. Microbiol.* **18**: 309–406.
10. Martin, J.F. and Liras, P. (1981) Biosynthetic pathways of secondary metabolites in industrial microorganisms. In: *Biotechnology, Vol. 1.* (Rehm, H.-J. and Reed, G., eds), Verlag Chemie, Weinheim, Germany, pp. 211–233.
11. Drew, S.W. and Demain, A.L. (1977) *Ann. Rev. Microbiol.* **31**: 343–356.
12. Kleinkauf, H. and von Döhren, H. (1987) *Ann. Rev. Microbiol.* **41**: 259–289.
13. Martin, J.F. and Liras, P. (1989) Enzymes involved in penicillin, cephalosporin and cephamycin biosynthesis. In: *Advances Biochemical Engineering/Biotechnology, Vol. 39.* (Fiechter, A., ed.), Springer Verlag, Berlin, Heidelberg, pp. 153–187.
14. Martin, J.F., Ingolia, T.D. and Queener, S.W. (1991) Molecular genetics of penicillin and cephalosporin antibiotic biosynthesis. In: *Molecular Industrial Mycology. Systems and Applications for Filamentous Fungi.* (Leong, S. and Berka, R. M., eds), Marcel Dekker, Inc., New York, pp. 149–196.
15. Beck, J., Ripka, S., Siegner, A., Schiltz, E. and Schweizer, E. (1990) *Eur. J. Biochem.* **192**: 487–498.
16. Campbell, I.M. (1984) *Adv. Microbial Physiol.* **25**: 2–60.
17. Martin, J.F. and Liras, P. (1989) *Ann. Rev. Microbiol.* **43**: 173–206.
18. Liras, P., Asturias, J.A. and Martin, J.F. (1990) *Trends Biotechnol.* **8**: 184–189.
19. Coque, J.J.R., Martin, J.F., Calzada, J.G. and Liras, P. (1991) *Mol. Microbiol.* **5**: 1125–1133.
20. Coque, J.J.R., Liras, P., Laiz, L. and Martin, J.F. (1991) *J. Bacteriol.* **173**: 6258–6264.
21. Aharonowitz, Y., Cohen, G. and Martin, J.F. (1992) *Ann. Rev. Microbiol.* **46** (in press).
22. Demain, A.L., Somkuti, G.A., Hunter-Cevera, J.C. and Rossmore, H.W. (1989) *Novel Microbial Products for Medicine and Agriculture.* Soc. Ind. Microbiol., Arlington, Va.
23. Turner, W.B. and Aldridge, D.C. (1983) *Fungal Metabolites II.* Academic Press, London, p. 630.
24. Steyn, P.S. (1980) *The Biosynthesis of Mycotoxins: A Study in Secondary Metabolism.* New York, Academic Press.
25. Bell, A.A. and Wheeler, M.H. (1986) *Ann. Rev. Phytopathol.* **24**: 411–451.
26. Dimroth, P., Walter, H. and Lynen, F. (1970) *Eur. J. Biochem.* **13**: 98–110.
27. Herbert, R.B. (1981) *The Biosynthesis of Secondary Metabolites*, Chapman and Hall, London, p. 178.
28. Wiesner, P., Beck, J., Beck, K.-F., Ripka, S., Müller, G., Lücke, S. and Schweizer, E. (1988) *Eur. J. Biochem.* **177**: 69–79.
29. Lynen, F. (1980) *Eur. J. Biochem.* **112**: 431–442.
30. Schweizer, E., Müller, G., Roberts, L.M., Schweizer, M., Rösch, J. *et al.* (1987) *Fat. Sci. Technol.* **89**: 570–577.
31. Sherman, D.H., Malpartida, F., Bibb, M.J., Kieser, H.M., Bibb, M.J. and Hopwood, D.A. (1989) *The EMBO Journal* **8**: 2717–2725.
32. Bibb, M.J., Biro, S., Motamedi, H., Collins, J.F. and Hutchinson, C.R. (1989) *The EMBO Journal* **8**: 2727–2736.

33. Wang, I.-K., Reeves, C. and Gaucher, G.M. (1991) *Can. J. Microbiol.* **37**: 86–95.
34. Mohr, H. and Kleinkauf, H. (1978) *Biochim. Biophys. Acta* **526**: 375–386.
35. Gerlach, M., Schwelle, N., Lerbs, W. and Luckner, M. (1985) *Phytochemistry* **24**: 1935–1939.
36. Lerbs, W. and Luckner, M. (1985) *J. Basic. Microbiol.* **25**: 387–391.
37. Hummel, W. and Diekmann, H. (1981) *Biochem. Biophys. Acta* **617**: 313–320.
38. Zocher, R., Keller, U. and Kleinkauf, H. (1982) *Biochemistry* **21**: 43–48.
39. Zocher, R., Keller, U. and Kleinkauf, H. (1983) *Biochem. Biophys. Res. Commun.* **110**: 292–299.
40. Zocher, R. (1986) *Biol. Chem. Hoppe-Seyler* **367**: 159 (suppl.).
41. Peeters, H., Zocher, R., Madry, N. and Kleinkauf, H. (1983) *Phytochemistry* **22**: 1719–1720.
42. Peeters, H., Zocher, R., Madry, N., Oelrichs, P.B., Kleinkauf, H. and Kraepelin, G. (1983) *J. Antibiot.* **36**: 1762–1766.
43. Lawen, A. and Zocher, R. (1990) *J. Biol. Chem.* **265**: 11355–11360.
44. Ueno, Y. (1980) *Adv. Nutr. Sci.* **3**: 301–353.
45. Cane, D.E., Ha, H., Paegellis, C., Waldmeir, F., Swanson, S. and Murthy, P.P.N. (1985) *Bioorg. Chem.* **13**: 246–265.
46. Beremand, M.N. and McCormick, S.P. (1991) Biosynthesis and regulation of trichothecene production by *Fusarium* species. In: *Mycotoxins in Ecological Systems.* (Bhatnagar, D., Lillehoj, E.B. and Arora, D.K., eds), Marcel Dekker, Inc., New York.
47. Croteau, R. and Cane, D.E. (1985) *Methods Enzymol.* **110**: 383–405.
48. Hohn, T.M. and Van Middlesworth, F. (1986) *Arch. Biochem. Biophys.* **251**: 756–761.
49. Hohn, T.M. and Beremand, P.D. (1989) *Gene* **79**: 131–138.
50. Bennet, J.W. and Papa, K.E. (1988) *Adv. Plant Pathol.* **6**: 263–280.
51. Keller, N.P., Cleveland, T.E. and Bhatnagar, D. (1991) Molecular approach towards understanding aflatoxin production. In: *Mycotoxins in Ecological Systems.* (Bhatnagar, D., Lillehoj, E.B. and Arora, D.K. eds), Marcel Dekker, Inc., New York.
52. Walton, J.D. (1991) Genetics and biochemistry of toxin synthesis in *Cochliobolus* (*Helminthosporium*). In: *Molecular Industrial Mycology.* (Leong, S.A. and Berka, R.M., eds), Marcel Dekker Inc., New York, pp. 225–249.
53. Hesseltine, C.W., Ellis, J.J. and Shotwell, O.L. (1971) *J. Agr. Food Chem.* **19**: 707–717.
54. Sugawara, F., Strobel, G., Strange, R.N., Siedow, J.N., Van Duyne, G.D. and Clardy, J. (1987) *Proc. Natl Sci. USA* **84**: 3081–3085.
55. Kono, Y. and Daly, J.M. (1979) *Bioorg. Chem.* **8**: 391–397.
56. Livingston, R.S. and Scheffer, R.P. (1984) *Plant Physiol.* **76**: 96–102.
57. Liesch, J.M., Sweeley, C.C., Staffeld, G.D., Anderson, M.S., Weber, D.J. and Scheffer, R.P. (1982) *Tetrahedron* **38**: 45–48.
58. Pope, M.R., Ciuffetti, L.M., Knoche, H.W., McCrery, D., Daly, J.M. and Dunkle, L.D. (1983) *Biochemistry* **22**: 3502–3506.
59. Walton, J.D. (1987) *Proc. Natl Acad. Sci. USA* **84**: 8444–8447.
60. Yoder, O.C. (1980) *Ann. Rev. Phytopathol.* **18**: 103–129.
61. Yoshida, M., Knomi, T., Kohsaka, M., Baldwin, J.E., Herlhen, S., Singh, P., Hunt, N.A. and Demain, A.L. (1978) *Proc. Natl Acad. Sci. USA* **75**: 6253–6257.
62. Mullin, P.G., Turgeon, B.G., Garber, R.C. and Yoder, O.C. (1988) *J. Cell. Biochem.* **12C**: 285.
63 Scheffer, R.P., Nelson, R.R. and Ullstrup, A.J. (1967) *Phytopathology* **57**: 1288–1291.
64. Scheffer, R.P. and Livingston, R.S. (1984) *Science* **223**: 17–21.
65. Walton, J.D., Earle, E.D., Stahelin, H., Grieder, A., Hirota, A. and Suzuki, A. (1985) *Experientia* **41**: 348–350.
66. Kawai, M., Pottorf, R.S. and Rich, D.H. (1986) *J. Am. Chem. Soc.* **29**: 2409–2411.
67. Walton, J.D. and Holden, F.R. (1988) *Mol. Plant-Microbe Interact.* **1**: 128–134.
68. Wessel, W.L., Clare, K.A. and Gibbons, W.A. (1987) *Biochem. Soc. Trans.* **15**: 917–918.
69. Wessel, W.L., Clare, K.A. and Gibbons, W.A. (1988) *Biochem. Soc. Trans.* **16**: 401–402.
70. Wessel, W.L., Clare, K.A. and Gibbons, W.A. (1988) *Biochem. Soc. Trans.* **16**: 402–403.
71. Takayama, S., Isogai, A., Nakata, M., Suzuki, H. and Suzuki, A. (1984) *Agric. Biol. Chem.* **48**: 839–842.
Turgeon, B.G., Garber, R.C. and Yoder, O.C. (1987) *Mol. Cell. Biol.* **7**: 3297–3305.
72. Closse, A. and Huguenin, R. (1974) *Helv. Chim. Acta* **57**: 533–545.
73. Schmidt, U., Beuttler, T., Lieberknecht, A. and Griesser, H. (1983) *Tetrahedron Lett.* **24**: 3573–3576.
74. Demain, A.L. (1983) Biosynthesis of β-lactam antibiotics. In: *Antibiotics Containing the*

β-lactam Structure, Vol. I. (Demain, A.L. and Salomon, N.A. eds), Springer-Verlag, Berlin, p. 189.
75. Martin, J.F. and Aharonowitz, Y. (1983) Regulation of biosynthesis of β-lactam antibiotics. In: *Antibiotics Containing the β-Lactam Structure, Vol. 1.* (Demain, A.L. and Solomon, N.A., eds), Springer-Verlag, Berlin, p. 229.
76. Queener, S.W., Wilkerson, S., Tunin, D.R., McDermott, J.R., Chapman, J.L., Nash, C., Platt, C. and Westpheling, J. (1984) Cephalosporin C fermentation: Biochemical and regulatory aspects of sulfur metabolism. In: *Biotechnology of Industrial Antibiotics.* (Vondamme, E. ed.), Marcel Dekker, Inc., New York, p. 141.
77. Baldwin, J.E. and Abraham, E. (1988) *Nat. Prod. Rep.* **1988**: 129–145.
78. Fawcett, P.A., Usher, J.J., Huddleston, J.A., Bleany, R.C., Nisbet, J.J. and Abraham, E.P. (1976) *Biochem. J.* **157**: 651–660.
79. Pang, C.P., Chakravarti, B., Adlington, R.M., Ting, H.H., White, R.L., Jayatilake, G.S., Baldwin, J.E. and Abraham, E.P. (1984) *Biochem. J.* **222**: 789–795.
80. Ramos, F.R., Lopez-Nieto, M.J. and Martin, J.F. (1985) *Antimicrob. Agents Chemother.* **27**: 380–387.
81. Pruess, D.L. and Johnson, M.J. (1967) *J. Bacteriol.* **94**: 1502–1508.
82. Martin, J.F., Diez, B., Alvarez, E., Barredo, J.L. and Cantoral, J.M. (1987) Development of a transformation system in *Penicillium chrysogenum.* Cloning of genes involved in penicillin biosynthesis. In: *Genetics of Industrial Microorganisms.* (Alacevic, M., Hranueli, D. and Toman, Z., eds), Pliva, Zagreb, p. 297.
83. Alvarez, E., Cantoral, J.M., Barredo, J.L., Diez, B. and Martin, J.F. (1987) *Antimicrob. Agent Chemother.* **31**: 1675–1682.
84. Hollander, I.J., Shen, Y.Q., Heim, J., Demain, A.L. and Wolfe, S. (1984) *Science* **224**: 610–612.
85. Jayatilake, G.S., Huddleston, J.A. and Abraham, E.P. (1981) *Biochem. J.* **194**: 645–648.
86. Kupka, J., Shen, Y.Q., Wolfe, S. and Demain, A.L. (1983) *Can. J. Microbiol.* **29**: 488–496.
87. Dotzlaf, J.E. and Yeh, W.K. (1987) *J. Bacteriol.* **169**: 1611–1618.
88. Scheidegger, A., Kuenzi, M.T. and Nuesch, J. (1984) *J. Antibiot.* **37**: 522–531.
89. Baldwin, J.E., Adlington, R.M., Cortes, J.B., Crabbe, J.C., Crouch, N.P., Keeping, J.W., Knight, G.C., Schofield, C.J., Ting, H.H., Vallejo, C.A., Thorniley, M. and Abraham, E.P. (1987) *Biochem. J.* **245**: 831–841.
90. Fujisawa, Y., Shirafuji, H., Kida, M., Nara, K., Yoneda, M. and Kanzaka, T. (1973) *Nature New Biol.* **246**: 154.
91. Fujisawa, Y., Shirafuji, H., Kida, M., Nara, K., Yoneda, M. and Kanazaki, T. (1975) *Agric. Biol. Chem.* **39**: 1295–1302.
92. Gutierrez, S., Velasco, J., Fernandez, F.J. and Martin, J.F. (1992) *J. Bacteriol.* (in press).
93. Xiao, X., Wolfe, S. and Demain, A.L. (1991) *Biochem. J.* **280**: 471–474.
94. Jensen, S.E. (1986) *CRC Crit. Rev. Biotechnol.* **3**: 277–310.
95. Nash, C.H., de la Higuera, N., Neuss, N. and Lemke, P.A. (1974) *Develop. Indust. Microbiol.* **15**: 114–123.
96. Normansell, P.J.M., Normansell, I.D. and Holt, G. (1979) *J. Gen. Microbiol.* **112**: 113–126.
97. Makins, J.F., Holt, G. and Macdonald, K.D. (1980) *J. Gen. Microbiol.* **119**: 397–404.
98. Makins, J.F., Allsop, A. and Holt, G. (1981) *J. Gen. Microbiol.* **122**: 339–343.
99. Arst, H.N. Jr. and Scazzocchio, C. (1985) Formal genetics and molecular biology of the control of gene expression in *Aspergillus nidulans.* In: *Gene Manipulations in Fungi* (Bennett, J.W. and Lasure, L.L., eds.), Academic Press, New York, pp. 309–343.
100. Edwards, G.F., Holt, G. and Macdonald, K.D. (1974) *J. Gen. Microbiol.* **84**: 420–422.
101. Makins, J.F., Holt, G. and MacDonald, K.D. (1983) *J. Gen. Microbiol.* **129**: 3027–3033.
102. Holt, G., Edwards, G.F. and Macdonald, K.D. (1976) The genetics of mutants impaired in the biosynthesis of penicillin. In: *Proceedings of the Second International Symposium on the Genetics of Industrial Microorganisms.* (Macdonald, K.D. ed.), Academic Press, London, p. 199.
103. Macdonald, K.D. (1983) Fungal genetics and antibiotic production. In: *Biochemistry and Genetic Regulation of Commercially Important Antibiotics.* (Vining, L.C. ed.), Addison-Wesley Pub. Co., Reading, Ma., p. 25.
104. MacCabe, A.P., Riach, M.B.R., Unkles, S.E. and Kinghorn, J.R. (1990) *The EMBO Journal* **9**: 279–287.

105. Lemke, P. and Nash, C. (1971) *Can. J. Antibiot.* **18**: 255–259.
106. Shirafuji, H., Fujisawa, Y., Kida, M., Kanzaki, T. and Yoneda, M. (1979) *Agric. Biol. Chem.* **43**: 155–160.
107. Ramos, F.R., Lopez-Nieto, M.J. and Martin, J.F. (1986) *FEMS Microbiol. Lett.* **35**: 123–127.
108. Ramsdem, M., McQuade, B.A., Saunders, K., Turner, M.K. and Harford, S. (1989) *Gene* **85**: 267–273.
109. Fujisawa, Y., Shirafuji, H. and Kanzaki, T. (1975) *Agric. Biol. Chem.* **39**: 1303–1305.
110. Queener, S.W., Capone, J.J., Radue, A.B. and Nagarajan, R. (1974) *Antimicrob. Agents Chemother.* **6**: 334–337.
111. Martin, J.F. (1987) *Trends Biotechnol.* **5**: 306–308.
112. Cortes, J., Liras, P., Castro, J.M. and Martin, J.F. (1986) *J. Gen. Microbiol.* **132**: 1805–1814.
113. Revilla, G., Ramos, F.R., Lopez-Nieto, M.J., Alvarez, E. and Martin, J.F. (1986) *J. Bacteriol.* **168**: 947–952.
114. Zhang, J.Y., Wolfe, S. and Demain, A.L. (1989) *Curr. Microbiol.* **18**: 361–367.
115. Zhang, J.Y., Wolfe, S. and Demain, A.L. (1989) *FEMS Microbiol. Lett.* **57**: 145–150.
116. Zhang, J.Y., Wolfe, S. and Demain, A.L. (1989) *Can. J. Microbiol.* **35**: 399–402.
117. Rollins, M.J., Jensen, S.W. and Westlake, D.W.S. (1991) *Appl. Microbiol. Biotechnol.* **35**: 83–88.
118. Lopez-Nieto, M.J., Ramos, F.R., Luengo, J.M. and Martin, J.F. (1985) *Appl. Microbiol. Biotechnol.* **22**: 343–351.
119. Banko, G., Wolfe, S. and Demain, A.L. (1986) *Biochem. Biophys. Res. Commun.* **137**: 528–535.
120. Banko, G., Demain, A.L. and Wolfe, S. (1987) *J. Am. Chem. Soc.* **109**: 2858–2860.
121. Jensen, S., Westlake, D.W.S. and Wolfe, S. (1988) *FEMS Microbiol. Lett.* **49**: 213–218.
122. Jensen, S.E., Wong, A., Rollins, M.J. and Westlake, D.W.S. (1990) *J. Bacteriol.* **172**: 7269–7271.
123. Van Liempt, H., von Döhren, H. and Kleinkauf, H. (1989) *J. Biol. Chem.* **264**: 3680–3684.
124. Baldwin, J.E., Bird, J.W., Field, R.A., O'Callaghan, N.M. and Schofield, C.J. (1990) *J. Antibiot.* **43**: 1055–1057.
125. Zhang, J.Y. and Demain, A.L. (1990) *Biotechnol. Lett.* **12**: 649–654.
126. MacCabe, A.P., van Liempt, H., Palissa, H., Unkles, S.E., Riach, M.B.R., Pfeifer, E., von Döhren, H. and Kinghorn, J.R. (1991) *J. Biol. Chem.* **266**: 12646–12654.
127. Diez, B., Barredo, J.L., Alvarez, E., Cantoral, J.M., van Solingen, P., Groenen, M.A.M., Veenstra, A.E. and Martín, J.F. (1989) *Mol. Gen. Genet.* **218**: 572–576.
128. Diez, B., Gutierrez, S., Barredo, J.L., van Solingen, P., van der Voort, L.H.M. and Martin, J.F. (1990) *J. Biol. Chem.* **265**: 16358–16365.
129. Smith, D.J., Earl, A.J. and Turner, G. (1990) *The EMBO Journal* **9**: 2743–2750.
130. Gutierrez, S., Diez, B., Montenegro, E. and Martin, J.F. (1991a) *J. Bacteriol.* **173**: 2354–2365.
131. Krätzschmar, J., Krause, M. and Marahiel, M.A. (1989) *J. Bacteriol.* **171**: 5422–5429.
132. Mittenhuber, G., Weckermann, R. and Marahiel, M.A. (1989) *J. Bacteriol.* **171**: 4881–4887.
133. Samson, S.M., Belagaje, R., Blankenship, D.T., Champman, J.L., Perry, D., Skatrud, P.L., Vanfrank, R.M., Abraham, E.P., Baldwin, J.E., Queener, S.W. and Ingolia, T.D. (1985) *Nature* **318**: 191–194.
134. Baldwin, J.E., Killin, S.J., Pratt, A.J., Sutherland, J.D., Turner, N.J., Crabbe, J.C., Abraham, E.P. and Willis, A.C. (1987) *J. Antibiot.* **40**: 652–659.
135. Chapman, J.L., Skatrud, P.L., Ingolia, T.D., Samson, S.M., Kaster, K.R. and Queener, S.W. (1987) *Dev. Ind. Microbiol.* **27**: 165–174.
136. Barredo, J.L., Cantoral, J.M., Alvarez, E., Diez, B. and Martin, J.F. (1989) *Mol. Gen. Genet.* **216**: 91–98.
137. Carr, L.G., Skatrud, P.L., Scheetz, M.E., Queener, S.W. and Ingolia, T.D. (1986) *Gene* **48**: 257–266.
138. Ramon, D., Carramolino, L., Patino, C., Sanchez, F. and Penalva, M.A. (1987) *Gene* **57**: 171–181.
139. Leskiw, B.K., Aharonowitz, Y., Mevarech, M., Wolfe, S., Vining, L.C., Westlake, D.W.S. and Jensen, S.E. (1988) *Gene* **62**: 187–196.
140. Garcia-Dominguez, M., Liras, P. and Martin, J.F. (1991) *Antimicrob. Agents Chemother.* **35**: 44–52.

141. Shiffman, D., Mevarech, M., Jensen, S.E., Cohen, G. and Aharonowitz, Y. (1988) *Mol. Gen. Genet.* **214**: 562–569.
142. Weigel, B.J., Bugett, S.G., Chen, V.J., Skatrud, P.L., Frolik, C.A., Queener, S.W. and Ingolia, T.D. (1988) *J. Bacteriol.* **170**: 3817–3826.
143. Shiffman, D., Cohen, G., Aharonowitz, Y., Palissa, H., von Döhren, H., Kleinkauf, H. and Mevarech, M. (1990) *Nuc. Acids Res.* **18**: 660.
144. Samson, S.M., Chapman, J.L., Belagaje, P., Queener, S.W. and Ingolia, T.D. (1987) *Proc. Natl Acad. Sci. USA* **84**: 5705–5709.
145. Castro, J.M., Liras, P., Laiz, L. and Martin, J.F. (1988) *J. Gen. Microbiol.* **134**: 133–141.
146. Chen, V.J. (1990) *Abstracts of the International Symposium on '50 years of Penicillin Utilization'*, Berlin 1990.
147. Gutierrez, S., Diez, B., Alvarez, E., Barredo, J.L. and Martin, J.F. (1991) *Mol. Gen. Genet.* **225**: 56–64.
148. Queener, S.W. and Neuss, N. (1982) The biosynthesis of beta-lactam antibiotics. In: *Chemistry and Biology of Beta-Lactam Antibiotics, Vol. 3.* (Morin, R.B. and Gorman, M., eds), Academic Press, New York, pp. 1–82.
149. Barredo, J.L., van Solingen, P., Diez, B., Alvarez, E., Cantoral, J.M., Kattevilder, A., Smaal, E.B., Groenen, M.A.M., Veenstra, A.E. and Martin, J.F. (1989) *Gene* **83**: 291–300.
150. Whiteman, P.A., Abraham, E.P., Baldwin, J.W., Fleming, M.D., Schofield, C.J., Sutherland, J.D. and Willis, A.C. (1990) *FEMS Lett.* **262**: 342–344.
151. Montenegro, E., Barredo, J.L., Gutierrez, S., Diez, B., Alvarez, E. and Martin, J.F. (1990) *Mol. Gen. Genet.* **221**: 322–330.
152. Ballance, D.J. (1986) *Yeast* **2**: 229–236.
153. Hoskins, J.A., O'Callaghan, N., Queener, S.W., Cantwell, C.A., Wood, J.S., Chen, V.J. and Skatrud, P.L. (1990) *Curr. Genet.* **18**: 523–530.
154. Samson, S.M., Dotzlaf, J.E., Slisz, M.L., Becker, G.W., Van Frank, R.M., Veal, L.E., Yeh, W.K., Miller, J.R., Queener, S.W. and Ingolia, T.D. (1987) *Bio/Technology* **5**: 1207–1214.
155. Skatrud, P.L. and Queener, S.W. (1989) *Gene* **79**: 331–338.
156. Smith, A.W., Collis, K., Ramsden, M., Fox, H.M. and Peberdy, J.E. (1991) *Curr. Genet.* **19**: 235–237.
157. Kovacevic, S., Weigel, B.J., Tobin, M.B., Ingolia, T.D. and Miller, J.R. (1989) *J. Bacteriol.* **171**: 754–760.
158. Kovacevic, S. and Miller, J.R. (1991) *J. Bacteriol.* **173**: 398–400.
159. Goodwin, T.W. (1980) *The Biochemistry of the Carotenoids*, 2nd ed., Vol. 1. Chapman and Hall, London.
160. Cerda-Olmedo, E. (1989) Production of carotenoids with fungi. In: *Biotechnology of Vitamin, Growth Factor and Pigment Production.* (Vadamme, E., ed.), Elsevier Applied Science, pp. 27–42.
161. Bramley, P.M. and Mackenzie, A. (1988) *Curr. Top. Cell. Regul.* **29**: 291–343.
162. Mitzka-Schnabel, U. and Rau, W. (1980) *Phytochemistry* **19**: 1409–1413.
163. Candau, R., Bejarano, E.R. and Cerda-Olmedo, E. (1991) *Proc. Natl Acad. Sci. USA* **88**: 4936–4940.
164. Goldie, A.H. and Subden, R.E. (1973) *Biochem. Genet.* **10**: 275–284.
165. Nelson, M.A., Morelli, G., Carottoli, A., Romano, N. and Macino, G. (1989) *Mol. Cell. Biol.* **9**: 1271–1276.
166. Schmidhauser, T.J., Lauter, F.R., Russo, V.E.A. and Yanofsky, C. (1990) *Mol. Cell. Biol.* **10**: 5064–5070.
167. Bartley, G., Viitanen, P.V., Pecker, I., Chamovitz, D., Hirchberg, J. and Scolnik, P.A. (1991) *Proc. Natl Acad. Sci. USA* **88**: 6532–6536.
168. Harding, R.W. and Turner, R.V. (1981) *Plant Physiol.* (*Bethesda*) **68**: 745–749.
169. Barredo, J.L., Diez, E., Alvarez, E. and Martin, J.F. (1989) *Curr. Genet.* **16**: 453–459.
170. Bhatnagar, D., Lillehoj, E.B. and Arora, D.K. (1991) *Mycotoxins in Ecological Systems.* Marcel Dekker, Inc., New York, pp. 464.
171. Fincham, J.R.S. (1989) *Microbiol. Rev.* **53**: 148–170.
172. Smith, D.J. Burnham, M.K.R., Edwards, J., Earl, A.J. and Turner, G. (1990) *Bio/Technology* **8**: 39–41.
173. Isogai, T., Fukagawa, M., Aramori, I., Iwami, M., Kojo, H., Ono, T., Ueda, Y., Kohsaka, M. and Imanaka, H. (1991) *Bio/Technology* **9**: 188–191.

11 Future prospects

J.R. KINGHORN and G. TURNER

The development of genetic manipulation techniques for the yeast *Saccharomyces cerevisiae* provided the impetus and many of the ideas for the development of similar techniques in the filamentous fungi. While these efforts began with the economically unimportant fungi *Neurospora crassa* and *Aspergillus nidulans* because of their well-developed genetic systems, the techniques have been rapidly transferred to other Ascomycetes, as well as the Deuteromycetes, Basidiomycetes, and Zygomycetes, where little or no classical genetic information, such as mutants and genetic maps, has been available. The reason for this transfer of technology has been the economic importance of a very large number of moulds, some useful to man, and some pests causing food spoilage and biodeterioration as well as animal and plant diseases. Molecular genetics has been used firstly to help understand processes and secondly to try to improve or modify these processes, and has already had a major impact in many areas.

Pharmaceutical companies continue to screen soil samples for novel compounds with biological activity, both antimicrobial and pharmacological. While the filamentous bacteria such as the Actinomycetes continue to provide the greatest variety of compounds, filamentous fungi are still important in such screens.

One important group of secondary metabolites are the polyketides, the biosynthesis and genetics of which have been the subject of intensive study in the Actinomycetes because of their potential to produce antibiotics and biologically active compounds [1]. Information on polyketide biosynthesis in fungi is at present rather limited. Polyketides are assembled from simple carboxylic acid units such as acetate, propionate and butyrate, and the early steps are similar in many organisms, resembling fatty acid biosynthesis. However, it is the huge variety of compounds generated by differences at later stages in biosynthesis that attracts interest in these pathways. Fungal polyketides include aflatoxins which are powerful carcinogens that can be produced in foodstuffs following fungal contamination.

One fungal polyketide of clinical importance is mevinolin, produced by a number of fungi including *Aspergillus terreus* and certain *Penicillium* and *Monascus* species. It is an inhibitor of hydroxymethylglutaryl-CoA reductase, which converts HMG CoA into mevalonic acid during the biosynthesis of cholesterol in mammals [2]. It is now used as a treatment for hypercholesterolaemia, a risk factor in atherosclerosis, and commercially produced as Lovastatin (Mevacor) by Merck Sharp & Dohme.

Recently, the gene encoding 6-methylsalicylic acid synthetase, an early step in the biosynthesis of the polyketide patulin, was isolated from *Penicillium patulum* [3]. Without doubt, a combination of chemistry, biochemistry and genetics will now be applied to the study of fungal polyketide biosynthesis, using similar techniques to those used so successfully to study β-lactam biosynthesis (chapter 10).

In principle, any new compound of interest produced by a microorganism can be synthesized by chemists, and whether or not fermentation is used in commercial production depends on the relative cost of chemical synthesis. If fermentation is to be used, then understanding of the biosynthetic pathway and the genes involved can lead to improvement or modification of the pathway. It may be possible to construct hybrid pathways, using genes derived from two or more organisms to generate new compounds, and this approach has great potential. Genetic manipulation of secondary metabolism in filamentous fungi is therefore likely to continue in importance alongside traditional mutagenesis/screening methods.

A number of fungal enzymes discussed in this book have been secreted fungal enzymes used in the food industry and more recently in detergents (chapter 3). However, fungal hydroxylation has been used for over 30 years to carry out biotransformation of sterols to corticosteroids [4]. Such biotransformations are stereospecific, and very expensive to carry out chemically. Clearly, there is potential for improving the activity of these enzyme preparations. Since they are generally membrane bound, purification has been difficult. Recently, a gene for benzoate hydroxylase has been cloned from *Aspergillus niger*, and peptides synthesized to raise antibodies to facilitate detection of the enzyme during purification [5]. This model system should lead to further developments with fungi such as *Rhizopus nigricans* and *Aspergillus ochraceus*, which hydroxylate progesterone.

Despite a number of exploratory attempts, expression of pharmaceutically important, high-value mammalian proteins has not yet proved its worth in filamentous fungi. Interest in filamentous fungi in this respect has been a result of their well-known ability to secrete native proteins in large amounts. However, moulds are also prodigious secretors of proteases, which can destroy the desired product, and although removal of some of these by mutagenesis or gene deletion is possible, it is still difficult to predict whether filamentous fungi will compete significantly with bacteria, yeasts and mammalian cells. The work by Genencor Inc. on the production of bovine chymosin by *Aspergillus awamori* for use in cheese manufacture (chapter 6) still represents the best example, to date, of mammalian protein production in a filamentous fungus.

Another interesting application of a fungal enzyme is the use of phytase from *Aspergillus* species to improve the quality of animal feed by releasing phosphate from phytate, which improves the phosphate availability [6]. A molecular approach is now being used for improvement of enzyme yield [7].

Filamentous fungi may also have a future role in decontamination of the environment. Some pathogens of cyanogenic plants are resistant to cyanide by means of a cyanide hydratase. An ICI fungal preparation Cyclear, for treatment of cyanide-containing waste [8], is prepared from *Fusarium lateritium* and progress is being made in the isolation of the cyanide-inducible gene encoding this enzyme from this organism [9].

The fungi provide a seemingly inexhaustible supply of variety and potential for innovation, and the application of molecular biology to their study and manipulation promises a healthy future for fungal biotechnology.

References

1. Hopwood, D.A. and Sherman, D.H. (1991) *Ann. Rev. Genet.* **24**: 37–66.
2. Alberts, A.W., Chen, J., Kuron, G., Hunt, V., Huff, J., Hoffman, C., Rothrock, J., Lopez, M., Joshua, H., Harris, E., Patchett, A., Monaghan, R., Currie, S., Stapley, E., Albers-Schonberg, G., Hensens, O., Hirshfield, J., Hoogsteen, K., Liesch, J. and Springer, J. (1980) *Proc. Natl Acad. Sci. USA* **77**: 3957–3961.
3. Beck, J., Ripka, S., Siegner, A., Schiltz, E. and Schweizer, E. (1990) *Eur. J. Biochem.* **192**: 487–498.
4. C. Vezina (1987) In: *Basic Biotechnology* (Bu'Lock, J. and Kristiansen, B. eds), Academic Press, London, p. 463–482.
5. Gerriste, K., van Gorcom, R., Fasbender, M., Lange, J., Schellekens, M., Zegers, N., Claasen, E. and Boersma, W. (1990) *Biochem. Biophys. Res. Commun.* **167**: 33–39.
6. Nelson, T. S., Shieh, T. R., Wodzinski, R. J. and Ware, J. H. (1968) *Poultry Science* **47**: 1842–1848.
7. van Hartingsveldt, C., van Zeijl, C., Hartefeld, M., Menke, H., Suykerbuyk, M., van Paridon, P., Selten, G., Veenstra, A., Van den Hondel, C. and van Gorcom, R. (1991) Poster Presentation EMBO Workshop on Molecular Biology of Filamentous Fungi, Berlin.
8. Harris, R. E., Bunch, A. W. and Knowles, C. J. (1987) *Science Progress* **71**: 293–304.
9. Brown, D., Cluness, M., Turner, P. and O'Reilly, C. (1991) Poster Presentation EMBO Workshop on Molecular Biology of Filamentous Fungi, Berlin.

Index

The manufacturer's authorised representative in the EU is Springer Nature Customer Service Centre GmbH, Europaplatz 3, 69115 Heidelberg, Germany. If you have any concerns regarding our products, please contact ProductSafety@springernature.com

Printed and bound by CPI Group (UK) Ltd, Croydon, CR0 4YY
22/09/2026
02220945-0006